Y0-BXQ-660

ISOTOPES IN ORGANIC CHEMISTRY

ELSEVIER SCIENTIFIC PUBLISHING COMPANY
335 JAN VAN GALENSTRAAT
P.O. BOX 211, AMSTERDAM, THE NETHERLANDS

AMERICAN ELSEVIER PUBLISHING COMPANY, INC.
52 VANDERBILT AVENUE
NEW YORK, NEW YORK 10017

Library of Congress Cataloging in Publication Data
Main entry under title:

Isotopes in molecular rearrangements.

(Isotopes in organic chemistry ; v. 1)
Includes bibliographical references and index.
1. Rearrangements (Chemistry) 2. Radioactive tracers.
I. Buncel, E. II. Lee, Choi Chuck, 1924-
III. Series.
QD281.R35I86 541'.388 74-10255
ISBN 0-444-41223-9

Library of Congress Card Number: 74-10255

ISBN: 0-444-41223-9

With 37 illustrations and 34 tables

Printed in The Netherlands

ISOTOPES IN ORGANIC CHEMISTRY

CONTRIBUTORS TO VOLUME 1

N.C. Deno	Department of Chemistry, The Pennsylvania State University, University Park, Pennsylvania, U.S.A.
W.R. Dolbier, Jr.	Department of Chemistry, University of Florida, Gainesville, Florida, U.S.A.
J.L. Holmes	Chemistry Department, University of Ottawa, Ottawa, Ontario, Canada
D.H. Hunter	Department of Chemistry, University of Western Ontario, London, Ontario, Canada
J.S. Swenton	Department of Chemistry, The Ohio State University, Columbus, Ohio, U.S.A.

ISOTOPES IN ORGANIC CHEMISTRY

Edited by

E. BUNCEL

Queen's University, Kingston, Ontario, Canada

and

C.C. LEE

University of Saskatchewan, Saskatoon, Saskatchewan, Canada

Volume 1

Isotopes in molecular rearrangements

ELSEVIER SCIENTIFIC PUBLISHING COMPANY

AMSTERDAM OXFORD NEW YORK

1975

ISOTOPES IN ORGANIC CHEMISTRY

Volume 2. Isotopes in hydrogen transfer processes

Maurice M. Kreevoy University of Minnesota	The effect of structure on proton-transfer isotope effects
Gérard Lamaty Université de Montpellier	Isotope exchange in carbonyl compounds
Kenneth T. Leffek Dalhousie University	Proton transfer in nitro compounds
Edward S. Lewis Rice University	Hydrogen atom transfer processes
Helmut Simon and Adolf Kraus Technische Universität Munich	Hydrogen isotope transfer in biochemical processes
Peter J. Smith University of Saskatchewan	Isotope effects and transition states in elimination reactions
Ross Stewart University of British Columbia	Isotopes in oxidation processes

Volume 3. Carbon-13 in organic chemistry

H.M. Buck Eindhoven University of Technology	Carbon-13 n.m.r. studies of carbonium ions
G.E. Dunn University of Manitoba	^{13}C Isotope effects in decarboxylation reactions
A. Fry, J. Hinton and M. Oka University of Arkansas	^{13}C n.m.r. methodology and mechanistic applications
G. Kunesch and C. Poupat Institut de Chimie des Substances Naturelles	Biosynthetic studies using ^{13}C precursors
G. Lukacs Institut de Chimie des Substances Naturelles	Structure elucidation of natural products and related compounds using ^{13}C n.m.r.
A.S. Perlin McGill University	Application of ^{13}C n.m.r. to problems of stereochemistry
A.V. Willi Columbia University	Kinetic carbon-13 and other isotope effects in cleavage and formation of bonds to carbon

FOREWORD

Organic chemistry is characterized by a vast variety of compounds, structures and reactions realized by a rather limited number of chemical elements. One and the same element is generally represented by a considerable number of atoms, playing several different roles. It is evident that a method enabling us to give the otherwise anonymous atom a kind of identity should be of particular value in this branch of chemistry.

Tracing by means of similar but still chemically discernible groups has been practised in organic chemistry for a long time, and has revealed that organic reactions are far more varied than expected. Isotopes, being chemically identical in a qualitative and usually also in an almost quantitative sense, are far more powerful as tracers, due to this similarity and to the fact that atoms rather than groups are labelled and can be traced as such.

A simple account of the molecular species involved, their structures and configurations, cannot be considered a complete description of a chemical system in equilibrium. We know from studies of non-equilibrium systems that opposite reactions balancing one another are generally taking place. As far as species of different molecular compositions are concerned, this has been realized for more than a century. It is only in the last few decades, however, that we have had at hand the means to measure the amounts of different isotopes and follow the behaviour of systems which are non-equilibrium ones with respect to isotopic composition. This has led to a still more vivid picture of most systems in equilibrium, with several exchange reactions taking place, sometimes at rates too large to be measured on the classical time scale of chemical reactions.

Even this is not enough for the true scientist who wants to go beyond the knowledge of which reactions actually take place and how fast they occur. He feels a desire to know also how the atomic nuclei and electrons behave in the transition called a chemical reaction. His questions come close to the fundamental limit set by the principle of uncertainty. At present the transition state of the rate-determining reaction step seems to be the most complete description attainable. In such studies it is not only the qualitative chemical similarity of isotopes, allowing the identity of atoms in the transition state to be revealed, but also their quantitative chemical dissimilarity which is of importance and allows a study of the force field and hence binding conditions in the transition state itself. Thanks to the fairly low atomic number of most atomic species of importance in organic chemistry, the relative mass differences between isotopes are sufficient to cause differences in quantitative behaviour, rather easily measurable with modern instruments.

Many scientists feel the flood of scientific publications as an encumbrance. The justification for the existence of a series like the one started by the present volume lies in the aid that the surveys it contains may offer the research worker and, perhaps more important, the stimulus for further research that may be provided. The application of isotope methods has undoubtedly a very important role in future research in organic chemistry. No attempt at a detailed prediction will be ventured here, however. It may suffice to refer to the development of the nuclear magnetic resonance technique. The studies of ordinary hydrogen nuclei, which have been of outstanding importance for the development of organic chemistry, can be considered as an application of isotope methods according to the ordinary usage of the concept only to the extent that deuterium has been used as a stand-in for protium. In the not-too-distant future, however, most laboratories will have equipment allowing routine studies of the less abundant carbon isotope ^{13}C, and then many chemists will be in possession of a sensitive probe in the centre of atoms of the most important element in organic chemistry. It will reveal not only details of molecular structure in the usual sense but also more subtle details about the electron distribution in the backbone of organic molecules. It is open to discussion, of course, whether this kind of work, which frequently makes use of the natural occurrence of heavy carbon, should be considered as an application of isotope methods. In any case, it utilizes a particular property of an isotope different from the most abundant one.

It is evident from the thoughts expressed in the last paragraph that the borders of the field "Isotopes in Organic Chemistry" are rather indeterminate. The editors' intention to apply as few restrictions as possible on subject matter seems wise, because then the interest taken in the present series by its future readers can be allowed to indicate the position of these borders in practice.

Göteborg Lars Melander

PREFACE

The publication, in the late 1940's, of a number of monographs on tracer methodology, particularly the authoritative volume on *Isotopic Carbon* by Calvin, Heidelberger, Reid, Tolbert and Yankwich, has given great impetus to tracer studies in organic chemistry. The utilization of kinetic isotope effects as a probe for the transition state also gained in momentum with the publication in 1960 of the monograph *Isotope Effects on Reaction Rates* by Melander. With these developments, and the more recent advent of techniques such as n.m.r. and mass spectrometry, applications of both radioactive and stable isotopes have become extremely useful in many areas of investigation.

It is the intent of this series to bring together information from diverse areas of organic chemistry under the common theme highlighting the use and value of isotopes. Since in the future one can look to increasingly wider opportunities for utilization of isotopic studies, it is our intention to place as few restrictions as possible on the subject matter to be covered. It may also be hoped thereby that the series will aid in providing a stimulus for further research.

It is our plan that each volume should have a central theme as a link for the various chapters. Thus the first volume contains contributions relating to the broad area of *Isotopes in Molecular Rearrangements,* and the second volume will deal with *Isotopes in Hydrogen Transfer Processes.* Other broad topics to be covered in future volumes will include *Carbon-13 in Organic Chemistry, Isotopes in Aromatic Systems, Isotopes in Structural Elucidations, Isotopic Sulfur in Organic Chemistry,* etc. Coverage of any one broad topic will not necessarily be limited to one volume. A second volume on molecular rearrangements, for example, is being planned.

We feel honoured in that Professor Lars Melander has consented to write a Foreword to this first volume in the series. Sincere appreciation is also extended to members of the Editorial Advisory Board for their valuable comments on various aspects of this undertaking, and most importantly, we would like to express our thanks to the contributing authors on whose efforts the success of this series will largely depend.

Kingston, Ontario — E.B.
Saskatoon, Saskatchewan — C.C.L.

CONTENTS

Chapter 1

DEUTERIUM LABELING IN CARBONIUM ION REARRANGEMENTS

N.C. DENO

Department of Chemistry, The Pennsylvania State University, University Park, Pennsylvania 16802 (U.S.A.)

I. INTRODUCTION

There are several reasons for reviewing the subject of deuterium labeling in carbonium ion rearrangements, the prime one being that in this area deuterium labeling has gone far beyond the classic technique of inserting a deuterium label in the reactant and noting its position in the product.

Deuterium has been introduced in either the reactant or the solvent or in both simultaneously. Multiple deuteration has been used. A mixture of deuterium-labeled and carbon-labeled substrate has been simultaneously monitored to relate the rates of hydrogen scrambling to those of carbon scrambling. Quenching experiments have been employed. Intensities of (P – Me) and (P – Et) mass spectrum bands have been used to determine the deuterium distribution in propyl and butyl groups.

Sometimes rearrangements were discovered where none had been expected, as in the addition of deuterated acids to cyclopropanes or in the mass spectroscopy of toluene and benzyl derivatives. Often complete equilibration of deuterium developed in chemically stable systems such as the *tert*-pentyl or methylcyclopentyl cations and deuterium labeling served to map out the sequence of events. Underlying it all is the propensity of carbonium ions to undergo complex multiple rearrangements with great rapidity which causes a fluidity of deuterium label not found in other types of organic reactions.

This review covers a selection of deuterium experiments which have been chosen to illustrate techniques that have been employed. No great effort has been made to secure complete coverage since many deuterium studies were auxillary to other work and are of import for their bearing on mechanistic problems rather than for illustrating techniques in deuterium labeling.

II. SEQUENTIAL INTRODUCTION OF DEUTERIUM LABEL FROM SOLVENT

A. Principles

Carbonium ions are in reversible equilibrium with their conjugate bases, the alkenes. This equilibrium serves to introduce deuterium from a deuterated solvent into the positions α to the positively charged carbon. This deuteration takes place not only in the reactant (which serves to label the substrate) and product (generally a bothersome complexity), but also on intermediates that might not otherwise be detected.

Work in this area has three features: (i) deuteration at positions other than α indicate reversible rearrangement(s) to cations which exist in amounts too small for direct observation or detection; (ii) any path must account for the positions of deuteriums in the product relative to the reactant as in classical deuterium labeling; (iii) appearance of excess deuterium in the product, above that possible from deuteration of reactant or product, signals a previously undetected intermediate which is undergoing exchange.

B. Addition of Deuterated Acids to Cyclopropane

Sulfuric acid adds to cyclopropane to form 1-propyl hydrogen sulfate as the exclusive initial product, eqn. (1) [1,2]. This is typical of additions of acids to cyclopropane. It is exemplified by the addition of hydrochloric acid, acetic acid, trifluoroacetic acid, and benzene, all of which have been shown to form exclusively 1-propyl products. This simplicity can be missed if long reaction times are used because the 1-propyl products undergo acid-catalyzed rearrangement to 2-propyl products as in the case of the hydrogen sulfate (eqn. (2))

$$H_2SO_4 + c\text{–}C_3H_6 \rightarrow CH_3CH_2CH_2OSO_3H \tag{1}$$

$$CH_3CH_2CH_2OSO_3H \rightarrow (CH_3)_2CHOSO_3H \tag{2}$$

An attractively simple mechanism for such additions is a concerted attack by proton plus a nucleophile, with ring opening and 1,3 addition. Such a mechanism with deuterated acids would form 1-propyl products containing a single deuterium and with the one deuterium entirely on C-3 as shown in eqn. (3)

$$D_2SO_4 + c\text{–}C_3H_6 \rightarrow CH_2DCH_2CH_2OSO_3H \tag{3}$$

An astonishing set of results was that in the addition of deuterosulphuric acid to cyclopropane: (i) the deuterium(s) were distributed over all three carbons of the 1-propyl system, (ii) more than one deuterium could be introduced and (iii) recovered cyclopropane was deuterated [1,3]. These results, first reported by Baird and Aboderin [1,3], firmly established the role of protonated cyclopropanes as intermediates. At the time, earlier experiments had been interpreted as disproving the intermediacy of protonated cyclopropanes in the reaction of propylamine with nitrous acid. These earlier studies were soon re-investigated and the results of three independent groups were found to be seriously in error [4,5]. The history of this has been reviewed [5].

Addition of D_2SO_4 to cyclopropane in 83%, 92%, and 99% D_2SO_4 gave 1-propyl hydrogen sulfate with the deuterium distributed in a random manner among the 7 hydrogens of the 1-propyl group [2]. This places 29% on C-1, 29% on C-2 and 42% on C-3 as shown in eqn. (4). This result was found for experiments in which only one deuterium was introduced (by continuously introducing fresh cyclopropane and removing slightly deuterated cyclopropane) as well as experiments in closed systems in which about 2 deuteriums were introduced [2]. It was true for the formation of 1-propyl acetate from cyclopropane plus 80% deutero sulfuric acid in CH_3COOD, the formation of 1-propyl trifluoroacetate from cyclopropane and CF_3COOD, and the formation of 1-propylbenzene from cyclopropane and deuterated benzene (c-C_6D_6) in the presence of 1–2% D_2O–$AlCl_3$ catalyst [2].

$$D_2SO_4 + c\text{–}C_3H_6 \rightarrow CH_3\text{–}CH_2\text{–}CH_2\text{–}OSO_3D \qquad (4)$$

$$42\% \quad 29\% \quad 29\% \qquad (\%D)$$

Under all of these conditions, the reaction is interpreted as occurring by the sequence

(i) addition of D^+ to cyclopropane forms c-$C_3H_6D^+$;

(ii) the hydrogen and deuterium rapidly scramble, this scrambling being a true rearrangement, since any other explanation implies that the seven hydrogens in c-$C_3H_7^+$ would have to be equivalent as in a regular heptagon. Although such a structure is geometrically possible, it has never been favored.

(iii) the equilibrating c-$C_3H_6D^+$ (structures unknown) either lose H^+ to form deuterated cyclopropane or add DSO_4^- to form deuterated 1-propyl hydrogen sulfate. The H^+ elimination and DSO_4^- addition occur at comparable rates.

Under a more limited set of conditions, the distribution of deuterium has been reported to deviate from random. Addition of deutero sulfuric acid to cyclopropane in 57% deutero sulfuric acid was reported to give ratios of 38% : 17% : 46% for deuterium on C-1, C-2, and C-3 [1]. Similar experiments in 60% and 79% acid using tritium labeling were reported to give ratios of 37% : 26% : 37% and 38% : 26% : 36% respectively [6]. Formation of 1-chloropropane by addition of hydrochloric acid to cyclopropane was reported to give ratios of 35% : 26% : 39% for deuterium chloride in the presence of 15% iron(III)chloride catalyst [2] and ratios of 38% : 19% : 43% for tritium chloride in the presence of zinc(II) chloride catalyst [7].

These deviations from random distribution have been interpreted as favoring edge-protonated structures for c-$C_3H_7^+$ intermediates [1,4–7]. The main feature of these arguments is that if products arose from corner-protonated cyclopropane, as shown in the following diagram, deuterium would appear first at C-3 and then equally at C-1 and C-2.

The alternative, products arising from edge-protonated cyclopropanes, accounts for the "apparent" sequential introduction of deuterium first into C-3, then into C-1, and finally into C-2 as shown below. All of these arguments presume that kinetic isotope effects are of the secondary type and can be neglected.

$CH_2D^{\oplus}$ / $CH_2\text{---}CH_2$ → $CH_3^{\oplus}$ / $CH_2\text{---}CHD$ corner-protonated cyclopropane

$\downarrow Y^{\ominus}$ $\downarrow Y^{\ominus}$

$CH_2DCH_2CH_2Y$ CH_3CHDCH_2Y + CH_3CH_2CHDY

$CH_2\text{---}D^{\oplus}$ / $CH_2\text{—}CH_2$ (1) → CH_2 / $CH_2\text{----}CHD$ / $H^{\oplus}$ (2) → $H^{\oplus}\text{---}CH_2$ / $CH_2\text{—}CHD$ edge-protonated cyclopropane

$\downarrow Y^{\ominus}$ $\downarrow Y^{\ominus}$ $\downarrow Y^{\ominus}$

$CH_2DCH_2CH_2Y$ $CH_2DCH_2CH_2Y$ + CH_3CH_2CHDY CH_3CHDCH_2Y

It does not appear to this reviewer that the deuterium was sequentially introduced. For sequential introduction, the ratio of deuterium (or tritium) label on C-3 to C-1 would always have to be greater than the random value of 1.5 (42:29). Actually, it ranges from 1.0 to 1.2 in the data reported. This would appear to remove the basis for favoring edge-protonated structures.

Since this conclusion is counter to several earlier papers, the statistical factors will be explained in detail. If (*1*) and (*2*) are completely equilibrated, the ratio of (*2*) to (*1*) would be 4 because there are four positions for deuterium on (*2*) and only one on (*1*). An additional statistical factor is that deuterium on C-3 can form in two equivalent ways from (*1*). The total deuterium on C-3 would be 2×1 (from (*1*)) plus 1×4 (from (*2*)) for a total of six. The deuterium on C-1 would be 1×4 (from (*2*)). The ratio would be 6:4 or 1.5. Any partial equilibration of (*1*) → (*2*) would increase the ratio above 1.5, the limit being infinite when all the product arises from (*1*). In other words, the limiting ratio for complete equilibration is 3:2 because there are three possible positions for the deuterium on C-3 and only two positions on C-1.

In fact, it is hard to imagine a mechanism that would not give 3:2 as the limiting ratio of deuterium on C-3 relative to C-1. Statistical arguments, entirely analogous to those above, show that the ratio of 3:2 is expected for corner protonated structures as well as the less likely structures such as regular heptagons of hydrogen or protonation on top of the cyclopropane ring. The ultimate conclusion is that the reports of 1.0–1.3 ratios either have a greater experimental error than previously supposed or that kinetic and equilibrium isotope effects are significant. This point was missed in several papers including a review, but then the field of carbonium ions has been a rocky road.

There are other reasons for de-emphasizing the reports of non-random distribution: (i) the deviations were not large and bordered on experimental error; (ii) the same degree of incomplete scrambling was reported in 60 and 79% sulfuric acid [6], which is unlikely in view of complete scrambling in 83% sulfuric acid [2]; (iii) the non-random values for the addition of cyclopropane to labeled sulfuric acid were obtained on 1-propanol, isolated by a long and arduous procedure.

Recent molecular orbital calculations on protonated cyclopropane have been in partial disagreement. One claimed that edge protonation was most stable and the other claimed that a structure between edge and corner protonation was most stable [8,8a]. Even if there were better agreement, such calculations do not take into account interaction energies with solvent and may not be relevant to the structure in solution, particularly in such delicately balanced situations as $c\text{-}C_3H_7^+$.

In summary, the deuterium studies were the key to proving the existence of $c\text{-}C_3H_7^+$ with its rapidly scrambling hydrogen. However, reports of a non-random distribution of deuterium (and tritium) label are open to question and structural assignments based on such reports are thus not convincing. Despite the skepticism regarding the experimental reports of non-random deuterium distribution, the interpretations by Baird and Aboderin were ingenious and illustrate an elegant potential technique in deuterium (or tritium) labeling.

The Baird and Aboderin type of argument was used recently in connection with studies on the trifluoroacetolysis of 1-propyl-1-^{14}C-mercuric perchlorate [8b,8c]. The result, incomplete scrambling with more carbon-14 on C-3 than C-2, was regarded as indicating that the product arose from edge-protonated cyclopropane prior to the formation of corner-protonated cyclopropanes. This conclusion seems marred only by the presence of mercury(II). This is well known to interact strongly with cyclopropanes and alkenes and may be involved in the mechanism.

The deuterium distributions were determined from n.m.r. band areas and a few comments on the technique may be helpful. Most n.m.r. spectra are taken under conditions of optimum resolution and wide sweep widths so that the spectra consist of sharp well-resolved spikes. Such spectra are not adapted to band area measurements since sharp spikes encompass little area and (in our experience) pen lag on sharp spikes can introduce up to 30% errors in band areas.

Two procedures have been used to optimize the accuracy of band area measurements. One was to use the smallest sweep width available (50 csec^{-1}). Although this broadened each band and allowed about a 100-fold increase in area to be fitted on the recording paper, it had the disadvantage that only small portions of the spectrum could be recorded on each sweep. This meant that several sweeps were necessary at different ranges of field strengths.

A second method was possible with the 1-propyl derivatives where the bands of hydrogens on C-1, C-2, and C-3 were well separated. The spectra were recorded on stationary samples with the sample spinner turned off. This converted each set of bands to broad envelopes. Again, about a 100-fold increase in area could be fitted onto the recording paper and pen lag problems were eliminated. These methods allowed band areas to be determined to 1% with a corresponding accuracy in hydrogen counts.

There are other sources of error in n.m.r. band area measurements such as saturation of field and Overhauser effects. Their nature and the appropriate precautions are described in books on n.m.r. spectroscopy.

C. Addition of Deuterated Acids to Substituted Cyclopropanes

There is growing evidence that with simple alkyl systems, protonated cyclopropanes are more stable than primary carbonium ions but less stable than secondary carbonium ions [2,9,10]. Thus, any alkyl or polyalkyl cyclopropane could add a deuteron to produce the secondary or tertiary alkyl cation directly and by-pass the protonated cyclopropane. The deuterium scrambling characteristic of intermediate protonated cyclopropanes would not be observed under such circumstances.

Accordingly, addition of deuterium chloride to methylcyclopropane gave exclusively 1,3 addition as shown in eqn. (5). No catalyst was necessary. In view of this result, it does not seem likely that protonated cyclopropane intermediates will play a significant role in the addition of acids to alkyl and polyalkyl cyclopropanes.

$$\text{(methylcyclopropane)} + DCl \longrightarrow CH_3CHClCH_2CH_2D \qquad (5)$$

Addition of deuterated acids to cyclopropanecarboxylic acids was investigated with the expectation that the carboxyl group might reduce the stability of secondary alkyl cation intermediates below that of protonated cyclopropane intermediates and allow the reappearance of the latter [11]. This was found for addition of deutero sulfuric acid to cyclopropanecarboxylic acid and 1-methylcyclopropanecarboxylic acid, but not for several other cyclopropanecarboxylic acids [11].

Addition of 98% deutero sulfuric acid to cyclopropanecarboxylic acid at 100° produced 24% butyrolactone and 76% of the hydrogen sulfate of 3-hydroxyisobutyric acid, eqn. (6). Analysis of the n.m.r. spectra of the reaction mixture showed that the added deuterium was statistically distributed in both products. The exact number of deuteriums introduced was not determined, but it must be at least one from the stoichiometry of eqn. (6).

$$\text{(cyclopropyl)}-COOH + D_2SO_4 \longrightarrow \underset{24\%}{\text{(butyrolactone)}} + \underset{76\%}{HO_3SOCH_2CH(CH_3)COOH} \qquad (6)$$

Addition of 98% deutero sulfuric acid to 1-methylcyclopropanecarboxylic acid at 100° gave 68% tiglic acid (*cis*-2-methyl-2-butenoic acid) and 32% 2-butanone, eqn. (7). The tiglic acid contained a total of 1.84 deuteriums. This is the multiple deuteration characteristic of an initial reversible addition of deuteron to form a protonated cyclopropane structure. Of the 1.84 deuteriums, 0.34 were on C-3 and the remaining 1.50

were probably on C-4, though the n.m.r. spectra did not permit a distinction between deuteriums on the C-4 methyl and the 2-methyl substituent. The deuterium introduction and distribution on the 2-butanone was not informative because of the rapid α-deuteration of 2-butanone under the reaction conditions.

(7)

In the two above examples, the deuterium scrambling and multiple introduction of deuterium indicated protonated cyclopropane intermediates.

D. Equilibration of Cyclohexenyl and Cyclopentenyl Cations

Generally, cyclohexenyl cations rearrange to form cyclopentenyl cations [12]. Both base-catalyzed and base-invariant mechanisms were demonstrated by showing that the ratios of cyclopentenyl cations formed underwent wide variation with acidity [13]. Sorensen discovered an example where the cyclohexenyl cation was the dominant form at equilibrium and went on to determine equilibrium constants for several interconverting cyclohexenyl–cyclopentenyl cation systems [14]. These interesting rearrangements clearly require intricate structural reorganization and so provided an opportunity to use deuterium labeling to study their course [13,15].

Equation (8) is typical of these rearrangements and is the one that has been studied by deuterium labeling. In one type of experiment [13], undeuterated 1,3,4,4-tetramethylcyclohexenyl cation (*3*) is generated in 96% deutero sulfuric acid. Based on studies of stable non-rearranging alkenyl cations [16], rearrangement of (*3*) to (*4*) and H–D exchange is believed to occur only on positions α to the allyl cation system. Not only do both (*3*) and (*4*) exchange, but intermediates in the rearrangement may also exchange and it is these added exchanges or lack of exchanges which are the most enlightening.

(8)

Table 1 contains the original data. The highlights are (i) the C-2 areas of (*3*) plus (*4*) remained constant which strongly implies that they were the same hydrogen and did not exchange (as expected for the vinyl hydrogen) [16]; (ii) the ratios of areas of three of the bands of (*3*) (gem dimethyls at C-4, C-5 and C-2) remained constant showing that they were not exchanging and that their disappearance rate repre-

TABLE 1

N.M.R. BAND AREAS FOR *(3)* → *(4)* IN 96% DEUTEROSULPHURIC ACID AT 35°

Position	Area under peaks at					
	4 min	12 min	19 min	64 min	104 min	297 min
Cation *(3)*[a]						
C-2	17	12	11	7.5	5	1
Me on C-1 and C-3	68	43	22	0	0	0
C-5	30	27	22	15	10	0
gem $(Me)_2$ on C-4	96	70	~55	*b*	*c*	0
Cation *(4)*						
C-2	0	5	6	11	12.5	16
C-4 and C-5	0	12	20	29	27	28
Me on C-3	0	0	16	20	16	16
tert-Butyl	0	28	~40	*b*	*c*	80

[a] The hydrogens at C-6 are not included.
[b] The sum was 95.
[c] The sum was 90.

sented the rate of disappearance of (*3*); (iii) the methyls at C-1 and C-3 of (*3*) disappeared faster than the other bands of (*3*) showing that H–D exchange is taking place (as expected for α-H's) and at a rate faster than rearrangement; (iv) all the hydrogens of (*4*) did not exchange at significant rates relative to the rearrangement rate.

These results are in accord with the path shown in eqn. (9) with the parentheses indicating partial deuteration. This path was particularly attractive because the gem dimethyl at C-4 of (*3*) was shown to facilitate greatly the rearrangement [12,13,16] in agreement with the formation of a tertiary alkyl cation (rather than the secondary or primary cation if the methyls were successively removed) in the intermediate (*5*). Also, the rearrangement could be formulated as the result of two successive 1,2 alkyl shifts of the classic type.

(3) —96% D_2SO_4→ (3d) → [(5)] → (4d) (9)

In a test of eqn. (9), Sorensen and Ranganayakulu [15] introduced (*6*) into DO_3SF at –80° and into 96% and 101% deutero sulfuric acid at 0°. The less stable cyclohexenyl cation (*3*) was predominantly formed (92%) relative to the 8% yield of the ultimate ground state, the cyclopentenyl cation (*4*). This allowed deuterium distribution in (*3*) to be examined with the results shown in eqn. (10). Only (*3a*) would have been expected from the mechanism of eqn. (9). The appearance of (*3b*) (and in a predominant amount) led to the proposal of a cyclopropylcarbinyl cation intermediate, (*7*).

The work in the paper [15] was extensive, but before eqn. (10) is accepted and eqn. (9) rejected, it is well to point out one discrepancy and a possible source of error. Equation (10) scrambles the gem dimethyls and the C-3 methyl before rearrangement so that in the experiment recorded in Table 1, the *tert*-butyl group in (*4*) should have had much more than the 3 deuteriums observed. The possible source of error is that the 1.35 δ band of (*3*) (methyl on C-4) and the 1.43 δ band of (*4*) (*tert*-butyl group) cannot be separated well enough (in our experience) for the accurate area measurements upon which eqn. (10) critically depends. Again, the logic of the method is correct and illustrates the potential technique.

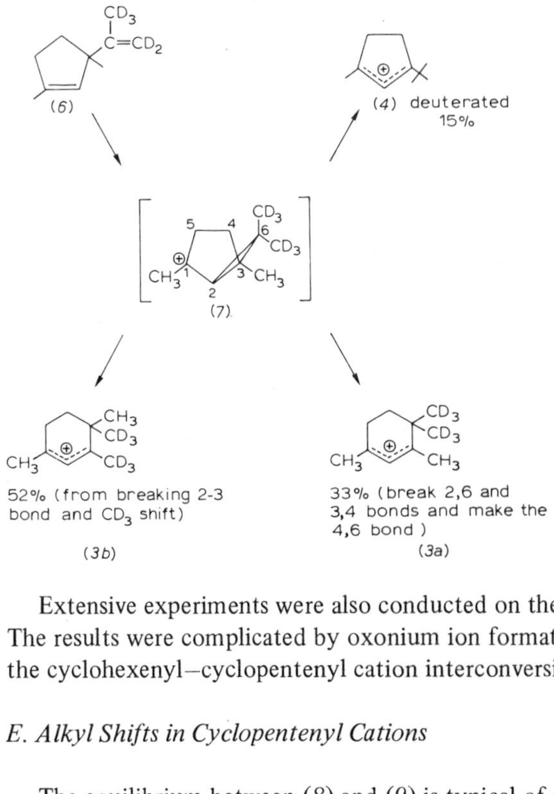

CD_3
$C{=}CD_2$
(*6*)

(*4*) deuterated
15%

CD_3 5 4 6 CD_3 CH_3 1 2 3 CH_3
(*7*)

(10)

CH_3 CD_3 CH_3 CD_3
52% (from breaking 2-3 bond and CD_3 shift)
(*3b*)

CD_3 CD_3 CH_3 CH_3
33% (break 2,6 and 3,4 bonds and make the 4,6 bond)
(*3a*)

Extensive experiments were also conducted on the alcohol corresponding to (*6*). The results were complicated by oxonium ion formation and do not apply directly to the cyclohexenyl–cyclopentenyl cation interconversion.

E. Alkyl Shifts in Cyclopentenyl Cations

The equilibrium between (*8*) and (*9*) is typical of a wide variety of 1,3,4-trialkylcyclopentenyl cations [16,17]. When conducted in 85% deutero sulfuric acid, the α hydrogens of (*8*) deuterate prior to rearrangement to form (*8d*) as shown in eqn. (11). As the rearrangement proceeds, all remaining positions deuterate except the two methyls on the isopropyl group. The disappearance of C-2 hydrogen on (*8*) is thus a measure of the rate of isomerization.

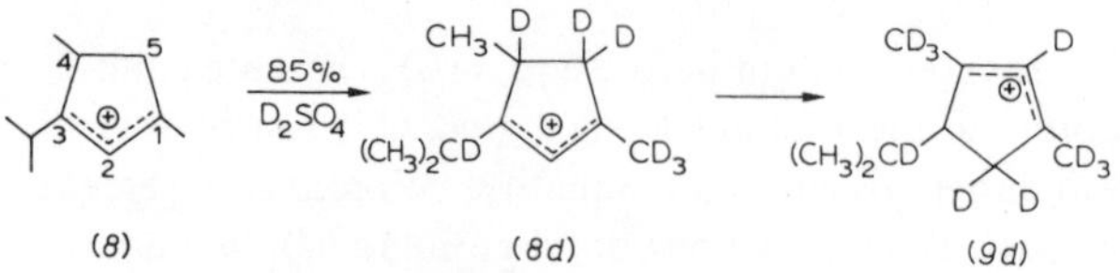

(11)

(*8*) (*8d*) (*9d*)

The conversion of (*8d*) to (*9d*) could occur by elimination–addition of deuteron; by two successive internal 1,2-hydride shifts; or by an internal 1,3-hydride shift. The hydride shifts would be insensitive to acid concentration whereas the elimination–addition would be sensitive to the large changes in the activity of water in 85–96% sulfuric acid [13,18]. It is thus very possible for the mechanism to change as the acid concentration is varied.

Work described under the rearrangement of dienyl cations to cyclopentenyl cations suggests that in 95% deutero sulfuric acid the internal hydride transfer path(s) prevail [17].

The isomerization shown in eqn. (12) is of the same type as eqn. (11) and is curious because product and reactant are identical in the absence of deuterium. In deutero sulfuric acid the α hydrogens deuterate first. The C-2 hydrogen and the C-4 methyl group deuterate simultaneously and much slower. The latter deuteration is indicative of the rearrangement shown in eqn. (12).

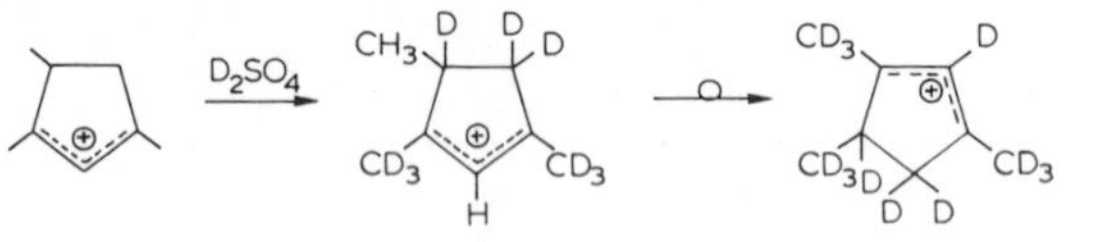

(12)

F. The Conversion of Dienyl Cations to Cyclopentenyl Cations

This rearrangement was first reported for dienyl cations [19] and was soon generalized to trienyl cations [20–23].

Introduction of triene (*10*) into 96% sulfuric acid gave a non-equilibrium mixture of two cyclopentenyl cations, eqn. (13) [17]. When the rearrangement was conducted in 96% deutero sulfuric acid, the two cyclopentenyl cations were initially formed in an undeuterated condition at C-5, eqn. (14).

4,7 - dimethyl - 1,3,5 - octatriene (*10*) (isomer mixture) $\xrightarrow[H_2SO_4]{96\%}$ (*11*) ⟶

55% (*12*) + 44% (*13*) (13)

(26% (*10*) and 74% (*11*) at equilibrium)

(*10*) $\xrightarrow{D_2SO_4}$ (*12*) + (*13*) (14)

This lack of deuteration at C-5 has an important consequence. If the dienyl cation (*11*) cyclized directly to a cyclopentenyl cation, it would form (*14*) (eqn. (15)). Cation (*14*) will stabilize by isomerization into (*12*) and (*13*). If this occurred by elimination of proton and addition of deuteron, C-5 would be deuterated. The fact that it is not (eqn. (14)) indicates that isomerization of (*14*) to (*12*) and (*13*) occurs by internal hydride transfer, by either 1, 3 or successive 1, 2 transfers [17]. This speculation rests on the assumption that (*14*) is the initial product of cyclization. Also, it may only hold for cyclizations in the stronger acids where elimination of proton (or deuteron) is retarded.

(*11*) in s-*cis*, s-*cis* form ⟶ (*14*) ⟶ (*12*) + (*13*) (15)

G. *Isomerization of Saturated Alkanes*

Over three decades ago it was shown that saturated alkanes containing tertiary hydrogens would exchange deuterium for hydrogen in deutero sulfuric acid [24,25]. Even the simplest member, isobutane, underwent exchange [26]. In 1951, a group at Shell Development Company began a detailed investigation of such exchanges [27,28]. Their first paper established that butane does not exchange and that isobutane (C_4H_{10}) exchanges up to a limit of 9 hydrogens, the 9 methyl hydrogens, but the tenth and unique tertiary hydrogen did not exchange. The interpretation of this in terms of intermediate *tert*-butyl cations which rapidly exchange α-hydrogens by reversible conversion to isobutene, the generation of a *tert*-butyl cation from isobutane by intermolecular hydride transfer, and the absence of interchange between primary and tertiary hydrogen, are now classic history.

A total of 15 simple alkanes were investigated with emphasis on 2-methylpentane. This alkane clearly isomerizes to 3-methylpentane in 95% sulfuric acid at 25°. Short residence times were used to insure that each isomerized molecule spent only one residence as an alkyl cation. The deuteration of 2-methylpentane in 95% deutero sulfuric acid was studied as well as the loss of deuterium from 2-methylpentane-2-*d*, -4-*d*, and -5-*d* in 95% sulfuric acid.

Some of the highlights of the extensive data on 2-methylpentane were that 3-methylpentane from 2-methylpentane-4-*d* always lost all the deuterium label, from which it was inferred that ten hydrogens (all the α-hydrogens in the 2-methyl-2-pentyl and 3-methyl-3-pentyl cations) had exchanged. About half of the 3-methylpentane from 2-methylpentane-5-*d* had lost the deuterium label indicating that movement of the carbonium ion center is sequential down the chain. This was confirmed by showing that undeuterated 2-methylpentane formed about twice as fast from 2-methylpentane-4-*d* as from 2-methylpentane-5-*d*. This is strong evidence that two successive 1,2 hydride shifts effect 1,3 hydride transfer rather than a direct 1,3 hydride shift.

The data on other alkanes also led to positive conclusions. The appearance of C_5HD_{11} from 2-methylbutane and C_7HD_{15} from 2,2,3-trimethylbutane showed that 1,2 methyl shifts were rapid and occurred in a single ionic residence time. The appearance of C_6HD_{13} from 2,3-dimethylbutane showed that the isothermic 1,2 hydride shift between the two *tert*-alkyl cations was rapid and occurred in a single ionic residence time.

All of these facts were to be abundantly confirmed a decade later by n.m.r. studies on stable alkyl cations. Yet, at the time, these studies were a landmark and it is remarkable that they were based entirely on relative intensities of mass spectra bands.

III. DOUBLE (DEUTERIUM AND CARBON-13) LABELING IN STABLE ALKYL CATIONS

The discovery that simple secondary and tertiary alkyl cations can be obtained as stable species in SbF_5-HO_3SF solutions ranks as one of the major recent advances in organic chemistry. The pioneering work was due to Olah followed shortly by work of Mackor and Brouwer. The most detailed studies have been by Saunders and it is this work that has utilized deuterium labeling.

The simplest alkyl cation for which n.m.r. spectra have been recorded is the 2-propyl or isopropyl cation. From the n.m.r. line broadening at low temperatures, it was known that the two kinds of hydrogen interchanged [9,10] and the two kinds of carbon also interchanged [9,29]. This set the background for an ingenious double labeling experiment which was designed to compare the rates of hydrogen exchange with the rate of carbon exchange.

An equimolar mixture of 1,1,1-trideutero-2-propyl cation and 2-^{13}C-2-propyl cation was prepared at −88°. The relative areas of methylene and methine peaks as well as the carbon-13 satellite bands could all be measured simultaneously on the same sample. The four deuterium-labeled isomers equilibrated and the two carbon-13 labeled isomers equilibrated. Hydrogen exchange was found to be slightly faster than carbon exchange by a factor of 1.5 ± 0.5. This was interpreted to mean that the 2-propyl cation was slowly converting to the 1-propyl cation which then either rapidly returned to the 2-propyl cation (eqn. (16)), effecting hydrogen exchange, or rapidly cyclized to a corner-protonated cyclopropane. Reopening of the corner-protonated cyclopropane effected the interchange of C-1 and C-2 and C-3 as shown in eqn. (17).

$$CH_3-CH^{\oplus}-CD_3 \longrightarrow CH_3-CHD-CD_2^{\oplus} \longrightarrow CH_3-CD^{\oplus}-CHD_2 \quad \text{etc} \qquad (16)$$

$$CH_3-CH^{\oplus}-CH_3 \longrightarrow CH_3-CH_2-CH_2^{\oplus} \longrightarrow \text{corner-protonated } CH_3\cdots CH_2\cdots CH_2 \longrightarrow$$

$$\overset{CH_2^{\oplus}}{CH_3 - CH_2} \longrightarrow \overset{CH_3}{CH_2 - CH^{\oplus}} \qquad (17)$$

The reaction mixtures were periodically quenched with methylcyclopentane which converted 2-propyl cation to propane. In principle, mass spectra analyses of these labeled propanes could show whether eqns. (16) and (17) adequately accounted for the relative rates of successive isomerizations or whether edge-protonated species were needed (eqn. (18)). Preliminary results were interpreted as suggesting that eqns. (16) and (17) are adequate [10].

$$^{\oplus}CH_3\cdots CH_2\cdots CH_2 \longrightarrow CH_2\cdots CH_2\cdots CH_2 \;(\text{edge-bridged by } H^{\oplus}) \longrightarrow CH_2\cdots CH_2\cdots CH_3^{\oplus} \qquad (18)$$

Saunders' novel use of simultaneous labeling was also applied to the methylcyclopentyl cation [30]. 1-Chloro-1-methylcyclopentane was prepared from equal amounts of $^{13}CH_3I$ (60% carbon-13) and CD_3I (95% deuterium). The corresponding methylcyclopentyl cations were formed in SbF_5-SO_2ClF. At about $-25°$, the carbon-13 methyl side bands rapidly decreased indicating equilibration of the carbon-13 of the methyl group with ring carbons. Simultaneously, deuterium moved out of the methyl group as indicated by the growth of the methyl n.m.r. band indicating H–D scrambling. The rates at $-33°$ were very close being $5 \pm 2 \times 10^{-4}$ sec^{-1} for methyl carbon mixing with ring carbons, and $3 \pm 2 \times 10^{-4}$ sec^{-1} for methyl deuteriums mixing with ring hydrogens [30].

The conclusion that eqns. (16) and (17), but not eqn. (18), is required [30] does not agree with the results of the addition of deuterated acids to cyclopropane. The latter showed that hydrogen scrambling was complete with each formation of $c\text{-}C_3H_7^+$ in 83–99% sulfuric acid and the same would be anticipated in the more acidic SbF_5-SO_2ClF system. The overview of the simultaneous labeling experiments on 2-propyl and methylcyclopentyl cations is that carbon and hydrogen mixing have very close rates and are probably the result of a single process. Until more details are available on the simultaneous labeling, it is attractive to describe the mixing process as hydrogen rearrangement in $c\text{-}C_3H_7^+$ (protonated cyclopropane) which scrambles both hydrogen and carbon simultaneously.

IV. THE CLASSICAL METHOD OF DEUTERIUM LABELING

A. Method

Deuterium labeling in its simplest form consists of introducing a deuterium label into a reactant and observing its position in the product. In carbonium chemistry, this method is largely restricted to cases where the carbonium ion is a fleeting intermediate and is trapped before the label is lost by exchange with solvent.

Keating and Skell [31] have separated methods for generating fleeting carbonium ion intermediates into two categories. In one, typified by solvolysis, the carbonium ion is strongly associated with counter ion or solvent and the distinction of this path with S_N2 displacement mechanisms becomes one of degree rather than kind. Carbonium ions formed in this way are less liable to rearrange and have been termed encumbered [31,32].

The contrasting category is comprised of reactions such as nitrous acid deamination of amines, deoxideation of alkoxides, and electrolytic oxidation of carboxylate anions where the carbonium ion is produced in a highly exothermic step. Such carbonium ions are more likely to rearrange and have been termed free or less aptly, hot [31].

The differences were originally rationalized in terms of carbonium ions in vibrational ground states contrasted with those in vibrationally excited states (hot). This view was championed by several noted chemists. It never seemed palatable to this reviewer because (i) nuclear movement between vibration states is much smaller than 1,2 shifts in carbonium ions and (ii) rates of vibrational transitions (10^{-12} to 10^{-13} sec half-lives) are faster than the fastest chemical reactions (10^{-10} sec) found by Eigen's temperature-jump methods.

Recently, the differences have been ascribed to carbonium ions in which solvent is oriented in the most stable (relaxed) arrangement in contrast to carbonium ions in which solvent is in less stable arrangements [31,32a–d]. These less stable arrangements could be energy minimums in which case the metastable arrangement might have an appreciable lifetime. The statement of Collins et al. [32a] is that differences are "...explained by counterion control in differently oriented ion pairs". The strongest line of experimental evidence for this view is the numerous examples [31,32a–d] in which carbonium ions from deamination, deoxideation, or electrolytic oxidation undergo several rearrangements and the resulting ion still shows more propensity to rearrange than the same ion produced by solvolysis. Surely, any excess vibrational energy would have been dissipated in several 1,2 shifts [32a].

The deuterium labeling experiments have been used to detect internal hydride shifts. Before discussing these, it will be helpful to summarize five facts about such shifts. (1) Exothermic and isothermic 1,2-hydride shifts have such fast rates that the resulting hybrid n.m.r. bands cannot be resolved at $-85°$ (ref. 33) and $-110°$ (refs. 34,35). (2) 1,3-Hydride transfers can also be extremely fast and proceed in a single step through an edge-protonated cyclopropane geometry [36]. It is not known whether this geometry is an intermediate or a transition state. (3) Closure of primary alkyl cations to protonated cyclopropanes followed by hydrogen scrambling and reopening can effect what appears to be 1,3-hydride transfer. (4) Hydride transfers more distant than 1,3 are possible but so far have been observed only in cyclic systems where the geometry is favorable for transannular hydride shifts [37–41]. (5) Some apparent hydride shifts are intermolecular processes and not internal rearrangements [42,43].

Classical deuterium labeling experiments have demonstrated transannular hydride shifts in solvolyses of medium sized (8–11) rings [37,38]. The subject has been recently reviewed [26–28,39] and will not be considered here.

B. Solvolyses

Dating from the innovation of $S_{N_1}-S_{N_2}$ terminology, it has been recognized that the transition state in solvolysis may resemble a carbonium ion in behavior. The closer the approach to a carbonium ion in structure, the more the appearance of rearrangements. This closeness of approach is dominantly governed by the nature of the alkyl group undergoing substitution (tertiary > secondary > primary). However, other factors can increase the carbonium character such as a better (more stable) leaving group and conducting the reaction in media of high polarity with a scarcity of nucleophilic species (high acidity).

Solvolysis of 1-propyl tosylate in acetic acid gave < 1% rearrangement to 2-propyl acetate and solvolysis in formic acid gave 1–2% 2-propyl formate [44]. Similarly, 1-butyl-1-*d*-4-nitrobenzenesulfonate gave no detectable rearrangement when solvolyzed under various conditions [45].

Three changes lead to the appearance of rearrangement. One is to use a system where the potential 1,2 shift (of hydrogen or alkyl) is the more exothermic conversion of a primary alkyl to a tertiary alkyl rather than a primary to a secondary. This is exemplified in eqn. (19) [46].

$$(CH_3)_2CHCH_2OSO_3^-Na^+ \xrightarrow[150^\circ]{H_2O,\ OH^-} \underset{87\%}{(CH_3)_2CHCH_2OH} + \underset{13\%}{(CH_3)_3COH} \qquad (19)$$

A second is to start with the more stable secondary alkyl system. For example, solvolysis of 3-methyl-2-butyl tosylate gives 97% rearrangement products [47]. By going to benzyl systems and other more stable carbonium ions, many examples could be found. In all of the above, rearrangement is evident without labeling.

The third change is to use more acidic systems. Here the need for deuterium labeling arises in order to detect or eliminate protonated cyclopropane intermediates (which scramble hydrogens) and alkene intermediates (which would lead to exchange of hydrogen with solvent). Three examples have been reported.

Solvolysis of $CH_3CH_2CD_2OTs$ in trifluoroacetic acid at 100° gave 84% 2-propyl trifluoroacetate and 16% of 1-propyl trifluoroacetate. No deuterium scrambling (n.m.r.) was observed for either product or either reactant. It was concluded that the 2-propyl ester is formed by a direct 1,2 hydride shift [48]. The result is in direct conflict with results reported from studies using ^{14}C labeling [49]. Scrambling of the label was reported to be 17% under identical conditions. However, in the presence of sodium trifluoroacetate, the extent of scrambling was only 2%. We can only speculate that either the deuterium experiments contained adventitious trifluoroacetate anions or that the reliability of labeling experiments continues to be fraught with pitfalls where minor products are concerned.

A similar result was found in the quantitative isomerization of 1-propyl hydrogen sulfate to 2-propyl hydrogen sulfate [2]. 1-Propyl hydrogen sulfate can be generated by addition of 1-propanol to sulfuric acid [50]. The half-life at 25° is 30 h in 95% sulfuric acid and less than one minute in 20% oleum. No rearrangement can be detected in 24 h in 80% sulfuric acid [2]. Conducting the rearrangement in 95% deutero sulfuric acid in deuterium oxide and monitoring the reaction mixture by n.m.r. showed that the 1-propyl moiety did not exchange hydrogen for deuterium before rearrangement and that the 2-propyl hydrogen sulfate still retained the hydrogen on C-2 after rearrangement (eqn. (20)). The complete deuteration of the two methyl groups, eqn. (20), is not significant because 2-propyl hydrogen sulfate (from 2-propanol) also undergoes complete H–D exchange on the methyl groups in less than 2 min at 25° (ref. 2).

$$CH_3CH_2CH_2OSO_3H \xrightarrow[25^\circ]{95\%\ D_2SO_4} (CD_3)_2CHOSO_3D \qquad (20)$$

These results show that the 1-propyl moiety, presumably as the cation, can rearrange to the 2-propyl cation without proceeding through the protonated cyclopropane. Intermediacy of the latter would have led to hydrogen scrambling and introduction of deuterium on C-2. The rapid deuteration of 2-propyl hydrogen sulfate at C-1 and C-3 indicates rapid reversible equilibrium with propene even though propene cannot be detected by n.m.r.

An interesting use of mass spectrometry was made in examining the isomerization of 1-bromopropane to 2-bromopropane, catalyzed by aluminium(III) bromide. Both $CH_3CH_2CD_2Br$ and $CH_3CD_2CH_2Br$ were used and both recovered 1-bromopropane and product 2-bromopropane were examined at spaced intervals. For the mass spectra, the bromides were hydrolyzed to the alcohols and the alcohols converted to the trimethylsilyl ethers [51].

For 1-bromopropane, the (P – Me) and (P – Et) bands were examined. The (P – Et) band is readily identified as arising from $^+CH_2OSiMe_3$ because $CH_3CH_2CD_2OH$ gave 99% d_2 for this band and $CH_3CD_2CH_2OH$ gave ∼ 1% d_2 [50]. The (P – Me) band was not clearly identified from the data and could have arisen by loss of methyl from the 1-propyl group or the trimethylsilyl group.

The (P – Me) band had 98–99% d_2 from the initial reactant, from the $CH_3CH_2CD_2Br$ recovered from 80% rearrangement, and from the $CH_3CD_2CH_2Br$ recovered from 65% rearrangement. This along with the absence of d_3 bands shows that no intermolecular exchanges occurred and served to eliminate the possibilities of intermolecular hydride transfers and reversible eliminations of hydrogen bromide. Fortunately, the conclusion is valid regardless of the identification of the (P – Me) band.

The (P – Et) band for either the recovered $CH_3CH_2CD_2Br$ or the recovered $CH_3CD_2CH_2Br$ exhibited leakage of deuterium to other positions. This leakage was slow and the percentage d_2 (for $CH_3CH_2CD_2Br$) had fallen only from 99% to 78% after 80% rearrangement. Correspondingly, the percentage d_0 in recovered $CH_3CD_2CH_2Br$ had fallen only from 99% to 91% after 65% rearrangement. Nevertheless, this slow scrambling does indicate a reversible conversion of 1-bromopropane to protonated cyclopropane, hydrogen scrambling, and a return to 1-bromopropane [51].

Of course, as scrambled 1-bromopropane accumulates, 2-bromopropane would increasingly form from such 1-bromopropane and would exhibit deuterium scrambling itself. Fortunately, the rearrangement rate is somewhat faster than deuterium scrambling so that 2-bromopropane was isolated with no label leakage and the conclusion could be drawn that the rearrangement occurred by an internal 1,2-hydride shift.

C. Deoxideation of Alkoxides

The reaction of an alkoxide with chloroform or bromoform is believed to proceed through a carbonium ion as shown in eqn. (21). The discovery and development of this reaction is due to Skell and co-workers [31]. Although these carbonium ions are produced in strongly alkaline media where their lifetime is of the order of 10^{-10} sec, they show great propensity for rearrangement. This has been attributed to their being formed in an exothermic process in which they arise in a free (non-solvated) state [31].

$$CHCl_3 + \text{base} \rightarrow CCl_2$$

$$RO^- + CCl_2 \rightarrow ROCCl_2^- \xrightarrow{-Cl^-} ROCCl \rightarrow R^+ + CO + Cl^- \qquad (21)$$

Bromoform was added slowly to a refluxing mixture of $CH_3CH_2CD_2OH$ and 50% aqueous potassium hydroxide. The cyclopropane product represents about 1% of the total product [31] (10% of the C_3H_6 fraction). The cyclopropane consisted of 94% of the expected $C_3H_4D_2$ but 5–6% of the unexpected C_3H_5D. This C_3H_5D would now be interpreted as arising from hydrogen scrambling in an intermediate dideuterated protonated cyclopropane, c-$C_3H_5D_2^+$ followed by elimination of either proton or deuteron.

Although the 6% of c-C_3H_5D superficially suggests little scrambling, this is not the case. This percentage would arise if the cyclopropane arose from completely scrambled

$c\text{-}C_3H_5D_2^+$ combined with the statistical factor of 2:7 and a $k_H{:}k_D$ ratio of 5 for deprotonation. Since 5 is within the range of primary kinetic isotope effects, it is concluded that scrambling is extensive and conceivably complete.

Deoxideation of 2-butanol-3,3-d_2 showed that 1,2 hydride shifts competed with deprotonation despite the strongly alkaline conditions [31]. From a consideration of the reversible nature of the hydride shift and statistical effects (but neglecting kinetic isotope effects), it was estimated that the ratio of rates of proton loss to 1,2 hydride shift was approximately 3. The experimental data are summarized in eqn. (22).

$$CH_3CD_2CHOHCH_3 \rightarrow CH_3CD_2CH^+CH_3 \rightarrow \underset{89}{CH_3CD=CHCH_3} + \underset{11}{CH_3CD=CDCH_3}$$

(plus other products) (22)

Skell [32c], Reichenbacher [52], and Hall [52a] devised a method (using deuterium) for determining the competition between 1,2-hydride shifts in a 2-butyl cation and the loss of H^+ from a 2-butyl cation to form 2-butene. Deoxideation of *dl-erythro*-2-butanol-3-*d* gave *cis*-2-butene containing 0.72 *d* and *trans*-2-butene containing 0.56 *d*. The intermediate $RO^+{=}C$ on loss of CO will directly form the three conformers shown below. Assuming *trans* elimination, the first conformer would lose deuteron to form undeuterated *trans*-2-butene. The second would lose a proton to form *cis*-2-butene-2-*d*. The third cannot directly form 2-butene. The fact that the *trans*-2-butene was partially deuterated and that the *cis*-2-butene was not fully monodeuterated shows that some rotation plus hydride shift (between C-2 and C-3) occurred before elimination to form 2-butene. The fact that the amounts of deuteration of *cis* and *trans*-2-butene were different showed that rotation and hydride shift equilibrations were not complete before 2-butene formation. The complimentary experiments were conducted on *dl-threo*-2-butanol-3-*d*. A more detailed interpretation is contained in the original reference.

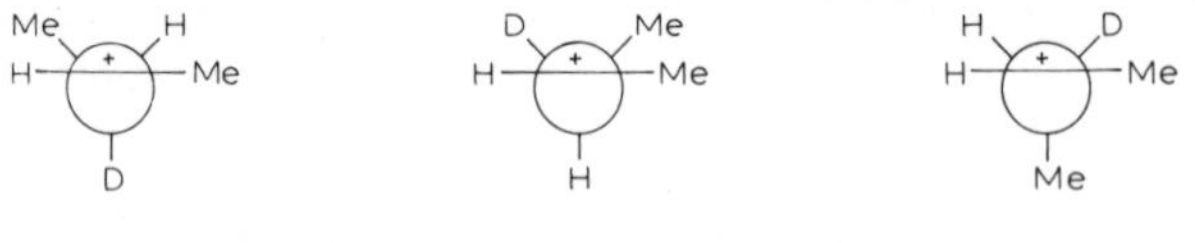

D. Nitrous Acid Deamination of Primary Amines

By showing that deoxideation of alcohols, anodic oxidation of carboxylate anions, and deamination of amines all gave similar ratios of rearranged products, Keating and Skell [31] concluded that free carbonium ions are responsible for the rearrangement products. By emphasizing experiments in nearly aprotic media (where ion pairing and polar aggregation effects are strong), Friedman [53] came to the contrary view that rearrangements were concerted and did not involve carbonium ions. The reader will recognize that it is largely a matter of convenience of description. The Keating–Skell terminology will be used here.

Nitrous acid deamination of amines has been extensively studied using deuterium, tritium, and carbon-13 labeling. However, two circumstances have arisen to cause the work to play a secondary role in establishing the fundamentals of protonated cyclopropane chemistry. One is that label migration in deamination of propylamine is limited to a minor product, 1-propanol (7% yield) [5], so that label scrambling must be conducted on a laboriously purified side product. The second circumstance stemming from the first is that the first three independent labeling studies on the deamination of propylamine contained errors in the data [5]. All these rejected the currently accepted protonated cyclopropane path for label scrambling. Collins has reviewed protonated cyclopropanes and provided a detailed analysis of the history of propylamine deaminations [5]. Keating and Skell in their review on free and encumbered carbonium ions have included some deamination work [31]. Friedman, in reviewing carbonium ion formation from diazonium ions, has covered most deaminations [53].

Results on deamination of variously deuterated propylamines [54–57] are now in accord with the authoritative carbon-14 and tritium labeling experiments [58,59]. Starting with a label on C-1, the yield of 1-propanol with a label on C-2 is 2% and C-3 is 2%. This 4% together with another 2% of 1-propanol (from label that is randomly scrambled but still on C-1) and the 1% yield of cyclopropane arise from a protonated cyclopropane intermediate [58,59]. The cyclopropane is extensively H–D exchanged with solvent [54].

Deamination of deuterium labeled butylamine and isobutylamine gave largely rearranged products but the only movement of the deuterium label was that associated with the 1,2 hydride shift required to convert the 1-butyl to the 2-butyl system and the isobutyl to the *tert*-butyl system [60,61]. These studies were unusually complete and used 1-butylamine which was dideuterated at C-1, C-2, or C-3 and isobutylamine which was dideuterated at C-1. The deuterium distributions in the products were determined by mass spectrometry using the intensity ratios of (P – Me), (P – Et), and (P – Pr) bands.

Indirectly related to labeling in deaminations is the fact that the intermediate diazonium ions can be in equilibrium with the diazoalkane through loss of proton [53]. This could introduce (or remove) label at C-1 prior to rearrangement and is a hazard in studies in this area. In fact, it has been speculated that this was a hidden source of error in the propylamine deaminations [5].

E. Decomposition of N-*Nitrosoamides*

The thermal decomposition of *N*-alkyl-*N*-nitrosoacetamides bears many similarities to the nitrous acid deamination of amines. These two reactions have been compared using deuterium labeling for the case where the alkyl group is isobutyl [62].

The deuterated isobutylamines $(CH_3)_2CHCD_2NH_2$, $(CH_3)_2CDCH_2NH_2$, and $(CD_3)_2CDCH_2NH_2$ were used. In a variety of conditions using protic and aprotic solvents and nitrous acid deamination and nitrosoamide decomposition, about 70% of the isobutylene showed the loss of label in accord with a 1,2 hydride shift and formation via the *tert*-butyl cation.

F. Bicycloalkyl Systems

A surfeit of studies has been conducted on bicyclo[2.2.1]heptyl (norbornyl)systems. Initially these studies were designed to illuminate questions regarding classical and non-classical carbonium ions. It is now apparent that the norbornyl cation is a substituted protonated cyclopropane [63] and thus a special case of the unsubstituted protonated cyclopropane, c-$C_3H_7^+$. The bicycloalkyl geometries provide additional stabilization for the protonated cyclopropane structure in some cases but are otherwise not unique.

Deuterium labeling has played a small role in this area. The rearrangements in eqn. (23) were argued to involve only 6,2 hydride shifts despite their complexity. The positions of deuterium label were essential to the argument [64].

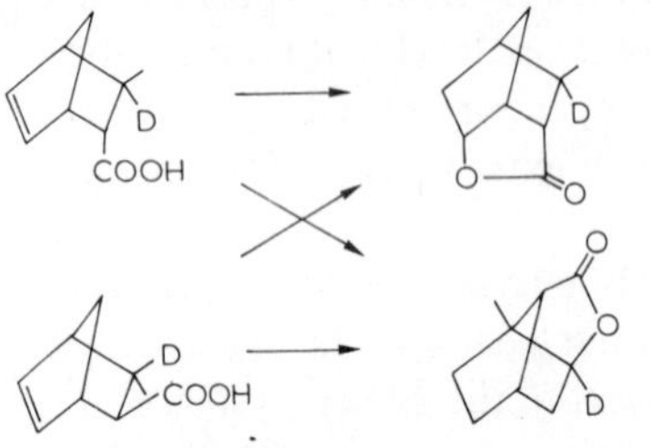

(23)

Collins and co-workers have conducted extensive investigations on nitrous acid deamination of norbornylamines containing phenyl and hydroxy substituents [32a,65,65a]. Labeling with deuterium, tritium and carbon-14 was used. The major conclusion was that carbonium ions derived by rearrangement still retain a greater propensity to rearrange than carbonium ions derived from solvolysis. This was discussed at the beginning of this section. This same point was shown by Skell and co-workers on the simplest alkyl systems [31,32c].

The Collins' work also established two other points. One was that 1,2 shifts do not always occur with inversion. A migrating group can shift to a carbonium center in a way that the carbonium ion carbon undergoes substitution with retention of configuration. What this means is that the carbonium ion is a true intermediate, that it rotates to a stable conformation, and that it preferentially reacts in one of the staggered conformations. The reacting conformation happens to be the one that results in substitution with retention. Related to this is a second conclusion that the products from nitrous acid deamination do not all arise by displacement on a diazonium ion, a view that had been proposed at one time.

Although much has been made of these two results [32a,65,65a], it seems to this reviewer that they are necessary and self-evident consequences of any system in which substituents render the carbonium ion stable enough to have an appreciable lifetime as an intermediate. The significance of the Collins' work is not in establishing such possibilities, but in delineating structural features that lead to them.

The work in this section has been treated lightly since the results are of more significance to individual mechanisms than to techniques and logic in the use of deuterium.

G. Hydride Shifts in Aldehydes

In strong acid systems, aliphatic aldehydes undergo a variety of condensations and rearrangements. Nevertheless, narrow temperature ranges were found in 5:1 HO_3SF–SbF_5 in which isobutyraldehyde-2-*d* and propionaldehyde-2-d_2 underwent internal 1,2 hydride shifts to scramble hydrogens and deuteriums on C-1 and C-2 prior to decomposition [66]. The scrambling with isobutyraldehyde was much faster than with propionaldehyde.

H. Cycloheptadienone Rearrangements

In DO_3SF (deuterated fluorosulfuric acid), 3,5-cycloheptadienone rearranges to 2,4-cycloheptadienone, eqn. (24), with incorporation of a single deuterium at C-6 [67]. This result uniquely shows that rearrangement proceeds by a single deuteration at C-6 and deprotonation at C-2. The authors [67] wrote this as taking place on the conjugate acid of the ketone. This is most probable since the equilibrium is highly in favor of the protonated ketone in fluorosulfuric acid solutions.

(24)

In sulfuric acid, 3,5-cycloheptadienone rearranged to nine parts 2,4-cycloheptadienone and one part 2,6-cycloheptadienone. Deuterium experiments were not helpful because in deutero sulfuric acid (relatively more basic than DO_3SF), complete deuteration at C-2 and C-7 took place before rearrangement and the products were extensively deuterated [67].

V. DEUTERIUM AS A STEREOCHEMICAL LABEL

Replacement of $-CH_2-$ by $-CHD-$ serves to introduce an asymmetric carbon without greatly altering its chemistry. This has been used in two ways. One is to investigate the stereochemistry of substitution at primary carbons. The second is to label *exo* and *endo* and *syn* and *anti* positions in bicycloalkyl systems stereochemically.

Nitrous acid deamination of (+)-butylamine-1-*d* in acetic acid gave 1-butyl acetate with 69 ± 7% net inversion [45]. The 2-butyl acetate and butene products would be optically inactive and were not examined.

Nitrous acid deamination of (+)-(R)-neopentylamine-1-*d* in acetic acid gave the products shown in eqn. (25) [68]. The dominant 1,2-methyl shift occurs with ⩾85% inversion.

$$(CH_3)_3C-CH(D)NH_2 \xrightarrow[AcOH]{BuONO} (CH_3)_3C-CH(D)N_2^{\oplus} \longrightarrow (CH_3)_2\overset{\oplus}{C}-CH(D)CH_3 \longrightarrow \qquad (25)$$

≥ 85% inverted

(–)-(R)-2-methyl-1-butene-3-*d*
2-methyl-2-butene-3-*d* (58% monodeuterated)
(–)-(R)-1, 1-dimethyl-1-propyl-3-*d* acetate (2-methyl-2-butyl-4-*d* acetate)
neopentyl acetate (a trace)

The neopentyl system has also been examined by deoxideation. (–)-(S)-Neopentyl alcohol formed (+)-(S)-2-methyl-1-butene-3-*d* using bromoform and potassium hydroxide [69]. Again the 1,2-methyl shift occurred with near complete or complete inversion at C-1.

It is attractive to interpret the rearrangement in the amine deaminations as proceeding through an intermediate diazonium ion and not through a free carbonium ion [5,69]. Similarly, in deoxideation, rearrangement occurs in a ROCCl carbene intermediate [69]. Such descriptions tend to overshadow the fact that energetics are very similar for deamination, deoxideation, and electrolytic oxidation as judged by remarkably similar product ratios [31]. The key to this dichotomy is that 1,2 hydride and 1,2 alkyl shifts have energy barriers comparable to rotational energy barriers [9,10,29, 33–36,51,70,71]. The transition states can thus be close to free carbonium ions in energy and yet produce highly stereospecific products in rearrangements.

A curious example involves introduction of deuterium label from the solvent, ethylene glycol-d_2. Decomposition of the tosylhydrazone anion of cyclopropanecarboxaldehyde gave bicyclobutane-2-*endo*-*d* along with cyclobutene and other products [72]. A consequence of this is that conducting the reaction on cyclopropanecarboxaldehyde-1-*d* tosylhydrazone in unlabeled ethylene glycol should yield bicyclobutane-2-*exo*-*d* and this was found [73,74].

Deamination of *exo,exo*-2,2-dideuterio-*anti*-7-norbornylamine in acetic acid gave a 95:5 ratio of the *anti* and *syn* acetates, eqn. (25). In this case the deuterium label introduced *syn–anti* isomerism [75].

H_2N (norbornyl, D, D) ⟶ AcO (norbornyl, D, D) + OAc (norbornyl, D, D) (26)

95% 5%

VI. FRAGMENTATION OF DEUTERATED SUBSTRATES IN MASS SPECTRA

The migration of deuterium or any other label cannot be detected from the parent bands in mass spectra. However, the intensities of fragment bands can reveal rearrangements. The relevance to carbonium ion rearrangements is that many and often the majority of bands in a mass spectrum of organic compounds are due to cleavage of the initially formed cation radical (substrate minus one electron) into a radical plus a carbonium ion. It is rearrangement within this carbonium ion followed by further fragmentation that reveals carbonium ion rearrangements. Only the positively charged fragments are recorded in the mass spectrum.

It has been debated whether the data from mass spectra are relevant to rearrangement of carbonium ions in solution. Certainly, the ions are generally produced in the mass spectrometer with excess energy so that rearrangements are possible (energetically at least) that would not be possible with carbonium ions in thermal equilibrium at ambient temperatures. The wonder is that the initial cation radicals cleave to carbonium ions (by loss of radical) and to simpler cation radicals (by loss of neutral moiety) and that the direction of these cleavages and their probability are largely in the order of their exothermicity as judged by stabilities in solution. Provisionally, it will be inferred that carbonium ion rearrangements will follow this same order.

Bursey and McLafferty have reviewed this field recently [76] so that only a few highlights that feature rearrangements will be summarized here.

Before deuterium labeling studies, the strong 91 band in the mass spectra of toluene, ethylbenzene, and many benzyl derivatives was thought to be due to the benzyl cations ($C_6H_5CH_2^+$). In 1957, studies were reported on toluene and five different deuterium-labeled toluenes and ethylbenzene and seven different deuterium-labeled ethylbenzenes. Attention was focused on the $C_5H_5^+$ band formed by the successive fragmentations shown in eqn. (27).

$$C_6H_5CH_2X \xrightarrow{-e^-} C_6H_5CH_2X^{+\cdot} \xrightarrow{-X^{\cdot}} C_7H_7^+ \xrightarrow{-C_2H_2} C_5H_5^+ \qquad (27)$$

For monodeuterated toluenes, the ratios of the 66 band ($C_5H_4D^+$) to the 65 band ($C_5H_5^+$) were 1.4 for *α-d*, 1.8 for *o-d*, 1.8 for *m-d*, and 1.8 for *p-d*. A similar set of ratios for monodeuterated ethylbenzenes were 1.6 for *α-d*, 1.7 for *o-d*, 2.1 for *m-d*, and 2.0 for *p-d*. All ratios were corrected for natural abundance of carbon-13 (ref. 77).

The values were felt to be close enough to each other and close enough to statistical values (2.5 assuming acetylene arises from contiguous carbons) to indicate that at least 90% of the 65 and 66 bands came from $C_7H_7^+$ in which all carbons are equivalent and all hydrogens are equivalent. This requires $>90\%$ of $C_7H_7^+$ to have the cycloheptatrienyl cation (tropylium) structure. Supporting this view were results on fragmentation of $C_7H_5D_2^+$. Statistically, relative ratios of bands at 67:66:65 would be 1:1:0.1. The observed values were 1:1:0.6 from $C_6H_5CD_2CD_3$, 1:1:0.4 from $C_6H_5CD_2CH_3$, and 1:0.7:0.5 from $C_6H_5CD_3$. None of these ratios were corrected for 4–5% of less deuterated material which partially accounts for the 65 band being much more intense than statistics would indicate.

These striking conclusions were subsequently confirmed by studies on toluene-α-^{13}C [78] and cycloheptatriene-7-*d* [79]. In particular, the scrambling of deuterium label in $C_7H_6D^+$ which had been deduced from the work on toluenes and ethylbenzenes was directly confirmed by the studies on cycloheptatriene-7-*d*.

A similar area involves the $C_3H_3^+$ band at *m/e* 39. It was a suspicious circumstance that allene, propyne, butyne, and 1,2 and 1,3-butadiene all gave an intense 39 band. This led to a study of $CD_3C{\equiv}CH$ [80]. The observed ratio of $C_3D_3^+$ (*m/e* 42) to $C_3D_2H^+$ (*m/e* 41) was the statistical 1:3 suggesting that all hydrogens and deuteriums occupied equivalent positions. This was supported by the fact that the appearance potentials were the same, again suggesting a common precursor. With the recognition of the great stability of the cyclopropenyl cation, it was evident that the *m/e* 39 band must be this cyclopropenyl cation and that it is a common feature of many mass spectra [81].

Hydrogen shifts have been detected in the mass spectra of menthene and camphor by deuterium labeling [82]. However, these shifts occur before fragmentation and are thus rearrangements in a cation radical and not in a carbonium ion.

REFERENCES

1 R.L. Baird and A. Aboderin, J. Amer. Chem. Soc., 86 (1964) 252.
2 N. Deno, D. LaVietes, J. Mockus and P.C. Scholl, J. Amer. Chem. Soc., 90 (1968) 6457.
3 R.L. Baird and A. Aboderin, Tetrahedron Lett., (1963) 235.
4 C.C. Lee, J.E. Kruger and E.W. Wong, J. Amer. Chem. Soc., 87 (1965) 3985, 3987.
4a C.C. Lee and J.E. Kruger, Can. J. Chem., 44 (1966) 2343.
5 C. Collins, Chem. Rev., 69 (1969) 543.
6 C.C. Lee and L. Gruber, J. Amer. Chem. Soc., 90 (1968) 3775.
7 C.C. Lee, W. Chwang and K. Wan, J. Amer. Chem. Soc., 90 (1968) 3778.
8 N. Bodor, M.J. Dewar and D.H. Lo, J. Amer. Chem. Soc., 94 (1972) 5305.
8a L. Radom, J.A. Pople, V. Buss and P.R. Schleyer, J. Amer. Chem. Soc., 94 (1972) 311.
8b C.C. Lee, S. Vassie and E.C.F. Ko, J. Amer. Chem. Soc., 94 (1972) 8931.
8c C.C. Lee, A.J. Cessna, E.C.F. Ko and S. Vassie, J. Amer. Chem. Soc., 95 (1973) 5688.
9 M. Saunders, E.L. Hagen and J. Rosenfeld, J. Amer. Chem. Soc., 90 (1968) 6882.
10 M. Saunders, P. Vogel, E.L. Hagen and J. Rosenfeld, Accounts Chem. Res., in press.
11 N. Deno, W.E. Billups, D. LaVietes, P.C. Scholl and S. Schneider, J. Amer. Chem. Soc., 92 (1970) 3700.
12 N. Deno and J.J. Houser, J. Amer. Chem. Soc., 86 (1964) 1741.
13 N. Deno and R.R. Lastomirsky, J. Amer. Chem. Soc., 90 (1968) 4085.
14 T.S. Sorensen, J. Amer. Chem. Soc., 91 (1969) 6398.
15 T.S. Sorensen and K. Ranganayakulu, J. Amer. Chem. Soc., 92 (1970) 6539.
16 N. Deno, N. Friedman, J.D. Hodge and J.J. Houser, J. Amer. Chem. Soc., 85 (1963) 2995.
17 T.S. Sorensen, J. Amer. Chem. Soc., 89 (1967) 3782, 3794.
18 N. Deno and R.W. Taft, Jr., J. Amer. Chem. Soc., 76 (1954) 244.
19 N. Deno and C.U. Pittman, Jr., J. Amer. Chem. Soc., 86 (1964) 1871.

20 N. Deno, C.U. Pittman, Jr. and J.O. Turner, J. Amer. Chem. Soc., 87 (1965) 2153.
21 T.S. Sorensen, Can. J. Chem., 42 (1964) 2768.
22 T.S. Sorensen, Can. J. Chem., 43 (1965) 2744.
23 G.A. Olah, C.U. Pittman, Jr. and T.S. Sorensen, J. Amer. Chem. Soc., 88 (1966) 2331.
24 C.K. Ingold, C.G. Raisin and C.L. Wilson, J. Chem. Soc., (1936) 1643.
25 R.L. Burwell, Jr. and G.S. Gordon, III, J. Amer. Chem. Soc., 70 (1948) 3128; 71 (1949) 2355;
26 T.D. Stewart and D. Harman, J. Amer. Chem. Soc., 68 (1946) 1135.
27 J.W. Otvos, D.P. Stevenson, C.D. Wagner and O. Beeck, J. Amer. Chem. Soc., 73 (1951) 5741.
28 D.P. Stevenson, C.D. Wagner, O. Beeck and J.W. Otvos, J. Amer. Chem. Soc. 74 (1952) 3269.
29 G.A. Olah and A.M. White, J. Amer. Chem. Soc., 91 (1969) 5801.
30 M. Saunders and J. Rosenfeld, J. Amer. Chem. Soc., 91 (1969) 7756.
31 J.T. Keating and P.S. Skell in G.A. Olah and P. Schleyer (Eds), Carbonium Ions, vol. II, Wiley-Interscience, New York, 1970, pp. 573–653.
32 R.H. Boyd, R.W. Taft, Jr., A.P. Wolf and D.R. Christman, J. Amer. Chem. Soc., 82 (1960) 4729.
32a C.J. Collins, I.T. Glover, M.D. Eckart, V.F. Raaen, B.M. Benjamin and B.S. Benjaminov, J. Amer. Chem. Soc., 94 (1972) 899.
32b C.J. Collins and B.M. Benjamin, J. Org. Chem., 37 (1972) 988.
32c P.S. Skell, Conference on Carbonium Ions, Cleveland, Ohio, October 1968.
32d J. Keating, Ph.D. Thesis, Pennsylvania State University, 1968.
33 D.M. Brouwer, C. MacLean and E.L. Mackor, Discuss. Faraday Soc., 39 (1965) 121.
34 M. Saunders, E.L. Hagen and J. Rosenfeld, J. Amer. Chem. Soc., 90 (1968) 6882.
35 G.A. Olah and J.A. Olah in G.A. Olah and P. Schleyer, (Eds), Carbonium Ions, vol. II, Wiley-Interscience, New York, 1970, pp. 715–782.
36 D.M. Brouwer and J.A. Van Doorn, Rec. Trav. Chim. Pays-Bas, 88 (1969) 573.
37 V. Prelog, J. Chem. Soc. (1950) 420.
37a R. Heck and V. Prelog, Helv. Chim. Acta, 38 (1955) 1541.
37b V. Prelog, W. Küng and T. Tomljenovic, Helv. Chim. Acta, 45 (1962) 1352.
38 H.C. Brown and C. Ham, J. Amer. Chem. Soc., 78 (1956) 2735.
39 V. Prelog and J.G. Traynham in P. deMayo (Ed.), Molecular Rearrangements, Part 1, Wiley-Interscience, New York, 1963, pp. 593–615.
40 V. Prelog, Rec. Chem. Progr., 18 (1957) 247.
41 Reprints and comments in P.D. Bartlett (Ed.), Nonclassical Ions, W.A. Benjamin, New York, 1965, pp. 197–210.
42 P. Schleyer, L.K. Lam, D.J. Raber, J.L. Fry, M.A. McKervey, J.R. Alford, B.D. Cuddy, V.G. Keizer, H.W. Geluk and J.L. Schlatmann, J. Amer. Chem. Soc., 92 (1970) 5246.
43 H. Van Bekkum, B. Van De Graaf, G. Van Minnen-Pathius, J.A. Peters, and B.M. Wepster, Rec. Trav. Chim. Pays-Bas, 89 (1970) 521.
44 C.C. Lee and J.E. Kruger, Can. J. Chem., 44 (1966) 2343.
45 A. Streitwieser, Jr. and W.D. Schaeffer, J. Amer. Chem. Soc., 79 (1957) 2888.
46 G.M. Calhoun and R.L. Burwell, Jr., J. Amer. Chem. Soc., 77 (1955) 6441.
47 S. Winstein and J. Takahishi, Tetrahedron, 2 (1958) 316.
48 I.L. Reich, A. Diaz and S. Winstein, J. Amer. Chem. Soc., 91 (1969) 5635.
49 C.C. Lee and W.K. Chwang, Can. J. Chem., 48 (1970) 1025.
50 N. Deno and M.S. Newman, J. Amer. Chem. Soc., 72 (1950) 3852; 73 (1951) 1920.
51 G.J. Karabatsos, J.L. Fry and S. Meyerson, Tetrahedron Lett., (1967) 3735.
52 P.H. Reichenbacher, Ph.D. Thesis, Penn. State Univ., 1967, pp. 188–196.
52a W.L. Hall, Ph.D. Thesis, Penn. State Univ., 1965, pp. 112–114.
53 L. Friedman in G.A. Olah and P. Schleyer (Eds), Carbonium Ions, vol. II, Wiley-Interscience, New York, 1970, pp. 655–713.

54 A.A. Aboderin and R.L. Baird, J. Amer. Chem. Soc., 86 (1964) 2300.
55 C.C. Lee and J.E. Kruger, Tetrahedron, 23 (1967) 2539.
56 G.J. Karabatsos and C.E. Orzech, Jr., J. Amer. Chem. Soc., 84 (1962) 2838.
57 G.J. Karabatsos, C.E. Orzech, Jr. and S. Meyerson, J. Amer. Chem. Soc., 87 (1965) 4394.
58 C.C. Lee, J.E. Kruger and E.W.C. Wong, J. Amer. Chem. Soc., 87 (1965) 3985.
59 C.C. Lee and J.E. Kruger, J. Amer. Chem. Soc., 87 (1965) 3986.
60 G.J. Karabatsos, N.Hsi and S. Meyerson, J. Amer. Chem. Soc., 88 (1966) 5649.
61 G.J. Karabatsos, R.A. Mount, D.O. Rickter and S. Meyerson, J. Amer. Chem. Soc., 88(1966) 5651.
62 J.H. Bayless, A.T. Jurewicz and L. Friedman, J. Amer. Chem. Soc., 90 (1968) 4466.
63 G.A. Olah, A.M. White, J.R. DeMember, A. Commeyras and C.Y. Lui, J. Amer. Chem. Soc., 92 (1970) 4627.
64 J.A. Berson and P.W. Grubb, J. Amer. Chem. Soc., 87 (1965) 4016.
65 C.J. Collins, in G.A. Olah and P. Schleyer (Eds), Carbonium Ions, vol. I, Wiley-Interscience, New York, 1968, p. 307–351.
65a C.J. Collins, Accounts Chem. Res., 4 (1971) 315.
66 D.M. Brouwer and J.A. Van Doorn, Rec. Trav. Chim. Pays-Bas, 90 (1971) 1010.
67 K.F. Hine and R.F. Childs, J. Amer. Chem. Soc., submitted 1972.
68 R.D. Guthrie, J. Amer. Chem. Soc., 89 (1967) 6718.
69 W.A. Sanderson and H.S. Mosher, J. Amer. Chem. Soc., 88 (1966) 4185.
70 D.M. Brouwer, C. MacLean and E.L. Mackor, Discuss. Faraday Soc., 39 (1965) 121.
71 D.M. Brouwer, Rec. Trav. Chim. Pays-Bas, 88 (1969) 9.
72 J.H. Bayless, L. Friedman, F.B. Cook and H. Shechter, J. Amer. Chem. Soc., 90 (1968) 531.
73 F. Cook, H. Shechter, J.H. Bayless, L. Friedman, L. Foltz and R. Randall, J. Amer. Chem. Soc., 88 (1966) 3870.
74 K. Wiberg and J. Lavanish, J. Amer. Chem. Soc., 88 (1966) 5272.
75 P.G. Gassman, J.M. Hornback and J.L. Marshall, J. Amer. Chem. Soc., 90 (1968) 6238.
76 M.M. Bursey and F.W. McLafferty in G.A. Olah and P. Schleyer (Eds.), Carbonium Ions, vol. I, Wiley-Interscience, New York, 1968, pp. 257–306.
77 P.N. Rylander, S. Meyerson and H.M. Grubb, J. Amer. Chem. Soc., 79 (1957) 842.
78 S. Meyerson and P. Rylander, J. Chem. Phys., 27 (1957) 901.
79 S. Meyerson, J. Amer. Chem. Soc., 85 (1963) 3340.
80 J. Collin and F.P. Lossing, J. Amer. Chem. Soc., 79 (1957) 5848; 80 (1958) 1568.
81 K.B. Wiberg, W.J. Bartley and F.P. Lossing, J. Amer. Chem. Soc., 84 (1962) 3980.
82 D. Weinberg and C. Djerassi, J. Org. Chem., 31 (1966) 115.

Chapter 2

ISOTOPE EFFECTS IN PERICYCLIC REACTIONS

W.R. DOLBIER, Jr.,

Department of Chemistry, University of Florida, Gainesville, Florida 32611 (U.S.A.)

I. INTRODUCTION

A determination of the timing in a reaction mechanism, i.e. how synchronous the various bond-breaking and bond-making processes are, has along with the determination of detailed transition state structures presented a constant challenge to the physical–organic chemist. This is particularly true for the class of reactions which are presently known as pericyclic or multi-centered reactions, which include cycloadditions, electrocyclic reactions and sigmatropic processes. Earlier such reactions were known, somewhat facetiously, as "no-mechanism" reactions because their detailed mechanisms had proved so inscrutible.

Although the advent of the orbital symmetry correlations of Woodward and Hoffmann along with other more recent theoretical innovations has provided workers with a theoretical basis upon which to design reaction systems, to make mechanistic predictions, and to rationalize experimental data, it still falls to the experimentalist to provide experimental data whereby mechanistic details can be rigorously proven.

All of the various physical–organic tools have been applied to the problem of pericyclic reactions, studies of product ratios, medium effects, substituent effects and perhaps most important, stereochemical probes. In general, kinetic probes have been very useful, and included among these is the use of kinetic isotope effects. Primary and secondary kinetic isotope effects are being increasingly utilized by workers to probe the simultaneity of mechanistic processes and to elucidate the nature of the transition state in multi-centered reactions.

This chapter will endeavor to present a current analysis of the progress being made in the area of the utilization of primary and secondary kinetic isotope effects to probe the mechanisms of pericyclic reactions. An emphasis will be made on the utilization of such effects somewhat empirically, that is without concomitant rigorous theoretical analysis. In some cases such analysis will seem to be worthwhile and perhaps even essential, but in many if not most cases, it will be seen that useful and sometimes definitive conclusions can be reached using simple theoretical models as guides to interpretation and prediction. Most of this chapter will be devoted to a discussion of secondary deuterium isotope effects, since they have proven to be of more general use and indeed have been demonstrated to be a most sensitive probe of transition state structure. Some primary heavy atom isotope effect studies will also be discussed and in those few cases where a C–H bond is broken, primary deuterium isotope effects will be discussed. This work, while not designed necessarily to be a comprehensive review of the literature, nevertheless endeavors to include a description and critical analysis of all pertinent isotope effect data related to the subject which was available as of April, 1973.

II. THEORY

What makes isotope effect studies unique, and different from substituent effect studies, is the fact that the potential energy surfaces for isotopic molecules are identical. Thus isotope effects have a vibrational, non-electronic origin. Moreover, it is found, at least for large molecules and for hydrogen isotopes, that zero point energy effects are the only major contributors to isotope effects. This comes about because of the relatively large vibrational frequencies and therefore large zero point energies for C–H vibrations and the resultant fact that all vibrations involving the hydrogen isotopes will remain, at ordinary temperatures, in the zero point energy levels.

The basic theory of isotope effects has been developed by Bigeleisen [1], Bigeleisen and Wolfsberg [2] and Stern and Wolfsberg [3]. Methods have been developed by these workers and others to make excellent computer calculations of kinetic isotope effects for even complex molecules. Thus by choosing a model "transition state" along with its appropriate force constants and geometries, one can calculate the isotope effect for the conversion of the starting material to this transition state. The main problems associated with such calculations are the choosing of the "correct" model and the assigning of appropriate force constants and molecular geometries. One often finds it is difficult to assign such parameters without ambiguity and also that there is usually more than one model which can give the experimental value for the isotope effect. Excellent discussions of the theory of isotope effects along with discussions of the empirical approximations which allow exact calculations of isotope effects may be found in the recent monograph by Collins and Bowman [4].

In this chapter we will be presenting theoretical models which the practicing organic chemist can use to reach conclusions. One will find that it is not always necessary to make theoretical calculations in order to make effective mechanistic use of kinetic isotope effects. Often, the choice between two logical mechanisms may be made by the single observation of whether or not there *is* a primary isotope effect, whether the secondary isotope effect is normal or inverse, or whether the intermolecular and intramolecular isotope effects are identical within experimental error.

Kinetic isotope effects can be divided into primary and secondary types. One often cannot completely distinguish these types and both very often act in the same system.

A. Primary Isotope Effects

A primary isotope effect may be observed when the bond to the isotopically labelled atom is broken during the reaction sequence. In a multi-step mechanism, where the rate of a "heavy" molecule is being compared with that of a "light" molecule, in order for an isotope effect to be observed, isotopic discrimination must occur during or before the rate-determining step. Such types of competitive studies are known as *intermolecular* isotope effect studies. An example would be a comparison of the hydrogen abstraction rates of methyl chloride with that of the deuterium abstraction of monodeuteriomethyl chloride.

$$CH_3 - Cl + X^{\cdot} \rightarrow {}^{\cdot}CH_2 - Cl + HX \quad (1)$$

$$CH_2DCl + X^{\cdot} \rightarrow {}^{\cdot}CH_2 - Cl + DX \quad (2)$$

Where the isotopic discrimination is made to occur within the same molecule, the competition is known as an *intramolecular* isotope effect study. In such competitions with a multi-step mechanism, an isotope effect should be observed regardless of the timing of the bond breakage. An example would be the abstraction of a hydrogen from monodeuteriomethylene chloride.

$$CHDCl_2 + X^{\cdot} \rightarrow {}^{\cdot}CDCl_2 + HX$$

$$\quad \llcorner\!\!\rightarrow {}^{\cdot}CHCl_2 + DX \quad (3)$$

Primary isotope effects can be approximated by assuming that one vibration of the isotopic atom in the reactant becomes the reaction-coordinate-motion of the transition state. In general, because of the lower zero point energy for the vibrations involving the heavier isotope, the lighter isotope is favored in bond breaking. Simple calculations [2] of maximum isotope effects, assuming that the useful vibration goes to zero frequency in the transition state, show $k_H/k_D = 18$, $k_H/k_T = 60$, $k^{C}{}_{12}/k^{C}{}_{13} = 1.25$, $k^{N}{}_{14}/k^{N}{}_{15} = 1.14$ and $k^{O}{}_{16}/k^{O}{}_{18} = 1.19$. Of course, force constants in actual transition states are never zero, thus observed primary isotope effects are always much smaller than these maximum values. The magnitude of the isotope effect is dependent on the temperature and the above values refer to 25°.

Primary Deuterium Isotope Effects

For hydrogen transfer reactions with open and probably linear transition states, the magnitude of k_H/k_D has generally been shown to be dependent upon the transition state structure, particularly with regard to its symmetry with respect to bond breaking and bond making. For reactant-like or product-like transition states, small isotope effects are generally observed, while reactions proceeding through symmetrical transition states produce relatively large effects. In the case of a linear transition state, moreover, k_H/k_D is determined principally by the zero-point energy change in the high frequency stretching mode in the reactants, and thus it is relatively large. Another factor which can give rise to an asymmetry of force constants in the transition state, whether it be reactant-like or product-like, is the involvement of a *hetero* atom in a transition state where a C–H bond is being broken.

One factor which is especially important in *pericyclic* reactions is the fact that generally a linear transition state for hydrogen transfer is not possible for such reactions. It has been demonstrated many times that a bent transition state geometry will give rise to relatively small primary isotope effects [5]. Simply stated, in a non-linear transition state k_H/k_D is determined principally by the zero-point energy change in a relatively low frequency bending mode of the reactants, thus producing a relatively small isotope effect.

In this chapter, essentially all primary isotope effect studies which will be discussed involve reactions where five- and six-membered ring transition states are involved. Thus the effect of transition state geometry will always be an important point to consider.

B. Secondary Deuterium Isotope Effects

Secondary deuterium isotope effects are observed in systems where the vibrational modes of the isotopically labelled site are "perturbed" during the transformation from reactant to product. As in systems involving primary effects, *intermolecular* competition studies can give information only about the rate-determining step, while intramolecular studies are not so restricted.

Secondary deuterium isotope effects are generally rationalized in terms of hybridization effects, steric effects, inductive effects, or hyperconjugative effects, the latter being applicable only to β deuterium isotope effects [4]. The use of these terms, which are usually relegated to discussions of substituent effects, can sometimes be misleading or confusing, since as mentioned earlier, isotopically labelled molecules have identical electronic potential energy surfaces. However, the use of such terms can perhaps be justified in that empirically they allow reasonable predictions and rationalizations to be made by the practicing chemist. Moreover they have some basis in fact. The steric effect of a C–D bond derives from the fact that it is effectively shorter than a C–H bond, mostly due to the isotope effect on the amplitude, purely a vibrational effect. The inductive effect derives from the C–D bond, being shorter, having a higher density and thus being more effectively electron supplying relative to the C–H bond. The hyperconjugation argument arises with regard to β substituted deuteriums, because CD_3 is effectively less delocalized than CH_3, and thus a radical, cation or π-system is less stabilized by an adjacent CD_3 than by a CH_3. The hybridization explanation is, I believe, completely encompassed by the steric argument and need not be invoked as a separate entity. In fact, since pericyclic reaction mechanisms generally involve little charge separation and usually proceed by either a concerted mechanism or one involving bi-radicals, one can generally rationalize most, if not all, of the data simply in terms of "steric" secondary deuterium isotope effects.

This steric isotope effect derives from the fact that the more sterically crowded a C–H bond is, the higher will be the vibrational frequencies associated with that bond, the out-of-plane bending frequencies generally being the more sensitive to crowding. Thus an sp^3 carbon bearing hydrogen will have significantly higher frequency bending vibrations associated with it than will an sp^2 carbon bearing hydrogen. Since the ratio of ν_{C-H}/ν_{C-D} = 1.35, a constant, it follows then that the secondary kinetic deuterium

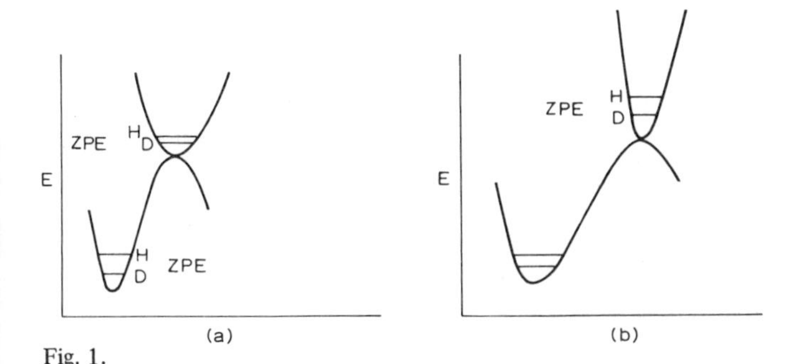

Fig. 1.

isotope effect for such a system as depicted in Fig. 1(a) would be expected to be normal, $k_H/k_D > 1$, and for an opposite situation as depicted in Fig. 1(b), the effect would be inverse, $k_H/k_D < 1$. A related factor should also be recognized as potentially giving rise to significant secondary isotope effects in pericyclic reactions, particularly those where a bond in a rigid cyclic system is broken (or formed) in a discriminatory step.

$$\text{methylenecyclopropane} \longrightarrow (\cdot CH_2)(\cdot CH_2)C{=}CH_2 \qquad (4)$$

Take, for example, the process of bond cleavage of methylenecyclopropane (eqn. (4)). The ring CH_2 groups of methylenecyclopropane are held rigidly with a relatively high twisting frequency (~1073cm^{-1}) [6]. On the other hand, as the bond is broken the CH_2 group becomes freer to rotate and the torsional frequency associated with this rotation should be *much* lower (~192cm^{-1}). These kinds of effects, where new torsional modes are created by bond breakage, have been hypothesized to contribute significantly to observed isotope effects [7]. On this basis, one would predict a $k_H/k_D > 1$ for the above cleavage process. In making predictions, we can utilize the general rule of thumb that the site with a deuterium label will generally accumulate at the more sterically congested or more rigid position in the transition state where discrimination occurs. In other words, if the isotopic site becomes "looser" in the discriminatory transition state, then the proton labelled species will pass through that transition state more rapidly. In intermolecular competitions this means that k_H/k_D will be normal for a rate-determining transition state which is looser and/or less rigid than the starting material. For intramolecular competitions, product ratios will be determined by the relative looseness of the transition state in which discrimination occurs, regardless of whether or not this step is rate determining.

The magnitude of the secondary deuterium isotope effect is generally of the order of 10–15% per deuterium, with smaller values usually being considered as indicative of less progress along the reaction coordinate.

III. CYCLOADDITIONS

A. [2 + 4] Cycloadditions

The earliest isotope effect studies involving pericyclic reactions were of the Diels–Alder reaction, the mechanism of which has challenged organic chemists for more than twenty years. A number of studies have been conducted which were designed to utilize knowledge and interpretation of the absolute magnitude of the heavy-atom primary and secondary deuterium isotope effects in the reaction, so as to conclude whether both new C–C bonds were formed simultaneously, or consecutively via an intermediate.

1. Primary Isotope Effects

There have been a limited number of studies of the Diels–Alder reaction using primary, heavy-atom isotope effects as a mechanistic probe. With very little data available for comparison purposes, largely because of the experimental difficulties involved in working with such small effects, this work is very much tied to interpretations of the absolute magnitude of the isotope effects.

Goldstein and co-workers have made the most significant effort in this area, their approach having been to utilize the Wolfsberg and Stern program and ascertain, through model calculations, the changes in force constants which are needed to reproduce their experimental isotope effects.

Goldstein and Thayer [8] first examined the retrodiene elimination of carbon dioxide from the adduct (*1*) from maleic anhydride and α-pyrone. In this reaction, they determined the carbon-13 and oxygen-18 isotope effects at

(*1*) ⟶ + CO_2 (5)

130° to be 1.030 and 1.014, respectively. Two basic models were used for their calculations: (a) simultaneous rupture of both bonds and (b) stepwise rupture. It was found that in order for simultaneous bond rupture to fit the data, C-1–C-2 bond cleavage must be much more substantial than C-4–O-3 cleavage. Good agreement can also be obtained for the stepwise mechanism if the rate-determining transition state involves C-1–C-2 bond cleavage concommitant with C-4–O-3 bond tightening. While this work cannot definitively distinguish between a one- or two-step mechanism, it certainly does point to the unsymmetrical nature of the rate-determining transition state and thus contributes significantly to our overall understanding of Diels–Alder reactions.

Next, Warren and Goldstein [9] examined a cycloaddition process involving a symmetrical diene and dienophile; the reaction of dicyanoacetylene with 1,3-cyclopentadiene.

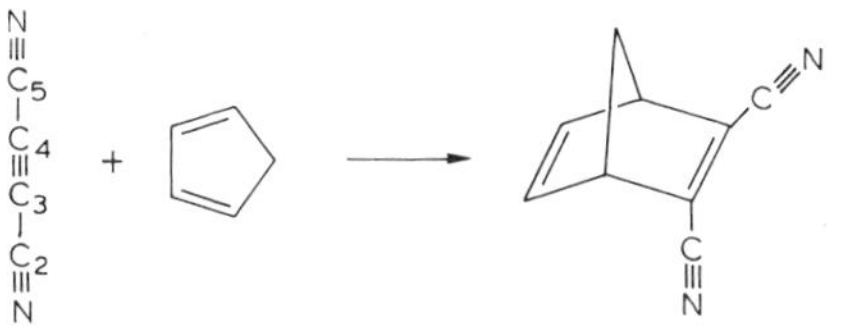

(6)

Using dicyanoacetylene they labelled (a) carbon-13 at one nitrile position, (b) carbon-13 at both nitrile positions, (c) carbon-13 at one acetylenic position and (d) carbon-13 at both acetylenic positions, and were able to observe (a) k_2/k_0, a single secondary isotope effect, (b) k_{25}/k_0, a double secondary isotope effect, (c) k_3/k_0, a single primary isotope effect and (d) k_{34}/k_0, a double primary isotope effect. The values for these isotope effects were found to be: $k_2/k_0 = 1.0004 \pm 0.0002$, $k_{25}/k_0 = 1.003 \pm 0.002$, $k_3/k_0 = 1.0204 \pm 0.0005$ and $k_{34}/k_0 = 1.046 \pm .006$. These isotope effects were, as in the earlier study, analyzed in terms of two models, one-center and two-center mechanisms. Using as an analogy a number of carbon-13 isotope effect studies in systems where C–C bonds were broken, and since in a biradical mechanism the net result in forming an intermediate is *one* bond broken, the observed k_{34}/k_0 value of 1.046 was found to be completely consistent with expectation for the "one-center" model. On the other hand, for the "two-center", concerted model, the observed isotope effect cannot be made consistent if bond making is equal to bond breaking in the transition state. In such a case, the k_{34}/k_0 should equal unity. If, however, the acetylenic bond weakens at a rate which is faster than the formation of the sigma bonds, then the isotope effect can be rationalized. While the observed isotope effects do not themselves distinguish between a biradical and a concerted mechanism, they contribute significant insight into the restraints imposed on the system in the transition state. It should also be stated that the authors qualitatively, through use of analogy, were able to reach the same basic mechanistic conclusions that were later verified by calculating the isotope effects using the γ technique [2].

2. Secondary Isotope Effects

Secondary deuterium isotope effects have also proved to be a valuable tool in elucidating the nature of the mechanism of various Diels–Alder reactions. Two early reports concerned themselves with isotope effects in the cycloaddition process while another examined the retrodiene process. Each of these works provided important groundwork for subsequent and future investigations.

Van Sickle and Rodin [10] found that deuterium substitution at the site of bond formation in either the diene or the dienophile gave rise to inverse deuterium isotope effects. In these studies, competitive kinetics in reactions of maleic anhydride and maleic anhydride-d_2 with butadiene, anthracene, or cyclopentadiene and of maleic anhydride with 1,3-butadiene-1,1,4,4-d_4 and anthracene-9,-10-d_2 were determined.

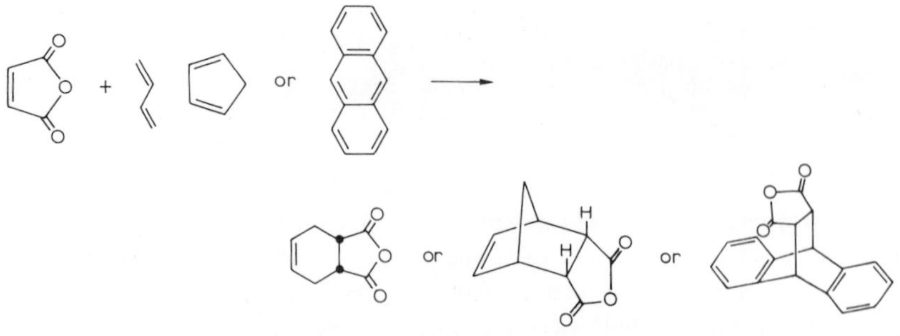

(7)

The values for these isotope effects are given in Table 1. Besides being consistent with the predictions of an inverse isotope effect, the results seem to indicate a transition state which has made little progress in bond making.

There seems to be no correlation possible regarding relative diene reactivity and magnitude of the isotope effect. Moreover, the larger isotope effect observed for the diene species seems to indicate that rehybridization is farther along in the transition state for the diene than for the dienophile. While consistent with a concerted single-transition-state mechanism, the results in the paper nevertheless provide no clear evidence that this will occur. A two-step process could also explain the results adequately, since one has no knowledge of the magnitude of the expected *absolute* isotope effect for stepwise versus simultaneous bond formation.

Similarly, Brown and Cookson [11] have examined the addition of various cyanoethylenes to anthracene and its 9-deuterio- and 9,10-dideuterio derivatives.

The reaction of tetracyanoethylene with 9,10-dideuterioanthracene resulted in a $k_H/k_D = 0.94$ while the reverse, *retro*diene process gave a larger, *normal* effect, $k_H/k_D = 1.09$. These results are consistent with the thermodynamics of the system, with Hammond's postulate, and with the law of microscopic reversibility, in that the cycloaddition apparently has a reactant-like transition state and the retrodiene process a product-like transition state.

The intramolecular competitions indicate a trend towards a non-synchronous

TABLE 1

SECONDARY DEUTERIUM ISOTOPE EFFECTS IN MALEIC ANHYDRIDE–DIENE ADDITIONS

Addition	k_H/k_D per D at 25°
Maleic anhydride-d_2 + butadiene	0.99
Maleic anhydride-d_2 + cyclopentadiene	0.97
Maleic anhydride-d_2 + anthracene	0.95
Maleic anhydride + anthracene-d_2	0.94
Maleic anhydride + butadiene-d_4	0.93

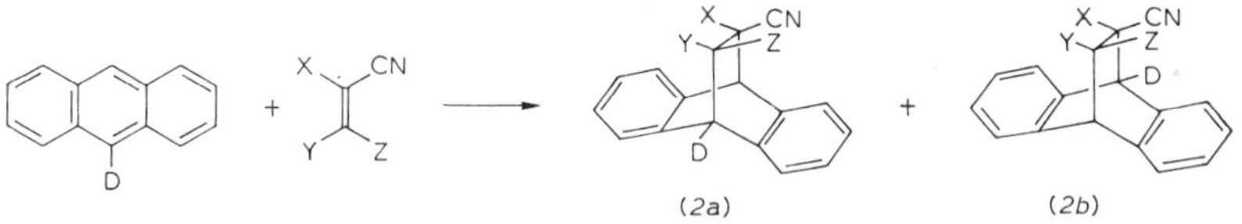

(8)

(2a) (2b)

mechanism with increasingly unsymmetrically substituted ethylenes. If the process were to be a two-step reaction, the compound (*2a*) would have been preferred since the more stable intermediate would result from initial bond formation to the *less* highly substituted ethylenic carbon. This is indeed what is observed. The data can thus be interpreted by either a two-step mechanism or a one-step mechanism which becomes more and more unsymmetrical or non-synchronous as the ethylene becomes more unsymmetrically substituted. In this study it would have been of interest to have had a comparison of the rate of reaction of tetracyanoethylene with both 9-deuterio- and 9,10-dideuterioanthracene. If a concerted reaction was occurring, the former isotope effect would have been the square root of the latter.

The retrodiene studies were significantly more precise in the conclusions which their data demanded. Seltzer [12] examined the retro Diels–Alder process for the exo adduct of 2-methylfuran and maleic anhydride, utilizing five different deuterated isomers in his study.

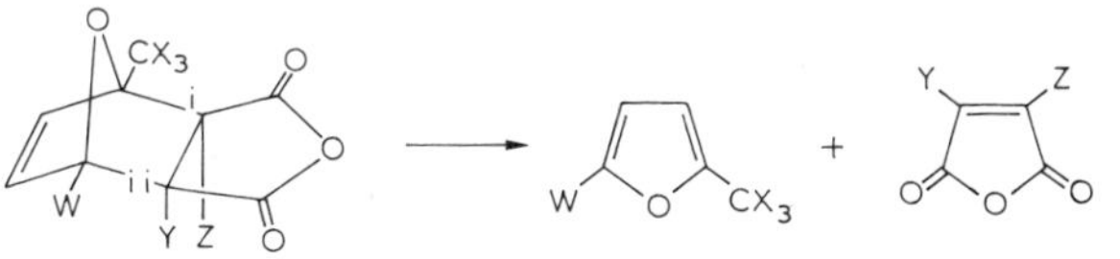

(9)

(*3a*) W = X = Y = Z = H
(*3b*) W = X = H, Y = Z = D
(*3c*) W = D, X = Y = Z = H
(*3d*) W = X = Y = H, Z = D
(*3e*) W = X = Z = H, Y = D
(*3f*) X = D, W = Y = Z = H

A number of aspects of this data should be noticed. First the isotope effect, per deuterium, with regard to the dienophile and with regard to the diene, as reflected in (*3a*)/(*3b*) and (*3a*)/(*3c*) respectively, are identical. This contrasts with Van Sickle's

TABLE 2

SECONDARY DEUTERIUM ISOTOPE EFFECTS IN THE ADDITION OF CYANOETHYLENES TO 9-DEUTERIOANTHRACENE

X	Y	Z	(*2a*)/(*2b*)
H	H	H	1.07
CN	H	H	1.07
CN	CN	H	1.03
CN	CN	Cl	1.02

TABLE 3

ISOTOPE EFFECTS IN THE RETRO-DIENE REACTION OF THE ADDUCT OF 2-METHYL-FURAN AND MALEIC ANHYDRIDE

Ratio of rates of compounds
(3a/b) = 1.16
(3a/c) = 1.08
(3a/f) = 1.03
(3d/e) = 1.00

results and indicates a similar extent of rehybridization in each species in the transition state. The relatively large isotope effect is consistent with Brown and Cookson's isotope effect for their retrodiene process and is indicative of a significant amount of bond breakage in the transition state. A value of key importance is the ratio which indicates that each bond of the unsymmetrical adduct is breaking to a similar extent in the rate-determining transition state. This result is strongly indicative of a concerted mechanism with simultaneous bond cleavage. The lack of a substantial β deuterium isotope effect (i.e. *(3a)*/*(3f)* = 1.03) also suggests the absence of an intermediate in the reaction. However, there is a remote possibility in this system that the rate equality for molecules *(3d)* and *(3e)* could have resulted from two different competing transition states each of which had bond breakage at only one site. Thus if bond i would be the easier to break, the resultant transition state would also be more reactant-like and the observed isotope effect smaller than for the breakage of bond ii. Thus, hypothetically, *(3d)*/*(3e)* could be unity for a two-step mechanism.

A study which allows distinction between these two alternatives and which leads to definitive conclusions independent of the absolute magnitude of the isotope effect, has recently been accomplished by Taagepera and Thornton [13]. They examined the retrodiene reaction of 9,10-dihydro-9,10-ethanoanthracene *(4)* and its -11,11-d_2 and -11,11,12,12-d_4 derivatives (eqn.(10)).

$$\text{(4)} \xrightarrow{\Delta} \text{anthracene} + CH_2{=}CH_2 \qquad (10)$$

The isotope effects observed at 219.84 ± 0.015° were $k_2/k_0 = 0.924 \pm 0.005$ and $k_4/k_0 = 0.852 \pm 0.007$ where k_0, k_2 and k_4 are the rates of the d_0 and d_2 and d_4 species, respectively. Theory predicts that for a concerted mechanism $k_4/k_0 = (k_2/k_0)^2$, while for a mechanism involving two equal two-step processes $k_4/k_0 = 2k_2/k_0 - 1$. With the very precise mass spectrometric measurements of the three labelled ethylenic species, a distinction between these two extreme mechanisms could be made and the conclusion was that the mechanism is concerted, with a symmetrical or nearly symmetrical transition state. This conclusion is independent of the absolute values of the observed isotope effects and is dependent only on a *comparison* of (k_4/k_0) and $(k_2/k_0)^2$.

Thus, elegant experiments by Thornton and by Seltzer each produced quite conclusive data that, at least in the two systems examined, a concerted mechanism with a symmetrical transition state was being followed. Brown and Cookson's results, on the other hand, indicated that as the systems became more and more unsymmetrical, the transition states also, predictably, became unsymmetrical. Whether in a particularly unsymmetrical system the reaction will become non-concerted is still a matter of conjecture. There is no isotope effect data available on such systems, if they indeed do exist. There are, however, published accounts of other types of evidence in support of such a mechanistic trend [14–18].

B. [2 + 3] Cycloadditions

There is a significant amount of kinetic and stereochemical evidence which indicates that 1,3-dipolar cycloadditions to alkenes proceed via concerted mechanisms. With regard to orbital symmetry considerations, 1,3-dipoles are analogous to 1,3-dienes in their symmetry properties.

Bayne and Snyder [19] have determined the intermolecular secondary deuterium isotope effect for the cyclo-addition reaction between tetracyanoethylene oxide and styrene. Linn and Benson [20], in a kinetic and stereochemical study, had earlier established the intermediacy of the 1,3-dipolar intermediate in this reaction and the probable concertedness of the intermediate's reaction with various alkenes. Bayne and Snyder utilized the three different monodeuterated styrenes for their competitive study and found the following values for k_H/k_D for the various competitions

ϕ CH=CH₂ (11)

0.96 H (α), H 0.97, H 0.97 (β)

With similar isotope effect values for the α and β positions the conclusion reached was that bond making occurs simultaneously at each site in the rate-determining step, although the authors could not rule out a two-step process with the second step being rate determining.

The intramolecular isotope effect study by Dolbier and Dai [21] of the addition of 1,1-dideuterioallene to tetracyanoethylene whereby $k_H/k_D = 0.97$ per deuterium can only, however, be rationalized in a consistent manner as being the result of a concerted process, since they have demonstrated that the intramolecular destruction of an intermediate allyl radical system results in a value of $k_H/k_D > 1$ (see Section III.F).

C. [2 + 2] Cycloadditions

The earliest determination of a kinetic secondary deuterium isotope effect in a

[2 + 2] cycloaddition was that of Katz and Dessau [22] in 1963. They examined the reaction of diphenylketene with 1-deuteriocyclohexene (eqn. (12)).

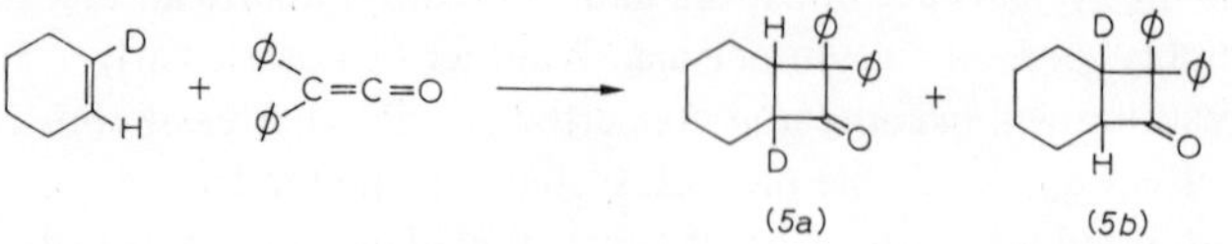

(12)

(5a) (5b)

In this reaction the ratio of adduct (*5b*)/(*5a*) averaged 0.87 ± 0.03. These results had to be interpreted before the authors had the benefit of stereochemical and kinetic evidence which later indicated that ketenes undergo concerted 2s + 2a cycloadditions with alkenes. They interpreted their results to mean that significantly more bonding is occurring in the transition state at the ketene carbonyl carbon atoms. The data did not allow any distinction between an unsymmetrical one step process or a multi-step process.

Baldwin and Kapecki [23] have, however, done an even more definitive study of ketene cycloadditions to alkenes, which allows more conclusive mechanistic statements to be made. These workers examined the intermolecular isotope effects for the cycloaddition of diphenylketene to styrene which was isotopically labelled on either the α or the β positions. Remarkably, the two isotope effects are completely opposite in nature.

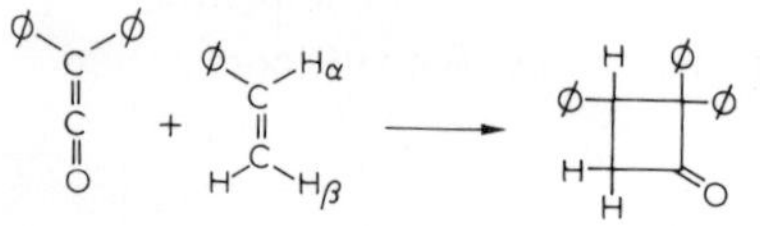

(13)

That is, the isotope effect at the β position was found to be inverse, $k_H/k_D = 0.91$, but that at the α position was large and normal, $k_H/k_D = 1.23$. These effects are completely consistent with Katz and Dessau's results which can be rationalized as being smaller in magnitude and a composite of both α and β effects. The β effect, which favors deuterium at the bond to the ketene carbonyl carbon atoms (as did the Katz system), is completely consistent with the isotope effect expected for a single bond-forming process, where the C–H bonds are being placed in a more crowded and thus higher frequency situation. The α effect however is completely without analogy in cycloaddition reaction. It certainly is not consistent with a biradical intermediate being formed, since this kind of a process should produce but a small isotope effect. Baldwin and Roy [24] have found, for example, that there is no observable kinetic isotope effect in the addition of α-deuterioacrylonitrile to allene.

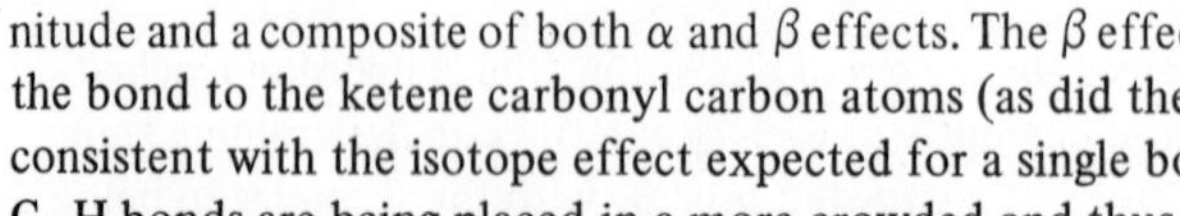

$$k_H/k_D = 1.00 \pm 0.02 \qquad (14)$$

Pryor and Henderson [25] have, however, found that a small, normal isotope effect ($k_H/k_D < 1.05$) is observed for the addition of radicals to styrene labelled at the α position. It does not seem likely, however, that a value as large as 1.23 can possibly be rationalized by such a mechanism. Baldwin and Kapecki have concluded that the reaction was indeed concerted, but that the α-C–H bond is considerably weakened in the transition state which they represented as structure (*6*). They propose that the orbital

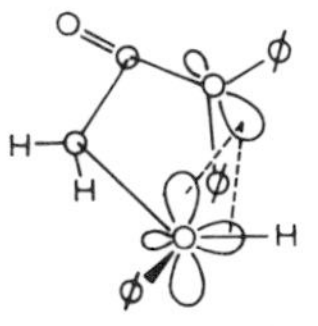

(6)

symmetry prohibition against a thermal suprafacial–suprafacial addition in this system leads to a geometry for the activated complex in which the *p* orbitals of the ketene carbon bearing the phenyls, which should be less electrophilic than the carbonyl carbon, and the α-C–H bond of styrene are not well disposed to overlap. The system is thus profitably lowered in energy through the interaction between the α-C–H bond of styrene and the *p* orbital of diphenylketene. This is effectively a hyperconjugative interaction in which a C–H bond should be more effective than a C–D bond.

It is apparent that there is not at present enough analogy regarding such effects to be able to reach definitive conclusions. Thus we must consider the source of the α effect to be a matter of conjecture at this time.

It is of interest that Koerner von Gustorf et al. [26] have observed a similar disparity in α and β effects for the addition of dimethylazodicarboxylate to α and β deuterated ethyl vinyl ether, and have concluded that the reaction proceeds *stepwise*.

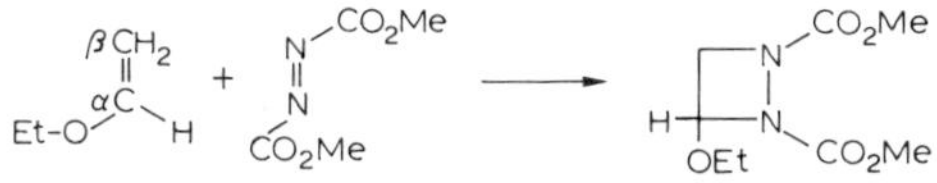

$\beta CH_2{=}\alpha CH{-}OEt$ + $MeO_2C{-}N{=}N{-}CO_2Me$ ⟶ (H, OEt-substituted diazetidine with two $N{-}CO_2Me$) (15)

In a series of intermolecular competitive studies, these workers determined for the β position $k_H/k_D = 0.83$, an inverse effect; while for the α position $k_H/k_D = 1.12$, a normal effect.

This normal effect is not as dramatically large as Baldwin's and possibly could be found consistent with a multi-step process. However, the results for these two reactions are close enough so that a common mechanistic pathway cannot be excluded. Azodicarboxylates have in the past been proposed as potential antarafacial agents in concerted [2 + 2] cyclo-additions because of the availability of the lone pairs on the nitrogens, orthogonal to the π-bond, but available for bonding.

It appears that there is a real need for additional data in the area of ketene and azodicarboxylate [2 + 2] cycloadditions. Because of the potential uniqueness of the mechanism for these cycloadditions they bear much interest to chemists who desire to use isotope effects as a diagnostic tool.

No studies other than in allene cycloadditions have been carried out regarding simple [2 + 2] systems. It will be seen in the discussion which follows on allene cycloadditions that, in general, [2 + 2] systems are clearly distinguishable as two-step processes by the utilization of both inter- and intramolecular secondary deuterium isotope effects.

D. [2 + 1] Cycloadditions

While there have regrettably been no definitive isotope effect investigations of carbene or nitrene additions to alkenes, there has been one related study; that of the addition of sulfur atoms to ethylene to form thiirane. It has been found by Strausz et al. [27] that whether the addition proceeds via excited state singlet $S(^1D)$ atoms or ground state triplet $S(^3P)$ atoms the episulfides are formed stereospecifically and while the singlet also gives insertion products, the triplet addition reaction is its only principal reaction pathway.

Strausz et al. [28] also measured the rate constants for the addition of $S(^3P)$ atoms to C_2D_4, CH_2CD_2 and *cis*-CHDCHD relative to ethylene. They found these intermolecular isotope effects to be k_H/k_D = 0.877, 0.93 and 0.96, respectively. These isotope effects are, as expected, inverse effects. They do not however allow distinction between a one or a two-step mechanism. An interesting aspect of this investigation resulted from some calculations done to test models for the transition state for addition.

$$S\ (^3P) + C_2H_4 \longrightarrow \text{(CH}_2\text{)}_2\text{S (thiirane)} \qquad (16)$$

Either a symmetrical or an asymmetric ring structure can be made to fit the data by the appropriate choice of structural parameters for the transition state. Interestingly, it was found that the single most important factor inducing a secondary hydrogen/deuterium isotope effect in addition reactions to olefinic double bonds, is not the out-of-plane bending motions of the C–H bonds but the gain in the isotopically sensitive vibrational degrees of freedom on going from reactant to activated complex.

The asymmetric model is comparable to that which prevails in the addition of monovalent atoms or radicals to olefins and as such, in an addition to a C=C bond, the C–C bond order is reduced to one, which results in the transformation of the terminal methylene torsional vibration into an internal rotation and thereby causes an increase in k_H/k_D. The authors conclude that the single most important contributing factor to the *inverse* isotope effect observed in the sulfur + ethylene reaction is the net gain in the isotope-sensitive vibrational modes during the activation process. The ethylene molecule has twelve normal modes of vibration while thiirane has fifteen. Of these, the asymmetric CH_2 twist in the symmetrical model and the twist of the CH_2 group which constitutes the reaction center towards the C–C–S plane in the asymmetric model, are particularly isotope sensitive and are the main source of the isotope effect.

The authors moreover generalize that the symmetrical model is a prototype for any cycloaddition reaction. In general, the addition of polyatomic reagent gives rise to a gain of six in the number of normal modes of vibration, one of which will correspond with the reaction coordinate. Of the remaining five, at least one, the CH_2 twist, will always be isotope sensitive and will generate a substantial inverse isotope effect.

Simons and Rabinovitch [29] examined the secondary deuterium isotope effect for the addition of methylene to perdeuterio-2-butene and found a $k_H/k_D = 1.07$. This value is rather difficult to interpret since it is a composite of α-deuterium isotope effects which should be inverse and β-deuterium isotope effects which should be normal.

$CD_3CD{=}CDCD_3$ vs. $CH_3CH{=}CHCH_3$ + CH_2: ⟶ (Me)C—C(Me) bridged by CH_2 (17)

E. [4 + 1] Cycloaddition

The extrusion of sulfur dioxide from sulfolenes is formally a retro [4 + 1] cycloaddition, but perhaps is better known now as an allowed linear cheletropic reaction [30]. The reaction has been demonstrated to proceed in a disrotatory manner with respect to the diene extruded or cyclo-added [31,32].

(18) 2,2-dideuterio-sulfolene —120°→ 1,1-dideuterio-butadiene + SO_2

(19) CH_3, D, D, H, CH_3 substituted sulfolene —105°→ D, D, CH_3 substituted diene + SO_2

Asperger et al. [33] have investigated intermolecular deuterium isotope effects for the reactions depicted in eqns. (18) and (19). In addition they examined the kinetic sulfur-34 isotope effect for reaction (18). The isotope effects were $k_H/k_D = 1.094 \pm 0.014$, $k_H/k_D = 1.05 \pm 0.019$ and ${}^{32}k/{}^{34}k = 1.009$. The results are consistent with a concerted mechanism, but a two-step mechanism cannot be excluded by the data. Such an interpretation would, however, be difficult because in reaction (19), a two step mechanism should result in prior cleavage of the C–S bond which would give rise to a secondary radical. In such a case, no significant intermolecular isotope effect would be expected, although a β isotope effect transmitted through a double bond could possibly be invoked. The sulfur primary isotope effect gives little information other than that the bonding to sulfur is diminished in the rate-determining transition state.

F. Allene Cycloadditions

Another approach toward probing the simultaneity of cycloaddition processes in general has evolved from Dai and Dolbier's broad examination of secondary deuterium isotope effects in allene cycloadditions [21,34–36]. Two approaches were taken by these workers: (a) an examination of intramolecular isotope effects for various cycloaddition reactions of 1,1-dideuterioallene, including [2 + 2], [2 + 3], and [2 + 4] processes, and (b) a comparison of these intramolecular isotope effects with intermolecular effects in a number of the same systems with the goal being to establish whether the rate-determining step and the product-forming step were one and the same.

One of the problems in interpreting most of the earlier work on secondary deuterium isotope effects was the necessity of creating theoretical models to rationalize experimental isotope effects, where often the model for the alternative mechanism

$CH_2{=}C{=}CH_2$ + Cl_6-cyclopentadiene $\xrightarrow{150°}$ Cl_6-methylenenorbornene (80%) (20)

$CH_2{=}C{=}CH_2$ + Cl_4-5,5-dimethoxycyclopentadiene $\xrightarrow{130°}$ Cl_4-7,7-dimethoxy-methylenenorbornene (75%) (21)

$CH_2{=}C{=}CH_2$ + tetracyanoethylene oxide $\xrightarrow{130°}$ tetracyano-methylenetetrahydrofuran (75%) (22)

$CH_2{=}C{=}CH_2$ + $CH_2{=}CHCN$ $\xrightarrow{210°}$ 3-methylenecyclobutanecarbonitrile (60%) (23)

$CH_2{=}C{=}CH_2$ + $CCl_2{=}CF_2$ $\xrightarrow{162°}$ 2,2-dichloro-3,3-difluoro-methylenecyclobutane (75%) (24)

2 $CH_2{=}C{=}CH_2$ $\xrightarrow{130°}$ 1,2-dimethylenecyclobutane (91%) (25)

2 cyclic allene $\xrightarrow{138°}$ dimer (95%) (26)

TABLE 4

SECONDARY DEUTERIUM ISOTOPE EFFECTS IN ALLENE CYCLOADDITIONS

Reaction (eqn. no.)	Cyclo-addition type	Intramol. k_H/k_D	Intermol. k_H/k_D	Equilibrium k_H/k_D
20	[2 + 4]	0.90 ± 0.02	0.90 ± 0.04	
21	[2 + 4]	0.92 ± 0.02		
22	[2 + 3]	0.93 ± 0.01		
23	[2 + 2]	1.13 – 1.21	1.04 ± 0.05	0.92 ± 0.01
24	[2 + 2]	1.17 ± 0.04		
25	[2 + 2]	1.14 ± 0.02	1.02 ± 0.03	
26	[2 + 2]	1.10 ± 0.03		

could be construed to fit the data. It was believed that by having data for both alternative mechanisms in allene cycloadditions that distinctions could more effectively be made for the mechanism involved in a particular sytem.

Dai and Dolbier examined a broad spectrum of cycloaddition reactions: two Diels–Alder reactions, the reactions of allene with hexachlorocyclopentadiene (eqn. (20)) and with 5,5-dimethoxytetrachlorocyclopentadiene (eqn. (21)); one [2 + 3] cycloaddition, the 1,3-dipolar cycloaddition of tetracyanoethylene oxide with allene (eqn. (22)); two [2 + 2] cycloadditions, the reaction of allene with acrylonitrile (eqn. (23)) and with 1,1-dichloro-2,2-difluoroethylene (eqn. (24)); and two dimerization processes formally [2 + 2] cycloadditions, that of allene itself (eqn. (25)) and of 1,2-cyclononadiene (eqn. (26)).

The isotope effects for the reactions are summarized in Table 4. Intramolecular isotope effects were obtained using 1,1-dideuterioallene in intramolecular competition. Intermolecular isotope effects were obtained through competition experiments using tetradeuterioallene and undeuterated allene. Each of the isotope effects determined was shown to be a kinetic isotope effect with one equilibrium isotope effect having been obtained for comparison purposes.

Where both new σ bonds are formed in a concerted reaction, an inverse intramolecular isotope effect would be expected, and since in this case the rate-determining step would also be product-forming, the intermolecular isotope effect should be of the same magnitude. As can be seen, these expectations were borne out for the [2 + 4] processes and for the [2 + 3] process reactions which according to Woodward–Hoffmann predictions should be concerted. On the other hand, a normal isotope effect was observed for all of the [2 + 2] processes, including the dimerizations. A normal effect is not consistent with expectations for a concerted reaction, but it can be rationalized for a two-step reaction, as will be discussed below. The non-identity of the intra- and intermolecular isotope effects in reactions (23) and (25) moreover indicates that the rate-determining and product-forming transition state *cannot* be identical; thus the mechanisms for these reactions must be *multi-step*.

While such comparisons clearly indicate the existence of an intermediate, the nature of the intramolecular isotope effect, i.e. the fact that it is normal, demands an explanation. Undoubtedly, the product-forming step, which gives rise to the intramolecular effect, involves a conversion of the isotopically labelled site from sp^2 to sp^3 hybridization, with a concomitant increase in steric crowding of the C–H bonds resulting in an *increase* in the C–H vibrational frequencies. Why then is a normal isotope effect observed?

To explain this one must first understand the nature of the product-forming step in these two-step processes. Examining the allene–dichlorodifluoroethylene reaction in detail (eqn. (27)) it can seen that the product-forming step involves the destruction of

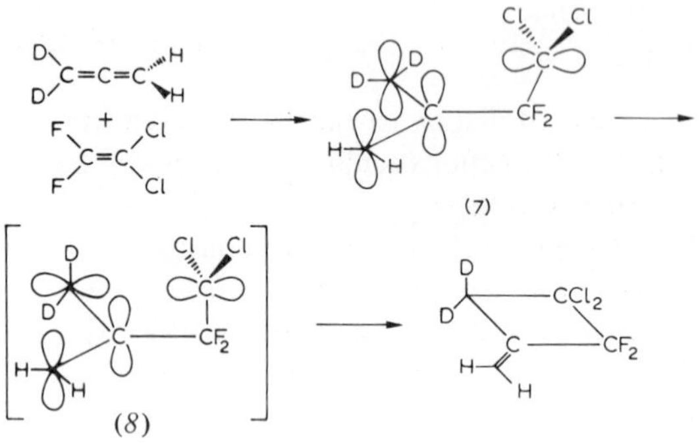

(7) (27)

(8)

a stabilized biradical (*7*). It has been established that the combination of two simple, non-stabilized radicals requires little or no activation energy. On the other hand [37], the destruction of a biradical such as (*7*) apparently requires an activation energy of approximately 13 kcal mol^{-1}. It is probable that the source of this activation energy is the energy required to rotate one of the methylenes of the allyl radical towards an orthogonal conformation, such as is depicted in a hypothetical transition state structure (*8*), which would have the correct geometry for bonding to occur. Thus there apparently need be little or no bonding in such a transition state and the isotope effect would derive from a loosening of the rigidity and the steric requirements of the planar allyl system. With this conception of the product-forming transition state it is not so surprising that a normal isotope effect is observed.

It is significant that earlier Crawford and Cameron [38] had observed a comparable isotope effect in the cyclization of a trimethylenemethane biradical. At that time the isotope effect was considered anomalous. Now it fits into the general picture nicely. The

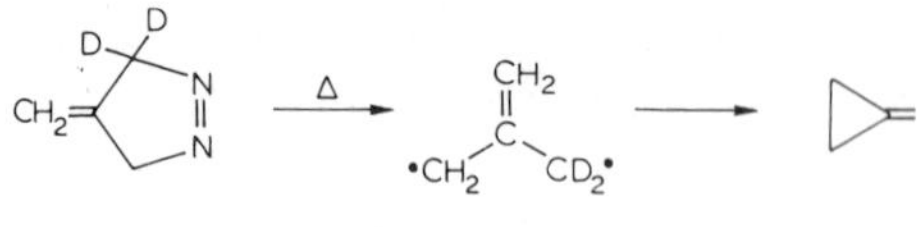

(28)

$k_H/k_D = 1.37 \pm 0.05$

fact that the isotope effect is so much larger in this cyclization might be rationalized on the basis that in order for a planar trimethylenemethane to convert to the product methylenecyclopropane, *two* CH_2 groups must rotate into an orthogonal situation, as opposed to *one* CH_2 group in the two-step cycloadditions. Thus Crawford's effect should be approximately the square of that observed in the allene systems, which is the case.

It was also of interest that under equilibrium conditions, the isotope effect in the acrylonitrile–allene system indicated an accumulation of an excess deuterium in the more sterically congested, sp^3, ring position.

Clearly, the study of such a broad range of reaction types in the allene investigation creates a situation in which comparison of the nature of the observed isotope effects in order to reach mechanistic conclusions becomes quite effective. In reaching these conclusions, one is not tied to an analysis of the absolute magnitude of the isotope effects. Thus in these systems, one has been able to gain significant insight into both the simultaneity of the cycloaddition bond-making process and the nature of the transition state for biradical destruction where biradicals are in fact involved.

Moore et al. [39] have extended the confidence of the conclusions reached above for allene dimerizations. They observed a k_H/k_D = 1.04 for the dimerization of 1-deuterio-1, 2-cyclohexadiene, a system which had independently been demonstrated

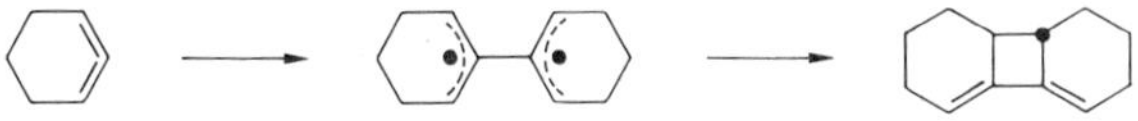

(29)

to involve biradicals as intermediates [40].

G. Electrocyclic Reactions

Electrocyclic reactions may of course be considered as intramolecular cycloadditions. While the literature abounds with stereochemical evidence to indicate that electrocyclic processes proceed in accordance with orbital symmetry predictions, there has been very little activity in the area of isotope effects in electrocyclic reactions.

One very elegant study was reported by Winter and Honig [41] in which they describe the "steric" intramolecular isotope effect in the conrotatory ring opening of *trans*-3, 4-dideuterio-1, 2-bis(trimethylsiloxy)cyclobutene to the two conrotatory diene products.

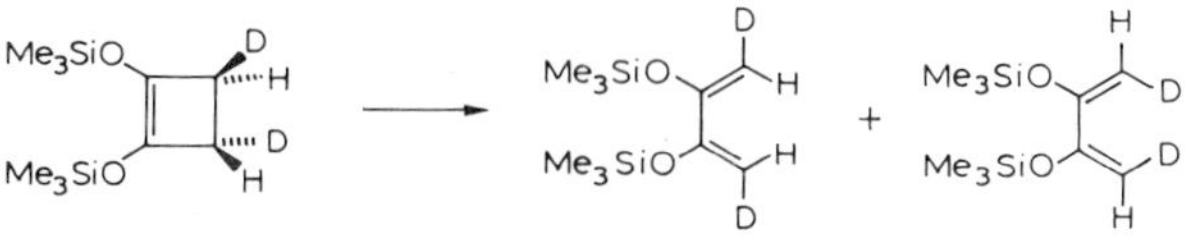

Me$_3$SiO, Me$_3$SiO, D, H, H, D → Me$_3$SiO, Me$_3$SiO, D, H, H, D (9a) + Me$_3$SiO, Me$_3$SiO, H, D, D, H (9b) (30)

It was found that isomer (*9b*) was favored kinetically over isomer (*9a*) with the ratio of the pair being 1.10. The authors interpret this selectivity in terms of a steric preference in the transition state for C–D, C–D interaction over C–H, C–H interaction. Moreover the authors assert that the models suggest that the necessary crowding in the transition state will only occur if the extent of C_3–C_4 bond cleavage is small. The interpretability of such an isotope effect as this is of course somewhat diminished by the lack of direct analogy with which to compare the effects. A rule of thumb prediction seems reasonable at this time, but only with more examples in many different types of cyclizations and ring opening processes will one be able to reach conclusions with any degree of confidence.

IV. SIGMATROPIC REARRANGEMENTS

The world of sigmatropic rearrangement can be construed to encompass essentially all thermal reorganization processes, be they orbital-symmetry-allowed or not. There has not been a great deal of application of isotope effect criterion to the investigations of the mechanisms of such reactions. In the few cases where they have been applied, however, the rewards have proven well worth the effort.

A. Cope-type Rearrangements

The Cope rearrangement has been probed in detail utilizing both thermodynamic and kinetic secondary deuterium isotope effects. Sunko and his co-workers [42] first examined the thermodynamic equilibrium between (*10a*) and (*10b*) and found the ratio of

CD_2 CD_2 (*10a*) ⇌ D D D D (*10b*) (31)

(*10b*)/(*10a*) = 1.23 at 200° (calculated to be ~1.41 at 25°). These results are completely consistent with expectations, since the deuterium would be expected to preferentially accumulate at the more sterically congested sp^3 position. Since in a thermodynamic equilibrium situation one is comparing the relative ground state energies of the two isomers, one can calculate the difference in the zero-point free energies between the two species, to be 204 cal mol^{-1}.

The same authors [43] have also determined the kinetic isotope effects for a Cope rearrangement, ascertaining the isotope effect both at the bond-breaking site and at the bond-making site (eqns. (32) and (33)).

CN Et Me NC D D $\xrightarrow{k_H/k_D\ =\ 1.19\pm0.03}$ NC Et NC Me D D (32)

$$\text{NC(CN)C(Et)=CHMe, CH}_2\text{CH=CD}_2 \xrightarrow{k_H/k_D = 0.94 \pm 0.02} \text{(NC)}_2\text{C=C(Et)CH(Me)CD}_2\text{CH=CH}_2 \quad (33)$$

The thermodynamics of this Cope rearrangement are known, the E_{act} = 26 kcal mol^{-1} while the ΔH° is approximately −5 kcal mol^{-1}. Thus one would expect, on the basis of Hammond's Postulate, that the transition state would lie slightly less than halfway along the reaction coordinate. Looking at the two secondary isotope effects, one might be tempted to assert that they provide evidence that bond breaking is much farther along than bond making in this reaction. As the authors point out, this kind of a conclusion may not be justified. Apparently the force constant in bond breaking or making is not a linear function of bond distance. In bond breaking the force constant decreases very rapidly initially upon a relatively small increase in bond distance. Likewise in bond formation the force constant only increases dramatically towards the end of the approach of the two atoms undergoing bonding.

It is also believed that the kinds of steric interactions which give rise to secondary deuterium isotope effects in bond-making or bond-breaking processes are also not a linear function of bond distance but maximize very early in bond breaking and very late in bond formation.

While the authors make an attempt to correlate the thermodynamic with the kinetic isotope effects, we will see that in general there need be no correlation. Recognizing the particular symmetry of the concerted process of the Cope rearrangement, such a correlation may be possible in this case, but one must be wary in general, since the structure of transition states which give rise to kinetic effects need not be related to product structures which give rise to thermodynamic effects.

McMichael [44] examined the kinetic secondary deuterium isotope effects in a related reaction, that of the conversion of allylthionbenzoate to allyl thiolbenzoate (eqn. (34)). The isotope effect for α,α-dideuterio species was k_H/k_D = 1.12 (1.06 per

$$\phi{-}\overset{\overset{\displaystyle S}{\|}}{C}{-}O{-}\overset{\alpha}{C}H_2CH{=}\overset{\gamma}{C}H_2 \longrightarrow CH_2{=}CHCH_2{-}S{-}\overset{\overset{\displaystyle O}{\|}}{C}{-}\phi \quad (34)$$

deuterium) while that for γ,γ-dideuterio was k_H/k_D = 0.94 (0.97 per deuterium). A large negative entropy of activation is strongly indicative of a cyclic transition state being involved and the isotope effect at the γ position indicates that it is involved to some extent in bond making in the transition state. Sunko and co-workers [45] had found earlier that for a simple ionization of α,α-dimethylallyl chloride, there was *no* isotope effect for deuterium substitution at the γ-position. Their results moreover are completely consistent with those on the Cope rearrangement as described above.

B. Vinylcyclopropane Rearrangements

In their study of allene cycloadditions, Dai and Dolbier apparently were able to detect the existence of π-delocalized biradical intermediates through the observation of a normal intramolecular deuterium isotope effect for the destruction of these biradicals. In an extension of this work, it was hoped that biradicals generated by other means, such as in thermal reorganizations, might also behave in a similar manner and thus be indirectly detected.

With this in mind, Dolbier and Alonso [46] examined the thermal reorganization

CD_2 CH_2 (11) —240°→ CD_2 CH_2 CH_2 → CD_2 H H + CH_2 D D (35)

of 1,1-divinylcyclopropane to 1-vinylcyclopentene. Utilizing the unsymmetrically deuterated species (*11*) they ascertained the intramolecular isotope effect to be k_H/k_D = 1.07 ± 0.02. Thus the expectation of a normal isotope effect for the destruction of the intermediate pentadienyl radical was fulfilled. The fact that the value was significantly smaller than the 1.15 value which was generally observed in the intermediate allyl radical systems, is not surprising since the pentadienyl system has more extended conjugation and should thus have a lower energy barrier for rotation out of planarity. This should result in less discrimination.

The intermolecular isotope effect which was ascertained from the ratio of rates of the tetradeuterio and undeuterated species was, however, also found to be significantly normal, k_H/k_D = 1.08±0.07. This led to ambiguity in the overall mechanistic interpretation. The essential identity of the two isotope effects certainly demands that a concerted mechanism be seriously considered. This interpretation was not favored, however, because of the *normal* nature of the effect. The preponderant weight of analogy insists that a rate-determining conversion of an sp^2 to an sp^3 carbon should be associated with an inverse kinetic secondary deuterium isotope effect.

On the other hand, with regard to the two-step process there is some analogy to expect a normal intermolecular isotope effect for the conversion sp^2 to radical [25] in which case the normal intermolecular effect would derive from the rate-determining formation of the biradical, while the normal intramolecular effect would derive from its product-forming destruction. If the second step were rate-determining, then the rate-determining and product-forming transition states would be identical as would be the isotope effects.

C. Methylenecyclopropane Rearrangements

The methylenecyclopropane rearrangement has elicited a tremendous amount of effort directed towards the elucidation of its detailed mechanism [47]. Because of the general reversibility of these reactions, kinetic isotope effects have not been easily examined.

Recently, Dolbier and Alonso [48] examined the irreversible conversion of dideuteriobiscyclopropylidene to three isomeric isotopically labelled methylenespiropentane products (eqn. (36)).

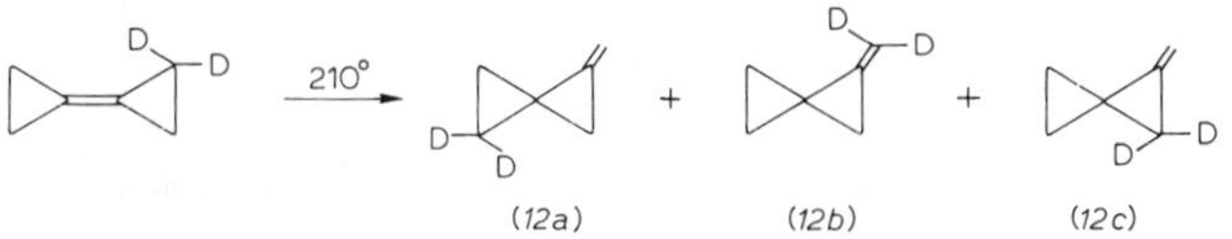

(12a) (12b) (12c) (36)

In this rearrangement, two distinct secondary isotope effects are operative as depicted in eqn. (37). Ratios of the three products will provide one with the values for the isotope effects.

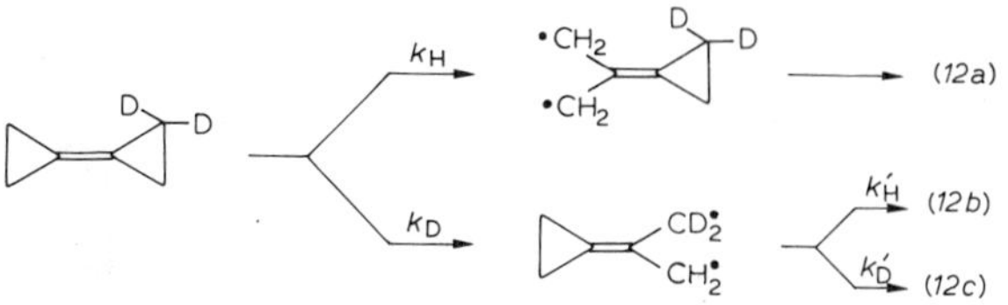

(37)

$$(12b)/(12c) = (k'_H/k'_D)_{cyclization} = 1.14 \pm 0.02 \quad (38)$$

$$(12a)/(12b) + (12c) = (k_H/k_D)_{cleavage} = 1.24 \pm 0.03 \quad (39)$$

The observed isotope effects are consistent with biradical mechanisms i and ii depicted in eqn. (40), and also with one possible concerted mechanism. It is believed that the large normal $(k_H/k_D)_{cleavage}$ arises from formation of transition state (*13*), where

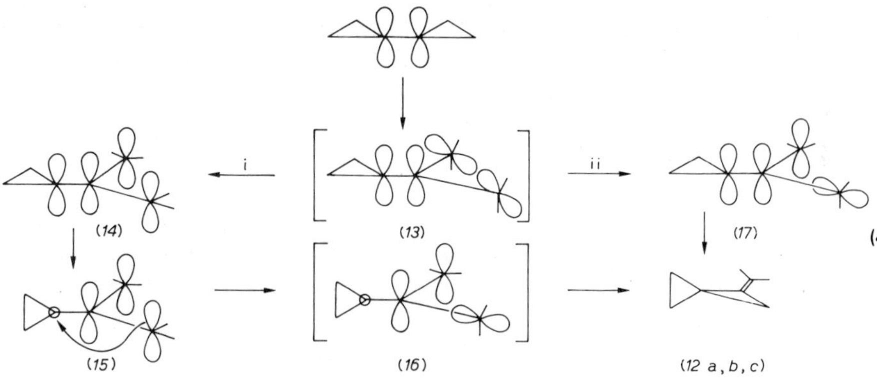

(40)

C–C bond breaking is occurring with some CH_2 twisting, to produce either the planar trimethylenemethane intermediate (*14*) and then (*15*), or the "pivot" intermediate (*17*), either of which can cyclize as shown to methylenespiropentane. The cleavage isotope effect can thus be rationalized in terms of a "steric" secondary deuterium isotope effect. That is, a normal isotope effect is expected if the transition state is "looser" or "less rigid" than the starting material. One very significant vibrational change undergone in proceeding to the transition state is the drastic lowering of the torsional force constant for the CH_2 twisting motion. This is a reflection of the reduction of the rotational barrier for the CH_2 twisting motion as the C–C bond is weakened.

The $(k'_H/k'_D)_{cyclization}$ is similar to those observed in other allyl radical cyclization processes. This normal effect may be rationalized via process i as arising from a reduction in "crowding" and/or rigidity of the species (*15*) in its procession to transition state (*16*). The effect can be rationalized for path ii if one considers that intermediate (*17a*) will be lower in energy than (*17b*).

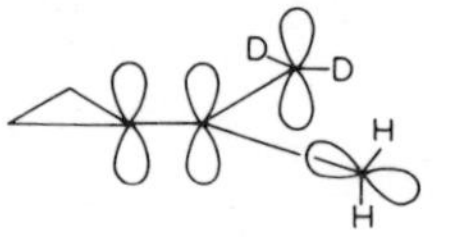

(17a) 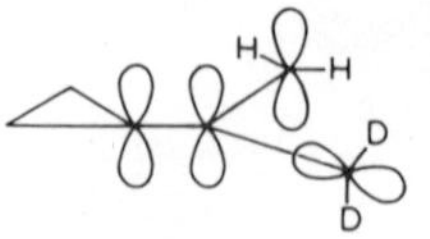 (17b)

In considering a concerted mechanism, the "pivot" type of transition state (transition state (*18*)) can rationalize the data if one assumes that the pivot methylene is the loosest CH_2 group in the transition state.

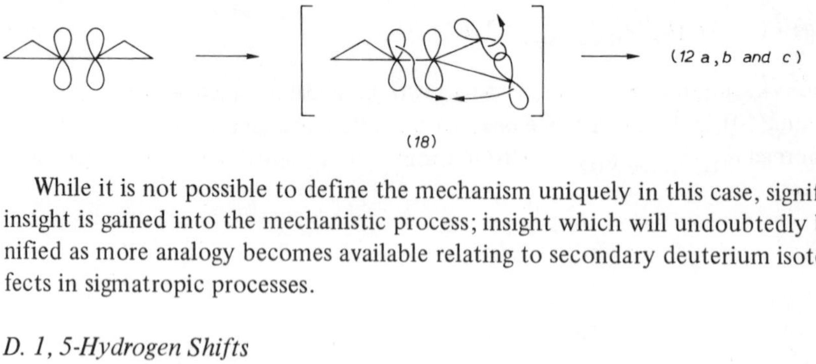

(12 a, b and c) (41)

(18)

While it is not possible to define the mechanism uniquely in this case, significant insight is gained into the mechanistic process; insight which will undoubtedly be magnified as more analogy becomes available relating to secondary deuterium isotope effects in sigmatropic processes.

D. 1, 5-Hydrogen Shifts

1,5-Hydrogen shifts in acyclic (eqn. (42)) and cyclic (eqn. (43)) 1,3-diene systems are very facile processes occurring with activation energies of approximately 35 and 20 kcal mol^{-1}, respectively [49].

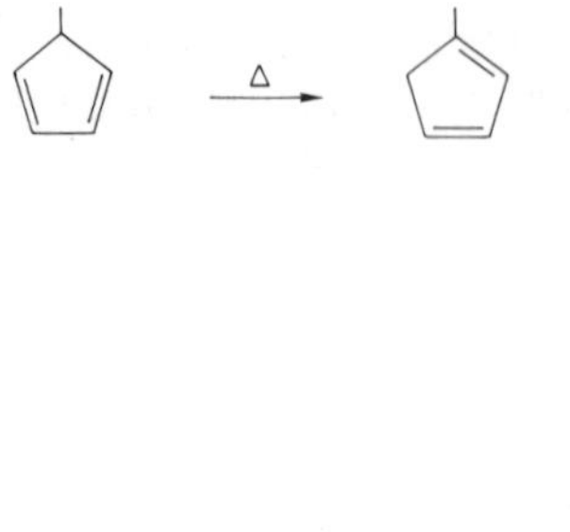

(42)

(43)

Orbital symmetry theory predicts that such rearrangements should be allowed, concerted suprafacial processes. Moreover, one would expect that since the thermodynamic driving force for such reactions is not overwhelming, the transition state should be relatively symmetrical and that primary kinetic hydrogen isotope effects should be relatively large.

Indeed, two isotope effect investigations have verified this expectation. Kloosterziel and Ter Borg [50] examined the thermal reorganization of the symmetrically deuterated 1,3,6-cyclooctatriene system (eqn. (44)), and found the primary intramolecular isotope effect to be $k_H/k_D = 5.0$ and the secondary isotope to be rather small ($k_H/k_D = 1.02 \pm 0.002$).

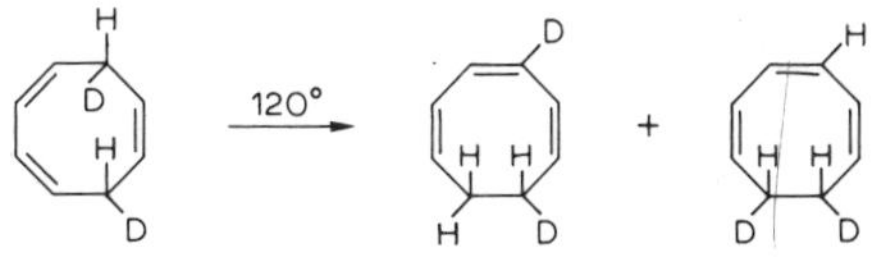

(44)

Roth and König [51] examined the even more symmetrical system of 1,3-pentadiene (eqns. (45) and (46)) and determined the intermolecular isotope effect to be $k_H/k_D = 12.2$ at 25°. They attributed this very large isotope effect to the highly symmetrical transition state which would be expected in this system. It should be mentioned that secondary isotope effects, both at the side of bond breakage and at the site of new C–H bond formation would tend to enhance the k_H/k_D ratio in this com-

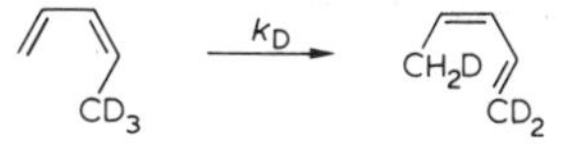

(45)

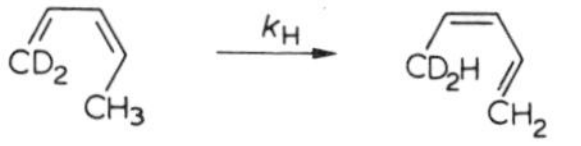

(46)

petitive system and while these effects may be relatively small, the authors have no way of estimating them from their data.

V. ENE REACTIONS

A process very similar mechanistically both to the Diels–Alder reaction and to the 1,5-hydrogen shift is the ene reaction which can be presented very simply by eqn. (47). Applying orbital symmetry principles to this system, one finds that it should be an

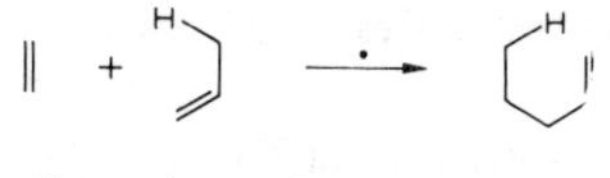

(47)

allowed, totally suprafacial concerted thermal process. Indeed there is much kinetic and stereochemical data which provide impressive evidence that the ene reaction is a synchronous process.

Ene reactions in systems which contain no hetero atoms have generally been very difficult to observe as clean reactions. On the other hand certain carbonyl compounds, azodicarboxylates and singlet oxygen have proven to be excellent ene components in their reactions with alkenes. It is generally in such systems that isotope effect studies have evolved as an effective mechanistic probe.

A single ene reaction with no hetero atom involved has been examined with respect to deuterium isotope effects. Foote and Mazur [52] studied the reaction of hexadeuterio-2,3-dimethyl-2-butene with dimethyl acetylenedicarboxylate.

$$(CH_3)_2C{=}C(CD_3)_2 + MeO_2C{-}C{\equiv}C{-}CO_2Me \xrightarrow{125^\circ} CH_2{=}C(CH_3){-}C(CD_3)_2{-}C(CO_2Me){=}CH(CO_2Me) + CD_2{=}C(CD_3){-}C(CH_3)_2{-}C(CO_2Me){=}CD(CO_2Me) \qquad (48)$$

k_H/k_D = 2.33±0.15

This intramolecular isotope effect is consistent with a concerted mechanism, but without the availability of intermolecular isotope effect data a non-concerted mechanism cannot be ruled out.

Dai and Dolbier [53] examined the very facile reaction between perfluorocyclobutanone (*19*) and various deuterated allenes (eqn. (49)).

$$F_6\text{-cyclobutanone } (19) + CH_2{=}C{=}CH_2 \xrightarrow{90^\circ} F_6\text{-cyclobutyl}(OH){-}CH_2{-}C{\equiv}C{-}H \qquad (49)$$

Intramolecular isotope effects were determined using 1,1-dideuterio- and 1,3-dideuterioallene while an intermolecular isotope effect was determined utilizing tetradeuterioallene.

The two intramolecular isotope effects, $(k_H/k_D)_{1,1} = 3.54 \pm 0.03$ and $(k_H/k_D)_{1,3} = 2.17 \pm 0.05$, and the intermolecular isotope effect, $(k_H/k_D)_{d_4} = 1.99 \pm 0.07$, are clearly non-identical in value. All can be understood as being different combinations of primary and secondary deuterium isotope effects, acting in a concerted mechanism.

In a concerted process one would expect a significant primary isotope effect due to proton transfer, as well as two secondary isotope effects due to changes in hybridization at both C-1 ($sp^2 \rightarrow sp^3$) and C-3 ($sp^2 \rightarrow sp$).

Letting X be the primary effect, Y be the former secondary effect, and Z be the latter secondary effect one sees that $(k_H/k_D)_{1,1} = XZ/Y^2$, $(k_H/k_D)_{1,3} = X/Z$, and $(k_H/k_D)_{d_4} = XY^2Z$. Solving the three simultaneous equations, we find that X = 2.41 Y = 0.868, and Z = 1.10, all this assuming a concerted mechanism.

A small primary isotope effect such as 2.41 might be expected for a reaction where the hydrogen transfer does not take place through a *linear* transition state. More O'Ferral [5] has found that primary isotope effects depend smoothly upon the β angles in the transition state, with a maximum being obtained where $\beta = 180°$ (see structure (*20*)). Using this method of calculation, an angle of 100° for β is obtained

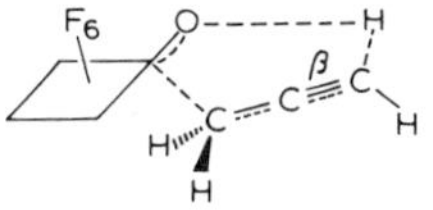

(*20*)

for the ene transition state, a value which certainly would be consistent with expectations.

The magnitude of the secondary deuterium isotope effect for $sp^2 \rightarrow sp^3$ rehybridization at C-1 was estimated to be $(k_H/k_D) = 0.868$. This value is comparable to those secondary effects mentioned earlier in systems where similar changes in bonding occurred, and the effect is consistent with substantial C–C bond formation in the concerted transition state.

Lastly, the value of the secondary deuterium isotope effect for $sp^2 \rightarrow sp$ rehybridization at C-3 was estimated to be $k_H/k_D = 1.10$. There are no other reported values for such a process although a negligibly small effect has been observed for the reaction of propargyl chloride with hexachlorocyclopentadiene [54]. Certainly a value of 1.10 is not inconsistent with a concerted mechanism.

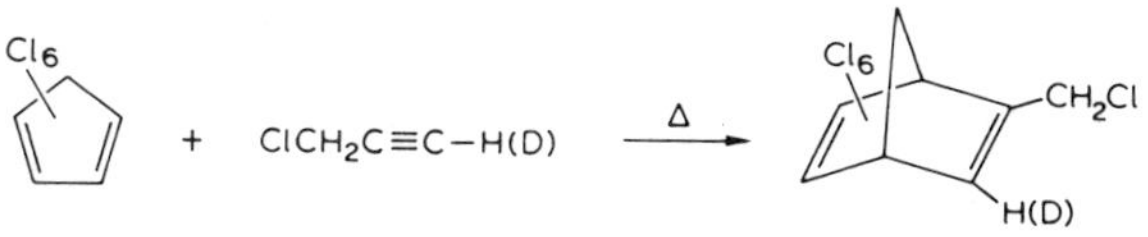

$$Cl_6\text{-cyclopentadiene} + ClCH_2C{\equiv}C{-}H(D) \xrightarrow{\Delta} \text{adduct } (CH_2Cl,\ H(D)) \qquad (50)$$

$k_H/k_D = 1.03 \pm 0.05$

It should also be mentioned that a mechanism involving a fully established equilibrium formation of an intermediate such as (*21*) would also be made consistent with the data. To the extent that such an equilibrium was *not* established, $(k_H/k_D)_{1,3}$ should have been greater than $(k_H/k_D)_{1,1}$. The opposite was actually observed.

Kinetic isotope effects in systems where azodicarboxylic esters are the enophile species have also been determined. Huisgen and Pohl [55] determined the intramolecular isotope effect in the reaction of diethyl azodicarboxylate with the labelled

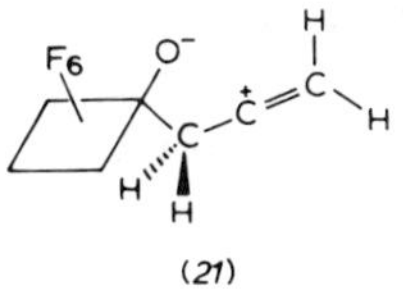

(*21*)

1,4-dihydronaphthalene species, (*22*).

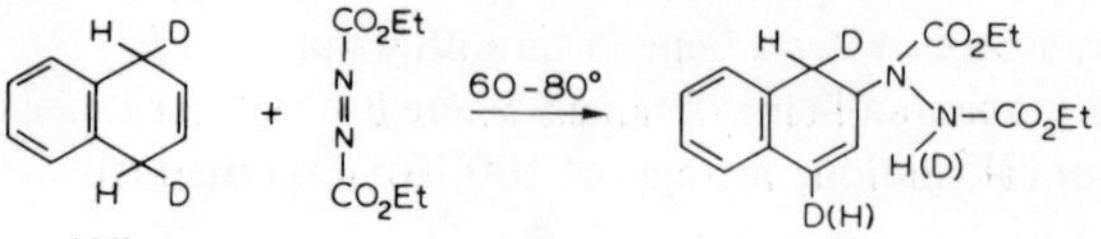

(51)

(*22*)

$k_H/k_D = 2.8$ at 80°

Mazur and Foote [52] have also examined the intramolecular isotope effect for a similar ene reaction, that of hexadeuterio-2,3-dimethyl-2-butene with dimethyl azodicarboxylate.

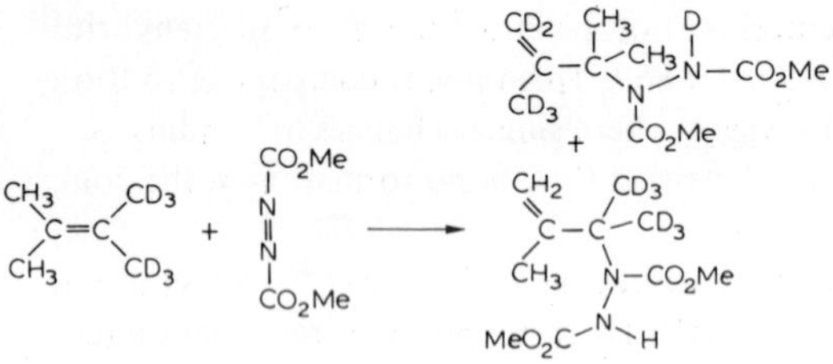

(52)

$k_H/k_D = 3.85$ at 78°
$= 2.97$ at 132°

There is also a report of an isotope effect for the reaction of diethyl azodicarboxylate with 7-deuteriocycloheptatriene [56].

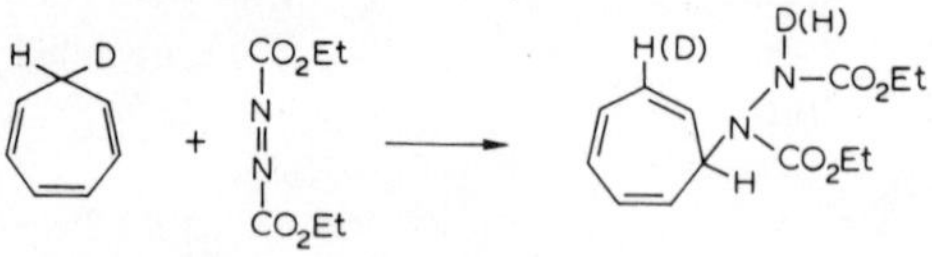

(53)

$k_H/k_D = 6.1$

These primary isotope effects while all being consistent with a concerted mechanism, do not provide conclusive evidence for such a pathway. They are all intramolecular effects and it would be worthwhile to have an analogous intermolecular isotope effect with which to make quantitative comparisons so as to gain knowledge of the simultaneity of the rate-determining and product-forming steps.

There is much evidence that the allylic hydroperoxidation of mono-olefins by singlet oxygen also proceeds by a cyclic ene mechanism. Primary deuterium isotope effects have been determined for such reactions in a number of systems. Nickon et al. [57] have, for example, determined the intramolecular isotope effect for the reaction of singlet oxygen with labelled molecules (*23*), (*24*) and (*25*) and found the isotope effects to be slightly dependent on the method of generating oxygen ($^1\Delta_g$) [57]. Struc-

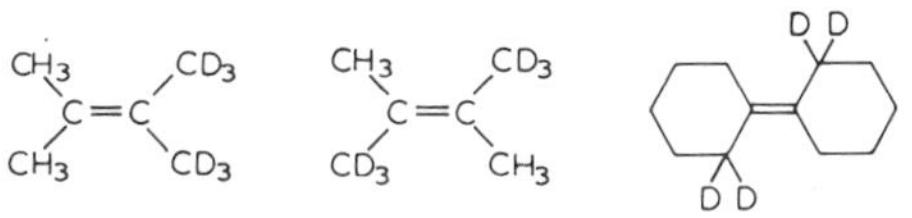

CH_3 CD_3 C=C CH_3 CD_3 — (23)

CH_3 CD_3 C=C CD_3 CH_3 — (24)

(25)

ture (*23*) gave a k_H/k_D = 1.78–2.42 depending on the source, structure (*24*) a k_H/k_D = 1.28–1.77 and structure (25) a k_H/k_D = 1.19. Such low isotope effects can be rationalized in terms of a very unsymmetrical transition state. Since singlet oxygen is far more reactive than the azodicarboxylates or acetylenedicarboxylates, one would expect the transition state geometry to resemble the starting materials and the C–H bond might thus be weakened very little in such a transition state. These conclusions are consistent with the finding that the thermodynamic stability of the final double bond has little effect on the orientation of the reaction and with the fact that the susceptibility of C–H to abstraction is not inherently related to whether this bond is primary, secondary or tertiary. Again a study of intermolecular effects would contribute much toward knowledge of the simultaneity of this type of ene reaction.

VI. RETROENE REACTIONS

There are many reactions of organic molecules which can be considered formally as retroene reactions. Most of these are six-membered-ring transition state fragmentation processes which are not generally thought of in this way. However, ester pyrolyses, Tschugaeff reactions, and other eliminations involving cyclic transition states may all be thought of, electronically, as formally retroene reactions. While not attempting to provide a complete review of the literature regarding studies of isotope effects in these systems, a number of such studies are presented to provide an overview of expectations for such reactions.

Curtin and Kellom [58] observed an intramolecular primary isotope effect in the ester pyrolysis shown in eqn. (54), while Blades and Gilderson [59] determined the intermolecular primary isotope effect for the system in eqn. (55). In another intramolecular competition Depuy et al. [60] found the isotope effect for the reaction depicted in eqn. (56).

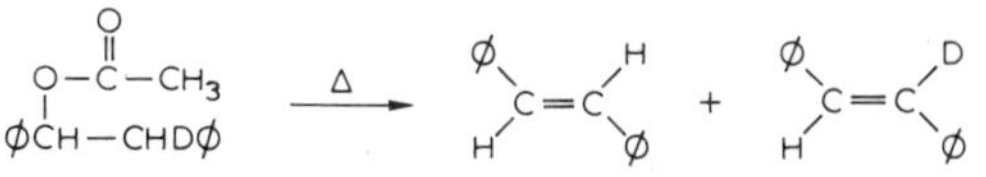

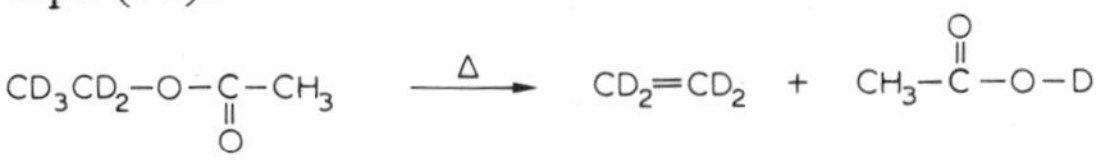

$$\phi CH(OCOCH_3)-CHD\phi \xrightarrow{\Delta} \phi HC{=}CH\phi + \phi HC{=}CD\phi \qquad (54)$$

k_H/k_D = 2.8±0.6 at 400°

$$CD_3CD_2-O-C(=O)-CH_3 \xrightarrow{\Delta} CD_2{=}CD_2 + CH_3-C(=O)-O-D \qquad (55)$$

k_H/k_D + 2.4 at 380°

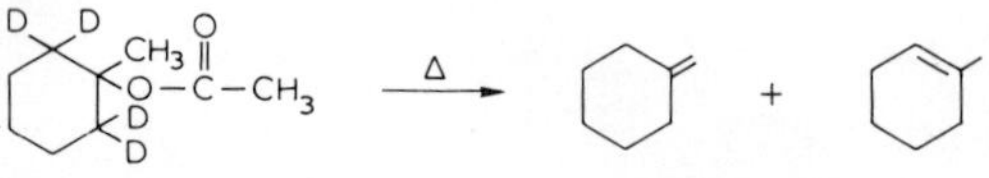

$$k_H/k_D = 1.9 \text{ at } 400°$$ (56)

The observation of both significant intermolecular and intramolecular isotope effects in these eliminations is strongly indicative of a cyclic, concerted process being involved.

A study of the heavy atom isotope effects in the Tschugaeff reaction was done by Bader and Bourns [61]. Three different kinetic isotope effects were measured: the

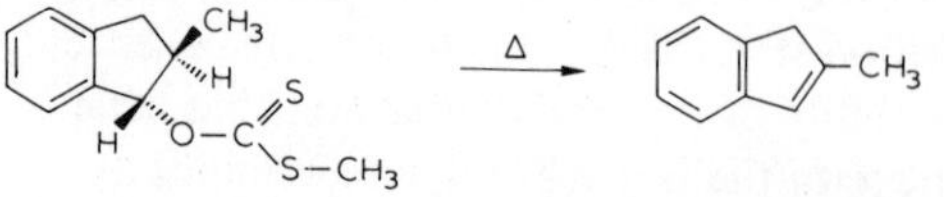

(57)

$^{32}S/^{34}S$ effect for the thioether sulfur linkage, the $^{32}S/^{34}S$ effect for the thion sulfur, and the $^{12}C/^{13}C$ effect for the carbonyl carbon. The isotope effects were respectively, 1.0021 ± 0.0007, 1.0086 ± 0.0016 and 1.0004 ± 0.0006.

Bader and Bourns sought to distinguish between two possible mechanisms depicted in eqns. (58) and (59).

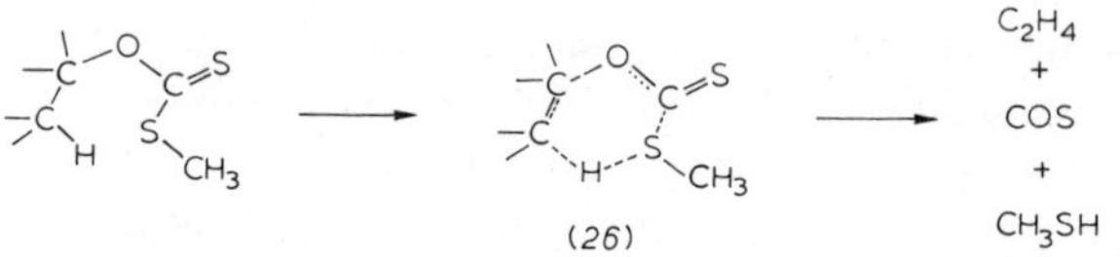

(26) → C_2H_4 + COS + CH_3SH (58)

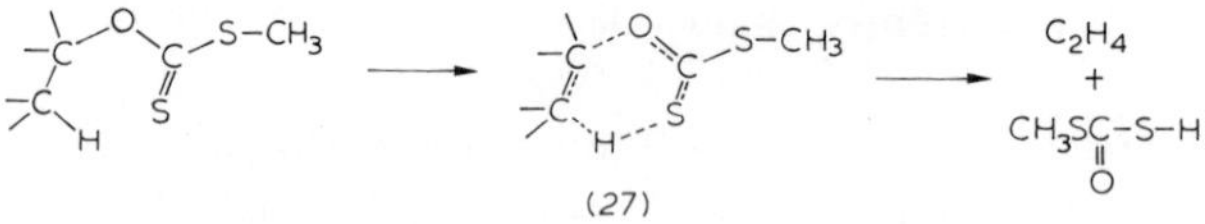

(27) → C_2H_4 + $CH_3SC(=O)-S-H$ (59)

It was predicted that passing through transition state (*26*) should give rise to significant isotope effects at the thioether sulfur and at the carbonyl carbon and negligible effect at the thion sulfur. In contrast, passing through transition state (*27*) should give rise to negligible isotope effect at the former two sites and a significant one at the thion sulfur.

In fact, of course, the results are more consistent with the mechanism in which the thion sulfur is the abstracting atom (eqn. (59)). Since the authors have considered only two such models one must ask if there are not other possibilities which might also explain the data. Indeed, a two-step mechanism, whereby the thion sulfur abstracts a proton in a rate-determining step, followed by a second rapid cleavage step should also explain the results. It does appear clear from these authors' work, however, that the abstracting atom is the thion sulfur and not the thioether sulfur.

VII. CONCLUSION

The utilization of isotope effects in the elucidation of mechanism in pericyclic processes appears to be gaining popularity and credibility as being a viable and effective tool in mechanistic investigations. In the past few years, increasing numbers of careful studies have been carried out, designed to demonstrate the effectiveness of this probe. Such effectiveness depends, to a great extent, upon the availability of data in the literature with which to compare new data. There are many as yet unexplained isotope effects which demand explanations, explanations which will come only after additional work with various systems carefully designed to probe the nature and source of such isotope effects. It now appears that secondary isotope effects have much potential for probing both the simultaneity of a mechanistic pathway and the structure of transition states in various steps along the way. In a field requiring analogy for comparisons, there is still not nearly enough data. Thus it is believed that future endeavors in this field will open new doors towards gaining detailed mechanistic insight into pericyclic processes.

It should be pointed out, however, that isotope effects do not, in general, provide a self-sufficient mechanistic probe for determining the mechanisms of the reactions under consideration in this account. They provide a unique insight but can never replace other mechanistic probes such as product analysis, stereochemical studies, kinetic studies, studies of substituent effects and medium effects, tracer studies, etc. Isotope effect studies should be recognized along with these others as having a unique role in effective mechanistic elucidation.

REFERENCES

1 J. Bigeleisen, J. Chem. Phys., 17 (1949) 675.
2 J. Bigeleisen and M. Wolfsberg, Advan. Chem. Phys., 1 (1958) 15.
3 M.J. Stern and M. Wolfsberg, J. Chem. Phys., 45 (1966) 15.
4 C.J. Collins and N.S. Bowman, Isotope Effects in Chemical Reactions, Van Nostrand Reinhold Company, New York, 1970.
5 R.A. More O'Ferral, J. Chem. Soc., B, (1970) 785.
6 J.E. Bertie and M.G. Norton, Can. J. Chem., 48 (1970) 3889.
7 I. Safarik and O.P. Strausz, J. Phys. Chem., 76 (1972) 3613.
8 M.J. Goldstein and G.L. Thayer, Jr., J. Amer. Chem. Soc., 87 (1965) 1933.
9 C.B. Warren, Ph.D. Thesis, Cornell University, 1970; Diss. Abstr., 70-14, 407.
10 D. Van Sickle and J.O. Rodin, J. Amer. Chem. Soc., 86 (1964) 3091.
11 P. Brown and R.C. Cookson, Tetrahedron, 21 (1965) 1993.
12 S. Seltzer, J. Amer. Chem. Soc., 87 (1965) 1534.
13 M. Taagepera and E.R. Thornton, J. Amer. Chem. Soc., 95 (1972) 1168.

14 J.C. Little, J. Amer. Chem. Soc., 87 (1965) 4020.
15 R.E. Banks, A.C. Harrison and R.N. Haszeldine, J. Chem. Soc. Chem. Commun., (1966) 338.
16 P.D. Bartlett, G.E.H. Wallbillich, A.S. Wingrove, J.S. Swenton, L.K. Montgomery and B.D. Kramer, J. Amer. Chem. Soc., 90 (1968) 2049.
17 J.S. Swenton and P.D. Bartlett, J. Amer. Chem. Soc., 90 (1968) 2056.
18 W. von E. Doering, M. Franck-Neumann, D. Hasselmann and R.L. Kaye, J. Amer. Chem. Soc., 94 (1972) 3833.
19 W.F. Bayne and E.I. Snyder, Tetrahedron Lett., (1970) 2263.
20 W.J. Linn and R.E. Benson, J. Amer. Chem. Soc., 87 (1965) 3657.
21 W.R. Dolbier, Jr. and S.-H. Dai, Tetrahedron Lett., (1970) 4645.
22 T.J. Katz and R. Dessau, J. Amer. Chem. Soc., 85 (1963) 2172.
23 J.E. Baldwin and J.A. Kapecki, J. Amer. Chem. Soc., 92 (1970) 4874.
24 J.E. Baldwin and U.V. Roy, unpublished work, referred to in J.E. Baldwin and R.H. Fleming, Fortsch. Chem. Forsch., 15 (1970) 281.
25 W.A. Pryor and L.W. Henderson, Int. J. Chem. Kinet., 4 (1972) 325.
26 E. Koerner von Gustorf, D.V. White, J. Leitich and D. Henneberg, Tetrahedron Lett., (1969) 3113.
27 O.P. Strausz, W.B. O'Callaghan, E.M. Lown and H.E. Gunning, J. Amer. Chem. Soc., 93 (1971) 559.
28 O.P. Strausz, I. Safarik, W.B. O'Callaghan and H.E. Gunning, J. Amer. Chem. Soc., 94 (1972) 1828.
29 J.W. Simons and B.S. Rabinovitch, J. Amer. Chem. Soc. 85 (1963) 1023.
30 R.B. Woodward and R. Hoffmann, The Conservation of Orbital Symmetry, Verlag Chemie, Academic Press, Weinheim/Bergstr., 1970.
31 S.D. McGregor and D.M. Lemal, J. Amer. Chem. Soc., 88 (1966) 2858.
32 W.L. Mock, J. Amer. Chem. Soc., 88 (1966) 2857.
33 S. Asperger, D. Hegedic, D. Pavlovic and S. Borcic, J. Org. Chem., 37 (1972) 1745.
34 W.R. Dolbier, Jr. and S.-H. Dai, J. Amer. Chem. Soc., 90 (1968) 5028.
35 S.-H. Dai and W.R. Dolbier, Jr., J. Amer. Chem. Soc., 92 (1970) 1774.
36 S.-H. Dai and W.R. Dolbier, Jr., J. Amer. Chem. Soc., 94 (1972) 3946.
37 J.J. Gajewski, J. Amer. Chem. Soc., 93 (1971) 4450.
38 R.J. Crawford and D.M. Cameron, J. Amer. Chem. Soc., 88 (1966) 2589.
39 W.R. Moore, P.D. Mogolesko and D.D. Traficante, J. Amer. Chem. Soc., 94 (1972) 4753.
40 W.R. Moore and W.R. Moser, J. Org. Chem., 35 (1970) 908.
41 R.E.K. Winter and M.L. Honig, J. Amer. Chem. Soc., 93 (1971) 4616.
42 R. Molojcic, K. Humski, S. Borcic and D.E. Sunko, Tetrahedron Lett., (1969) 2003.
43 K. Humski, R. Malojcic, S. Borcic and D.E. Sunko, J. Amer. Chem. Soc., 92 (1970) 6534.
44 K.D. McMichael, J. Amer. Chem. Soc., 89 (1967) 2943.
45 V. Berlanic-Lipovac, S. Borcic and D.E. Sunko, Croat. Chem. Acta, 37 (1965) 61.
46 W.R. Dolbier, Jr. and J.H. Alonso, J. Amer. Chem. Soc., 94 (1972) 2544.
47 W. von E. Doering and L. Buladeanu, Tetrahedron, 29 (1973) 499 and references therein.
48 W.R. Dolbier, Jr. and J.H. Alonso, J. Amer. Chem. Soc., 95 (1973) 4421.
49 S. McLean and P. Haynes, Tetrahedron, 21 (1965) 2329.
50 H. Kloosterziel and A.P. Ter Borg, Rec. Trav. Chim. Pays-Bas, 84 (1965) 1305.
51 W.R. Roth and J. König, Ann. Chem., 699 (1966) 24.
52 C.S. Foote and S. Mazur, unpublished communication; see S. Mazur, Ph.D. Thesis, University of California, 1971; Diss. Abstr. 72-5848.
53 S.-H. Dai and W.R. Dolbier, Jr., J. Amer. Chem. Soc., 94 (1972) 3953.
54 S.-H. Dai and W.R. Dolbier, Jr., unpublished communication.
55 R. Huisgen and H. Pohl, Chem. Ber., 93 (1960) 527.

56 M.J. Goldstein, unpublished results.
57 A. Nickon, V.T. Chuang, P.J.L. Daniels, R.W. Denny, J.B. DiGiorgio, J. Tsunetsugu, H.G. Vilhuber and E. Werstiuk, J. Amer. Chem. Soc., 94 (1972) 5517.
58 D.Y. Curtin and D.B. Kellom, J. Amer. Chem. Soc., 75 (1953) 6011.
59 A.T. Blades and P.W. Gilderson, Can. J. Chem., 38 (1960) 1407.
60 C.H. Depuy, R.W. King and D.H. Froemsdorf, Tetrahedron, 7 (1959) 123.
61 R.F.W. Bader and A.N. Bourns, Can. J. Chem., 39 (1961) 348.

Chapter 3

THE ELUCIDATION OF MASS SPECTRAL FRAGMENTATION MECHANISMS BY ISOTOPIC LABELLING

J.L. HOLMES

Chemistry Department, University of Ottawa, Ontario (Canada)

I. INTRODUCTION AND METHODOLOGY

This review will be concerned with the use of isotopic labelling as an aid to determining the fragmentation mechanisms of positively charged ions generated by electron impact in a mass spectrometer. In this branch of physical–organic chemistry the commonly used isotopes are deuterium and carbon-13. Occasionally, oxygen-18 and nitrogen-15 have been employed, but molecules containing many hetero-atoms are uncommon and the need to identify specific hetero-atoms is infrequent. The emphasis will be upon recent work involving small organic compounds (of molecular weight < 300 a.m.u.), the study of which provides the principles for understanding the behaviour of larger and more complex molecules. It will be assumed that readers are reasonably familiar with conventional single- and double-focussing instruments and the separation of metastable ions by techniques that are now a routine procedure.

The general aim of isotopic substitution experiments is to discover therefrom which atoms are involved in a given fragmentation of an ion. Such studies often permit one to describe the transition state for the dissociation but alone they provide little information as to the ground state structure of the precursor ion, the daughter ion or the neutral fragment. It has become customary to study the distribution of label among daughter ions at two or three time intervals after the generation of their precursors. This kinetic approach to mechanistic problem solving is a relatively recent advance and has increased greatly the power and scope of labelling experiments. The significance of such observations will be considered next.

A. Ion Observed in a "Normal" Mass Spectrum

These are the ions which are generated in the ion source and are recorded after mass separation (single-focussing instrument) or velocity and mass separation (double-focussing instrument.) Such ions have resulted from all dissociations of molecular and fragment ions which have taken place in up to 0.5–5.0 μ sec after the primary ionisation of the neutral molecule. Ions which fragment more slowly may be observed as metastable ion peaks; these are discussed later. In order for ions to dissociate in the ion source they must often contain energy appreciably in excess of the threshold requirement for the fragmentation in question. Thus the ions observed in the normal mass spectrum are generated from precursors having a very wide range of energies and with lifetimes of one to about a million vibrations. In general, it is to be expected that rapid dissociations will be accompanied by less rearrangement and loss of original positional identity of label than will slow dissociations. However, in practice, the multitude of adjacent peaks in the mass spectra of many (unlabelled) molecules, renders the adequate interpretation of the mass spectra of their isotopically labelled analogues very difficult. This problem arises most commonly when an ion loses, for example, fragments corresponding to A, (A + H), (A + 2H) etc. Nevertheless, as will be seen below, many fragmentation mechanisms have been successfully elucidated by observing normal mass spectra alone and unless stated otherwise it will be assumed that results refer thereto.

B. Metastable Ion Peaks

The diffuse signals which appear at non-integral mass in a normal mass spectrum arise from the fragmentation of ions in the field-free region of the instrument preceding the (magnetic field) mass analyser. In a double-focussing instrument this is generally the *second* field-free region of the apparatus. The relationship between the apparent mass of the metastable ion peak, m^*, and the masses of the precursor (m_1) and daughter (m_2) ions is well known

$$m^* = m_2^2/m_1$$

Thus the appearance of a metastable ion peak permits the almost unequivocal identification of the fragmentation sequence

$$m_1^+ \xrightarrow{*} m_2^+ + \text{neutral fragment}$$

The use of metastable ion peaks is then of obvious help in following the participation of an isotope in a given fragmentation. Two points should, however, be emphasised. Metastable ion peaks observed in a normal mass spectrum are weak signals and are often partially obscured by the neighbouring daughter ion peaks; if they occur at integral mass they may be missed altogether. Their abundance relative to daughter ion peaks may be improved by reducing the ionising electron energy and/or increasing the repeller potential [1]. A small additional problem concerns consecutive fragmentations. It must be remembered that the equation for m^* involves only precursor and daughter ion masses and does not include the possibility that the fragmentation might be a two-step process. For example, the apparent losses of C_2O_2 (2CO) (ref. 2), H_4O_2 ($2H_2O$) (ref. 3), and H_3O_2 (H_2O, OH) (ref. 4) have been reported. All that is needed for such an observation is that *both* dissociations take place within the time spent passing through the field-free region.

Ions which generate metastable ion peaks possess little energy above the threshold for dissociation; they are long-lived species and therefore may have undergone extensive internal rearrangement prior to dissociation. Indeed, much of our knowledge of such rearrangements has come from observations of metastable ion peaks. In the limit, it is often observed that the daughter ions retain the isotopic label in abundances which must have resulted from a complete loss of positional identity of the labelled and unlabelled atoms within the precursor ion. Such processes will be referred to as involving complete randomisation or "scrambling" of atoms. Metastable ion peaks examined in the normal mass spectrum may subsequently be referred to as $^2m^*$.

In double-focussing mass spectrometers where the electric sector precedes the magnetic sector, there is another field-free region, between the ion source and the electric sector. This is referred to as the *first* field-free region and metastable dissociations observed therein are represented as $^1m^*$. The methods of observing and separating the first field-free region generated daughter ions from the main ion beam are well known and need not be discussed in detail here [5]. However, they form an intermediate life-time region, between the source and second field-free region generated ions and they can be observed, completely separated from the normal ions, as signals of adequate intensity. The most frequently used method of separating first field-free region metastable ions from the main ion beam is that due to Barber and Elliott [6]. This essentially involves adjusting the instrument in normal operating conditions such that ions m/e m_2 are transmitted. The ion acceleration potential is then increased; when the acceleration potential has been increased by the ratio m_1/m_2, daughter ions of mass m_2, generated *after full* acceleration and during passage through the first field-free region, will now possess sufficient energy to be transmitted through the instrument and will therefore be detected. Thus, determination of the required acceleration potential ratio permits identification of the precursor (m_1) of m_2. Of course, daughter ions of low mass may have several precursors and thus signals will be observed at acceleration potentials corresponding to the several appropriate m_1/m_2 values.

Isotope effects will not be singled out for special attention but will be discussed for specific cases. Indeed, in many studies it is implicit that there *is no* isotope effect! This is a reasonable assumption where it can be expected to be small as, for example, where methyl or CD_3 in similar environments are competitively being lost from an ion. However, where loss of H/D or $H_2/HD/D_2$, or specific hydrogen atom transfer processes are involved, such an assumption will or may be an oversimplification.

In the following sections compounds have been divided according to the presence of particular atoms or functional groups.

II. ALIPHATIC HYDROCARBONS

In this section we will deal with dissociations not only of molecular ions, but also of ionised hydrocarbon radicals which frequently appear as prominent daughter ions in the mass spectra of many organic compounds. Indeed, the former are of relatively little interest to the organic chemist who is more familiar with molecules containing a heteroatom.

A. Methane and the Methyl Radical

The effect of deuterium on the mass spectrometric behaviour of methane has been studied by several workers with particular regard to metastable ion peaks. Dibeler and Rosenstock [7], repeating some earlier experiments, reported that only one truly unimolecular (i.e. not collision-induced) metastable ion could be found in the mass spectrum of methane and methane-d_4, namely that resulting from loss of an atom of hydrogen or deuterium. Ottinger [8] and recently Beynon [9], have identified more metastable ions. There is a large isotope effect for H vs. D loss in partially labelled methanes, of at least 30 for CH_3D, $>$ 80 for CH_2D_2 and $>$ 70 for CHD_3 (ref. 8).

B. Ethane and the Ethyl Cation

A detailed description of the normal mass spectra of six deuterated ethanes has been presented by Amenomiya and Pottie [10]. The purpose of the study was to obtain relative ion abundances for quantitative mass spectrometric analysis, but this work also provides some evidence in favour of partial atom scrambling prior to dissociation into a methyl radical and methyl cation. Löhle and Ottinger [11] and Lifshitz and Sternberg [12] have examined metastable ion peaks in the mass spectra of ethane and CH_3CD_3. Unfortunately, as is described in some detail elsewhere [13], there is considerable disagreement between the two sets of experimental results. It is at present uncertain as to whether the loss of a hydrogen molecule from the ethyl cation is preceded by slight, partial or complete statistical randomisation of the hydrogen atoms. However, it *is* agreed that the ethane moleculai ion does not suffer hydrogen scrambling prior to dissociation by loss of a hydrogen molecule. Inconclusive evidence that the ethyl cation may exist in the gas phase as $CH_3CH_2^+$ rather than a symmetrical protonated ethylene has come from i.c.r. (ion cyclotron resonance) studies [14], but a recent report [15] on proton and hydride transfer reactions of deuterium-labelled ethyl cations requires that statistical atom scrambling precedes their reaction.

C. C_3 Species

The behaviour of deuterium-labelled propanes in the fragmentation

$$C_3H_8^{+\cdot} \rightarrow C_2H_4^{+\cdot} + CH_4$$

was studied by Lifshitz and Shapiro [16,17] for metastable dissociations. In the first paper, observations on propane and C_3D_8 were presented; the results, however, were inconclusive, because it was realised that two competing mechanisms must be distinguished, a 1,3 and a 1,2 elimination

$$CH_3CH_2CH_3^{+\cdot} \rightarrow CH_2CH_2^{+\cdot} + CH_4$$

$$CH_3CH_2CH_3^{+\cdot} \rightarrow CH_3CH^{+\cdot} + CH_4$$

The second report [17] contains results for $CH_3CD_2CH_3$ and $CD_3CH_2CD_3$. Unfortunately, only a metastable ion peak for methane loss from $CH_3CD_2CH_3$ could be found, the remaining three (namely CH_3D from $CH_3CD_2CH_3$ and CD_4 and CD_3H from $CD_3CH_2CD_3$) being below the limit of detection. Ottinger [18] also examined two labelled propanes, namely $CH_3CD_2CH_3$ and CD_3CHDCD_3. He, too, failed to observe all the above four metastable ion peaks, and instead directed his attention to hydrogen-loss processes. For the molecular ions, the major metastable ion peaks were for loss of a deuterium atom from $CH_3CD_2CH_3$ and loss of a hydrogen atom from CD_3CHDCD_3, as expected from these low-energy ions (the heat of formation of iso-propyl$^+$ is 192 kcal. mol.$^{-1}$ and is appreciably lower than that of *n*-propyl$^+$ which is 208 kcal. mol.$^{-1}$) [19]. For the competing processes, an isotope effect H/D $\sim$ 300 was observed.

$$CD_3CHDCD_3^{+\cdot} \rightarrow CD_3\overset{+}{C}DCD_3 + H^\cdot$$

$$\rightarrow CD_3\overset{+}{C}HCD_3 + D^\cdot$$

For the propyl cations thus produced, $CH_3\overset{+}{C}DCH_3$ lost a hydrogen molecule and HD whilst $CD_3\overset{+}{C}HCD_3$ lost HD and a molecule of deuterium.

The system was examined again in 1970 by Vestal and Futrell [20]. They showed that so far as metastable ions were concerned, the 1,2 elimination discussed by Lifshitz and Shapiro [16,17] is of negligible importance. There is thus no evidence to support hydrogen randomisation prior to dissociation of the molecular ion. For the propyl cation, however, metastable ion studies of hydrogen molecule loss are presented in Table 1. With the exception of the third and the last results, the values shown are compatible with dissociation taking place from a propyl cation in which the hydrogen and deuterium atoms have lost their identity and in which an average isotope effect of 2.3 is operating, favouring hydrogen over deuterium. As will be seen in subsequent sections, an isotope effect of this magnitude is typical for such competitive (metastable) fragmentations involving loss of hydrogen.

McAdoo et al. [21] have described i.c.r. experiments performed on propyl cations. They found little evidence for positional isomerisation among low-energy (non-decomposing) deuterium-labelled iso-propyl cations in the time required for i.c.r. reactions ($\sim 10^{-3}$ sec). On the same time scale, virtually all *n*-propyl cations had isomerised to iso-propyl ions. The $C_3H_7^+$ ions produced by protonation of cyclopropane were identical with iso-propyl ions in so far as their reactivity towards methanol was concerned.

To summarise the above results, there is no evidence that propane molecular ions undergo hydrogen randomisation prior to fragmentation. Propyl cations with sufficient energy to dissociate and generate metastable ion peaks, do so with extensive hydrogen randomisation; such dissociations take place from an iso-propyl cation irrespective of their precursor molecules' structure. Very long-lived iso-propyl cations ($> 10^{-3}$ sec) retain the positional integrity of hydrogen atoms.

TABLE 1

METASTABLE DISSOCIATIONS OF DEUTERIUM-LABELLED PROPYL CATIONS (STATISTICAL RATIOS IN PARENTHESES)

Propyl cation	[−HD]/[−H2]	[−HD]/[−D2]	Ref.
$CH_3CH_2\overset{+}{C}HD$ [a]	0.15 (0.4)		21
$CH_3\overset{+}{C}DCH_3$ [b]	0.14 (0.4)		21
$CH_3\overset{+}{C}D\,CH_3$ [d]	<0.04 (0.4)		18
$CD_3\overset{+}{C}H\,CD_3$ [c]		0.86 (2.5)	21
$CD_3\overset{+}{C}H\,CD_3$ [f]		0.6 (2.5)	20
$CD_3\overset{+}{C}H\,CD_3$ [e]		1.0 (2.5)	18
$CD_3CD_2\overset{+}{C}H_2$ [g]	0.46 (1.0)		20
$CD_3CH_2\overset{+}{C}D_2$ [f]	0.47 (1.0)		20

[a] Generated from CH_3CH_2CHDBr.
[b] Generated from $CH_3CDBrCH_3$.
[c] Generated from $CD_3CHBrCD_3$.
[d] Generated from $CH_3CD_2CH_3$.
[e] Generated from CD_3CHDCD_3.
[f] Generated from $CD_3CH_2CD_3$; $CD_3CH_2CD_2^+$ not unequivocally identified.
[g] Presumed structure; generated from $CH_3CD_2CH_3$.

(Conflicting results have been reported for the behaviour of labelled propyl cations generated as fragment ions in the mass spectra of deuterium-labelled *n*-hexanes [22].)

D. C_4 Species. Butanes and Butenes

The fragmentation behaviour of the isomeric butenes has received considerable attention. Early experiments clearly showed that for source-generated allyl cations at 70eV and near the appearance potential, positional identity of hydrogen atoms was completely lost.

$$C_4H_8^{+\cdot} \rightarrow C_3H_5^+ + CH_3^\cdot$$

(For $CH_2{=}CHCH_2CD_3$ (ref. 23) and for $CH_2DCH{=}CHCH_2D$ (ref. 24) as labelled butene precursors.) The mechanism for hydrogen randomisation can be represented as involving a sequence of 1,2 and/or 1,3 shifts. It is, however, possible that carbon atom migration is also involved and with this point in mind, Meisels et al. [25] investigated the fragmentation behaviour of $CH_2{=}CHCH_2CD_3$ and $CH_2{=}CHCH_2{}^{13}CH_3$ for nominal ionising electron energies between 9.0 and 15.0 eV. For the deuterated compound, the earlier results [23] were confirmed and, furthermore, for the carbon-13 analogue, the allyl daughter cation contained 75% of the label. Thus for source-generated ions, the hydrogen and carbon atoms are randomised prior to dissociation by methyl loss.

This result is compatible with some recent observations by Lossing [26], who found a unique heat of formation for the allyl cation generated from the isomeric butene molecules, showing that the precursor molecular ions had all rearranged to a common intermediate prior to dissociation at threshold.

Lias and Ausloos [27] have shown that $C_4H_8^{+\cdot}$ ions generated in the (radiolytic) fragmentation of cycloalkane molecular ions consist of a mixture of the three double bond isomers. It should be noted that these ions are produced with little rearrangement of the cycloalkane prior to its dissociation and that they are ions with insufficient internal energy to fragment further. In conventional mass spectrometry, therefore, they are $C_4H_8^{+\cdot}$ ions which would be observed in the normal mass spectrum. The fragmentation behaviours of carbon-13 and deuterium-labelled $C_4H_8^{+\cdot}$ and *n*-butyl cations were examined by Davis et al. [28]. The cations all were generated from compounds of type $CH_3CH_2CH_2CH_2X$. Also studied were the butyl cations generated from *n*-, iso-, *sec*- and *tert*-butyl structures. For the fragmentations listed below, complete atom randomisation was observed among daughter ions generated in the first and second field-free regions.

$$CH_3CH_2CH_2CH_2X^{+\cdot} \xrightarrow{-HX} C_4H_8^{+\cdot} \xrightarrow[\text{randomisation}]{\text{complete atom}} C_3H_5^+ + CH_3^\cdot$$

$$\left.\begin{array}{l} CH_3CH_2CH_2CH_2X^{+\cdot} \\ CH_3CH_2CHXCH_3^{+\cdot} \\ (CH_3)_2CHCH_2X^{+\cdot} \\ (CH_3)_3CX^{+\cdot} \end{array}\right\} \xrightarrow{-X^\cdot} C_4H_9^+ \xrightarrow[\text{randomisation}]{\text{complete atom}} \begin{cases} C_3H_5^+ + CH_4 \\ C_2H_5^+ + C_2H_4 \end{cases}$$

The energetics of the methane-generating dissociation of the butyl cation were discussed shortly afterwards by Yeo and Williams [29]. They measured the appearance potentials for the metastable ion peak for $C_4H_9^+ \rightarrow C_3H_5^+ + CH_4$. They found that for the *tert*-butyl cation, the energy required for fragmentation was 1.9 ± 0.3 eV while for the *n*-butyl cation, only 0.5 ± 0.3 eV was necessary. The energy diagram presented by Yeo and Williams, modified to include more recent thermochemical data, is shown in Fig. 1. Note that for all butyl cations the dissociation is endothermic; the complete randomisation of atoms in these fragmentations indicates a common structure (or structures) being generated before or at the transition state. Thus the transition state lies between 3 and 10 kcal. $mol.^{-1}$ above the products' energy level. It is worth noting here that the metastable ion peak for this fragmentation is a composite species. The energy release for the major component has been estimated [30] as not exceeding 7 kcal. $mol.^{-1}$. Much, or indeed all of the reverse activation energy may therefore be partitioned as translational kinetic energy of the products.

Liardon and Gäumann [31] have also examined the behaviour of labelled butyl cations produced from simple halide precursor molecules. In particular they wished to extend observations to lifetimes shorter than those studied by Davis et al. [28] who had only considered field-free region dissociations. For *n*-, *sec*- and *tert*-butyl cations,

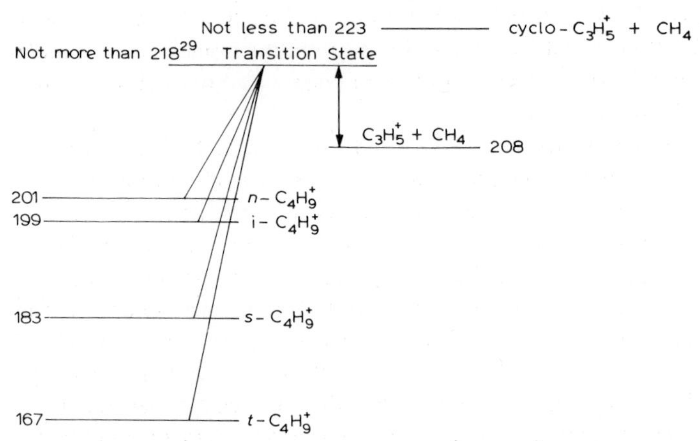

Fig. 1. Energy diagram for the dissociation $C_4H_9^+ \rightarrow C_3H_5^+ + CH_4$ showing an activation energy leading to a common transition state. (Heats of formation (in kcal mol^{-1}) from refs. 19, 26.)

labelled with carbon-13 and deuterium, the authors concluded that the carbon and hydrogen atoms had become essentially positionally equivalent prior to dissociation by elimination of methane and ethylene *outside* the ion source. This is in agreement with Davis et al. [28]. However, all ions produced *in* the source showed significantly less mixing of label (particularly deuterium) than that required for complete randomisation. The observations could be satisfactorily explained by proposing that protonated cyclobutane is the intermediate species in which carbon and hydrogen randomisation takes place. Such a cyclic intermediate immediately permits a single carbon-13 label to lose its identity (with respect to methane and ethylene loss) without further rearrangement; small deviations from statistical loss of carbon-13 were explained as arising from fast cleavages prior to cyclisation. (Note that an earlier and less satisfactory proposal for carbon-13-labelled *tert*-butyl cations [32] considered that methylated cyclopropane was the intermediate.) Hydrogen/deuterium atom mixing was proposed as resulting from successive atom interchanges within the protonated cyclobutane and this is more time-dependent than carbon atom mixing if the dissociating butyl ion contains only one carbon-13 carbon. The possibility that carbon atoms rearrange in the cyclobutane intermediate could, in principle, be examined, e.g. by generating a $^{13}CH_3CH_2CH_2{}^{13}CH_2^+$ cation and observing ethylene, $^{13}CCH_4$ and $^{13}C_2H_4$ losses. (However, such an approach necessarily *assumes* that the ring is produced by the formation of a bond between C-1 and C-4.)

Shaw et al. [33] performed a series of measurements on ions of the general formulae $C_nH_{2n-1}^+$ and $C_nH_{2n-3}^+$. They found that the relative metastable ion peak abundances for competing dissociations of (unlabelled) $C_4H_7^+$, $C_5H_9^+$, $C_6H_9^+$ and $C_6H_{11}^+$ were independent of the structure of their precursor molecules and concluded, therefore, that common intermediates are produced prior to fragmentation. The hypothesis is that if a given ion observed in the mass spectra of a variety of molecules displays the same metastable ion peaks having the same relative abundances, then the said given ion has a common structure and/or energy content irrespective of the structure of its precursor ion. (For discussion thereof see refs. 33, 108, 109 and 110.) For $C_5H_9^+$, the d_4 analogue was generated from 2,2,5,5 - d_4-cyclopentyl bromide, and it was observed that prior to dissociation by loss of ethylene, randomisation of hydrogen/deuterium atoms had taken place. (The labelled $C_5H_9^+$ result has been confirmed by Tomer et al. [34].)

E. C_5 and Larger Species

The mass spectrum of neopentane -2-^{13}C was reported and discussed a decade ago [32]. The molecular ion dissociates without rearrangement to yield a labelled *tert*-butyl cation (see above). The neopentyl cation fragments chiefly by loss of ethylene; deuterium and carbon-13 labelling experiments [35] have shown that for the metastable ion-peak-generating dissociation all atoms have lost their positional identity and are randomly eliminated.

The mass spectral fragmentation behaviour of *n*-hexane has been extensively examined by Liardon and Gäumann [22,36]. In their earlier publication they examined the loss of a methyl radical, as an ion source process only, using twelve deuterium-labelled *n*-hexanes. They concluded that the results were adequately explained by the participation of two reaction pathways, one being the loss of a terminal methyl group and the other involving the loss of an internal carbon atom (shown below).

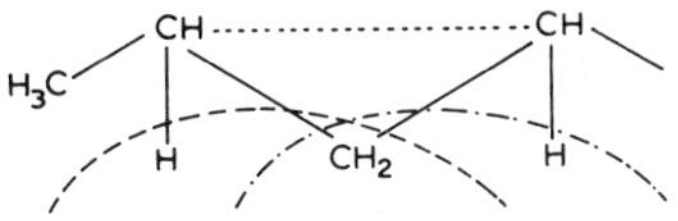

(For longer-chain alkanes they considered that the second mode of methyl loss correspondingly increased in importance.) Labelled $C_5H_{11}^+$ daughter ions having sufficient internal energy to dissociate in the first field-free region were observed to do so (by ethylene loss) with complete randomisation of hydrogen/deuterium atoms.

Weinberg and Scoggins [37] studied a series of deuterium-labelled isomeric octanes, produced in each case by catalytic deuteration of the corresponding octene. Their results showed that loss of methyl, ethyl and propyl groups from the molecular ions could very well be explained using a set of simple rules, namely

(a) no scrambling in molecular ions; no isotope effect on C–C bond cleavages

(b) *n*-alkanes-d_2 randomly cleave C–C bonds followed by loss of hydrogen atoms

(c) branched alkanes-d_2 cleave exclusively at points of branching, followed by loss of hydrogen atoms.

The isotope content of C_4 ions could not, however, be simulated by application of the above rules, indicating that they were mainly produced as secondary fragments, few if any coming directly from cleavage of molecular ions.

Goldenfeld and Korostyshevsky [38] examined the behaviour of *n*-octane-2-d_1 and *n*-nonane-5-^{13}C in a field ionisation mass spectrometer. The use of such an instrument permits the observation of ions at much shorter time intervals after generation than in a conventional electron impact mass spectrometer. When these two molecules were observed under electron impact conditions [39,40] it was clear that the methyl, ethyl, propyl and butyl daughter cations resulted from rearranged species (and of course by more than one route). The distribution of the labels in the field ionisation spectra, however, could be explained by invoking single C–C bond cleavages *alone.*

Relatively few mass spectrometric investigations involving labelled molecules have been reported of larger, non-aromatic hydrocarbons. Among these, the mass spectra of the C_7H_{12} isomers, 1-methylcyclohexene and methylenecyclohexane have been compared [41]. These compounds display very similar mass spectra, the base peak being $C_6H_9^+$ in each case. Only normal mass spectra were examined; deuterium labelling showed no evidence for complete atom randomisation in fragmentations of the molecular ions but the results were too complex to permit firm mechanistic conclusions to be reached. 7-^{13}C-labelled methylenecyclohexane was observed [42] to lose a methyl radical with approximately 50% retention of label; normal ion peak measurements on deuterium-labelled analogues agreed with the earlier findings.

Holmes and McGillivray [43] recently completed a study of the C_7H_{10} isomers, norbornene and nortricyclene, using extensive deuterium labelling, separated metastable ion peaks and appearance potential measurements, to elucidate fragmentation pathways. The molecular ions do not rearrange to a common intermediate prior to fragmentation by ethylene loss which for norbornene is a simple retro Diels–Alder reaction. Methyl loss from the molecular ions was complex but there was evidence for hydrogen atom scrambling in the nortricyclene fragmentation.

Kraft and Spiteller [44] showed by means of deuterium labelling, that olefins of form II suffered no isomerisation before fragmentation, unlike those of type I.

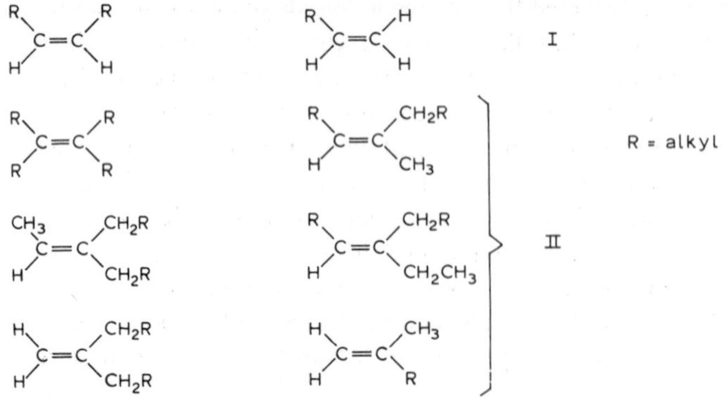

I

R = alkyl

II

A recent field ionisation and electron impact study of deuterium-labelled cyclohexene [45] has shown that hydrogen/deuterium atom randomisation is extensive in the molecular ion ranging from partial at times below 10^{-9} sec, to essentially complete at times in excess of 10^{-9} sec. However, the loss of ethylene by a retro Diels–Alder reaction predominates (over methyl radical loss) at very short times and it is this process which shows greatest deviation from statistical loss of label. It was proposed that the randomisation of hydrogen/deuterium atoms resulted from successive 1,3-allylic rearrangements, viz.

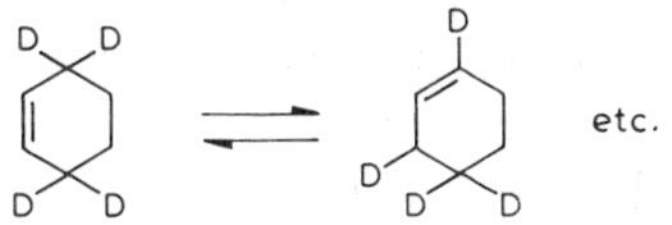

The elimination of benzene from 1-phenyltetralin has been shown by deuterium labelling to proceed mainly by a 1,4 elimination [46], somewhat similar to the loss of water from the molecular ion of cyclohexanol (see pp. 110–112)

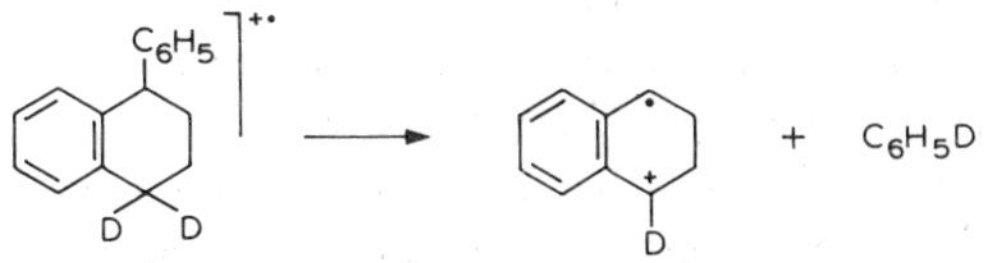

It is, however, uncertain whether this benzene elimination is a stereospecific *cis*-1,4 elimination.

The fragmentation behaviour of some isomeric deuterium-labelled nonynes and decynes has recently been described [47]. It is evident from this and related earlier studies [48,49] that few dissociations of the molecular ions are not preceded by some triple bond and atom migrations, thus making uncertain any exact mechanistic description.

III. AROMATIC HYDROCARBONS

A great deal of work has been directed to furthering our understanding of the fragmentation behaviour of aromatic hydrocarbons. In this section the results of labelling experiments on benzene (and phenyl) and toluene (benzyl and tropyl) will be discussed in detail; for other and related aromatic molecules only the major conclusions will be outlined.

A. Benzene

The first studies of deuterium-labelled benzenes [50,51] were marred by incomplete labelling; nevertheless the results were best explained by proposing the complete randomisation of hydrogen and deuterium atoms prior to fragmentation of the molecular ion. Jennings [52] studied metastable ion peaks ($^{2}m^{*}$) in the mass spectra of benzene-1,4-d_2 and 1,3,5-d_3; their relative abundances for the loss of acetylenes (C_2H_2, C_2HD, C_2D_2) from the molecular ion and for the loss of C_3H_3 and labelled analogues from the molecular ion indicated that complete randomisation of hydrogen/deuterium atoms had taken place prior to fragmentation. Such loss of positional identity could result from processes analogous to those photochemically induced, e.g. via the intermediate involvement of benzvalene, prismane and 'Dewar' benzenes [53]. The point which now had to be established was whether carbon atoms also lost their positional identity. Williams and co-workers [54] reported that in the molecular ion of 1,3,5-$^{13}C_3$-benzene, when dissociating by loss of acetylene in the first field-free region of a double-focussing mass spectrometer, all six carbon atoms were completely randomised. Later in the same year Beynon and co-workers [55] briefly described the behaviour of benzene - 1,2 - $^{13}C_2$ - 3,4,5,6-d_4. They studied the formation of $C_3H_3^+$ and $C_4H_4^+$ ions (and their labelled analogues) in the normal mass spectrum (using high resolution to separate some of the isobaric species). Their principal aim was to distinguish between three postulated mechanisms: (a) scrambling of the six C atoms without breaking C–H bonds, (b) scrambling of hydrogens without loss of positional identity of the carbons and (c) independent scrambling of carbons and hydrogens. These preliminary results were not conclusive but a comprehensive study [56] was reported in 1972. In this work, first field-free region dissociations were also examined. It was concluded that for the latter, long-lived ions, complete randomisation of all atoms had taken place prior to dissociation of the molecular ion. For ions produced in the source, the results indicated that total randomisation was not complete, some hydrogens having scrambled without carbon scrambling. Dickinson and Williams [57] also showed that for the same metastable ion peak generating dissociations in benzene-1-^{13}C-1-d_1, complete loss of atom identity occurred. In addition it was shown that the phenyl cation fragmented with complete loss of identity of carbon and hydrogen atoms. (See also ref. 58.)

B. Toluene and the $C_7H_7^+$ Ion

The ubiquitous nature and great stability of the $C_7H_7^+$ ion in the mass spectra of toluene and its isomers, and many benzylic and related compounds, was explained [59] in 1957 as being due to its existence as the symmetrical tropyl cation rather than as the benzyl cation. In recent years, interest in the fragmentation behaviour of the toluene molecular ion and the $C_7H_7^+$ species has revived, particularly with the advent of metastable ion separation techniques. Meyerson et al. [60,61] showed, via deuterium labelling studies, that for the dissociation

$$C_7H_7^+ \rightarrow C_5H_5^+ + C_2H_2$$

observed in the normal mass spectrum, all the hydrogen atoms in $C_7H_7^+$ had lost their positional identity when the precursor molecules were benzyl alcohol, ethylbenzene, and bibenzyl. (Observations on diphenylmethane-α-d_2 were inconclusive, but were interpreted as showing the formation of tropylium-1,2-d_2 which retained positional integrity of label before fragmentation by loss of acetylene.) These results provided no information as to the possible mechanism for formation of tropyl from benzyl (e.g. insertion of the methylene carbon on either side of C-1, or random insertion, followed by hydrogen atom migration). Rinehart et al. [62] prepared toluene-α, 1-$^{13}C_2$, toluene-1-^{13}C and toluene-α-^{13}C and measured, under high resolution, peaks in the normal mass spectrum in the region *m/e* 65–67, corresponding to labelled $C_5H_5^+$ daughter ions. The intensities of the $C_5H_5^+$, $^{13}CC_4H_5^+$ and $^{13}C_2C_3H_5^+$ ions were close (within experimental error) to the distribution expected for complete loss of positional identity of the carbon atoms in $C_7H_7^+$ prior to fragmentation. Similar results were reported by Siegel [63] for toluene-2-^{13}C and toluene-2,6-$^{13}C_2$. Note that this again gives no insight into the mechanism of the formation of $C_7H_7^+$, assuming that it indeed has the tropyl structure. Howe and McLafferty [64,65] performed a series of experiments on deuterium-labelled toluenes, studying both normal and metastable ion peak relative abundances. The effect of internal energy on the fragmentation behaviour of the molecular ions was estimated as follows, using toluene-α-d_3 and toluene-ring-d_5. For the loss of a hydrogen or deuterium atom, four ranges of increasing internal energy can be defined.

(1) Those molecular ions which do not dissociate until they have entered the first field-free region and are thus observed as generating metastable ion peaks.

(2) Slightly higher energy molecular ions may be produced by admitting a collision gas (argon) to the first field-free region. Collision-induced metastable dissociations may then be observed.

(3) Daughter ions produced in the ion source provide another range of internal energies (i.e. these have insufficient energy to fragment further after their formation).

(4) Molecular ions which yield metastable ion peaks for the secondary fragmentation (yielding $C_5H_5^+$); such ions have (on average) higher energies than those of (3).

The results for toluene and cycloheptatriene and the estimated average total energies of ions dissociating in regions 1–4 are shown in Table 2. The deuterium isotope effect decreases with increasing internal energy and it appears likely that the toluene molecular ion isomerises to that of cycloheptatriene prior to dissociation, certainly for the low internal energy species. Differences in the results for the higher energy toluene

TABLE 2

ISOTOPE EFFECT (I) AND DEGREE OF HYDROGEN ATOM RANDOMISATION (R) FOR $C_7H_8^{+\cdot} \rightarrow C_7H_7^+ + H^\cdot$ IN TOLUENE AND CYLOHEPTATRIENE

Energy region (see text)		Toluene		Cycloheptatriene	
		I	R	I	R
1	(11.8eV)	2.8	1.0	2.6	0.99
2	(12.4eV)	2.3	0.98	2.1	0.95
3	(14.0eV)	1.5	0.89	1.5	0.92
4	(15.8eV)	1.4	0.72	1.2	0.88

ions may indicate that direct decomposition from the unscrambled molecular ions may be taking place.

A somewhat similar study by Beynon et al. [66] appeared shortly afterwards. They only studied first field-free region metastable ion dissociations, but agreed with Howe and McLafferty that for such low internal energy ions, scrambling was complete prior to fragmentation. Their isotope effect was reported to be 3.50, but this result was later amended [67] to 2.4, which is in satisfactory agreement with that of Howe and McLafferty. A "preference factor" of 1.0 was deduced from the experiments; this factor compares the probability of loss of a ring hydrogen (or deuterium) atom with that for a side-chain hydrogen (or deuterium) atom; thus deviations from 1.0 indicate incomplete scrambling. A value of 1.32 had earlier been reported by Meyer and Harrison [68], together with an isotope effect of 1.58, for ion source generated $(M-H)^+$ ions (in good agreement with the values in Table 2).

Observations of doubly charged ions in the mass spectra of labelled toluenes have been described by Ast et al. [69,70]. They concluded that for the $C_7H_8{}^{2+} \rightarrow C_7H_7{}^{2+\cdot} + H^\cdot$ fragmentation occurring in the ion source, there is a preference factor of 1:3.3 and an isotope effect of 1.3; for reactions involving loss of molecules of hydrogen, HD and deuterium, randomisation is essentially complete prior to fragmentation. Similarly, for slow reactions involving loss of methyl cations and $C_3H_3{}^+$ ions, randomisation of hydrogen/deuterium is nearly complete.

From time to time, new experimental results appear which question the assumption that the $C_7H_7{}^+$ ion from toluene should be represented as the symmetrical tropyl cation. For example, Yamamoto et al. [71] found that in the gas-phase radiolysis of toluene-α-d_3, the ratio of mixed labelled methyldiphenylmethanes produced could not be explained solely on the basis that a randomised tropyl cation was the reactive species. They concluded that two thirds of their reacting labelled $C_7H_7{}^+$ ions should be represented as benzylic. The problem remains one of considerable interest and should occupy mass spectrometrists' ingenuity for some time to come.

Bruins et al. [72] briefly reported that there was a marked difference in fragmentation behaviour between the $(M-H)^+$ ion from benzylamine and that from p-aminotoluene. They showed first that amino hydrogen was not lost in either case; the benzylamine-ND_2 generated $C_7H_6ND_2{}^+$ ion lost exclusively hydrogen cyanide whilst the same ion from p-aminotoluene-ND_2 lost both hydrogen cyanide and deuterium cyanide in the ratio 1:0.7. This result indicates that the two $(M-H)^+$ ions probably do *not* have the same structure, i.e. that the sequence shown below is incorrect.

CH₂NH₂]⁺˙ , CH₃]⁺˙ with NH₂ —(−H)→ tropylium–NH₂

TABLE 3

DISSOCIATIONS OF AROMATIC IONS FOR WHICH COMPLETE ATOM RANDOMISATION HAS BEEN IDENTIFIED

Precursor molecule	Fragmentation	Label	Energy region			Ref.
			s	$^1m^*$	$^2m^*$	
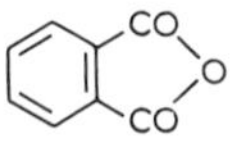 CO, O, CO	$C_6H_4^{+\cdot} \rightarrow C_4H_2^{+\cdot} + C_2H_2$	D	X			88
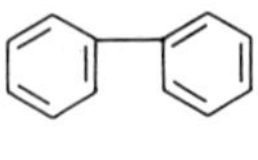	$M^{+\cdot} \rightarrow C_{11}H_7^{+} + CH_3^{\cdot}$	D	X			88
	$\rightarrow C_{10}H_8^{+\cdot} + C_2H_2$	D	X			88
	$\rightarrow C_9H_7^{+} + C_3H_3^{\cdot}$	D	X			88
C≡C, CH_3	$C_{15}H_{11}^{+} \rightarrow C_{13}H_9^{+} + C_2H_2$	D	X[a]			84
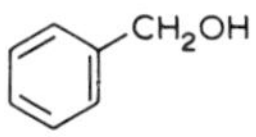 CH_2OH	$C_7H_7^{+} \rightarrow C_5H_5^{+} + C_2H_2$	D	X			89
HC≡CCH=CHCH=CHCH$_2$OH	$C_7H_7^{+} \rightarrow C_5H_5^{+} + C_2H_2$	D	X			89
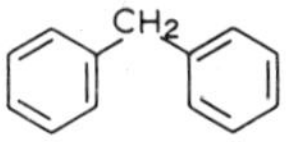 CH_2	$M^{+\cdot} \rightarrow C_{12}H_9^{+} + CH_3^{\cdot}$	D,^{13}C	X		X	60,90
	$C_{13}H_{11}^{+} \rightarrow C_{12}H_8^{+\cdot} + CH_3^{\cdot}$	D,^{13}C	X		X	60,90
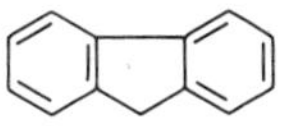 and 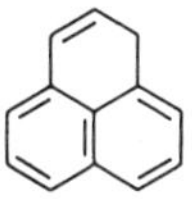	$M^{+\cdot} \rightarrow C_{13}H_9^{+}$[b] $+ H^{\cdot}$	D		X	X	91
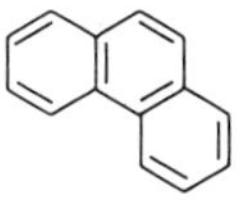	$M^{+\cdot} \rightarrow C_{12}H_8^{+\cdot} + C_2H_2$	D	X			92
CH=CH	$M^{+\cdot} \rightarrow C_{14}H_{11}^{+} + H^{\cdot}$	D	X		X	93, 94, 95
	$M^{+\cdot} \rightarrow C_{13}H_9^{+} + CH_3^{\cdot}$	D	X[c]		X[c]	
	$(M^{+\cdot} \rightarrow C_7H_6^{+\cdot} + C_6H_6)$	D	(note this is not a random process)			
	$M^{2+} \rightarrow C_8H_6^{2+} + C_6H_6$	D	X			96
	$M^{2+} \rightarrow C_{14}H_{10}^{2+} + H_2$	D	X			96
	$M^{2+} \rightarrow C_{13}H_9^{2+\cdot} + CH_3^{\cdot}$	D	X			96

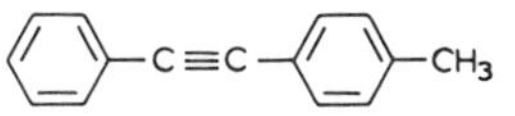

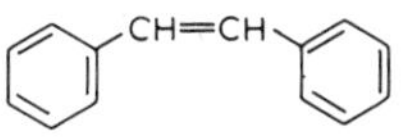

TABLE 3 (continued)

DISSOCIATIONS OF AROMATIC IONS FOR WHICH COMPLETE ATOM RANDOMISATION HAS BEEN IDENTIFIED

Precursor molecule	Fragmentation	Label	Energy region			Ref.
			s	$^{1}m^*$	$^{2}m^*$	
CH_2CH_2Br						
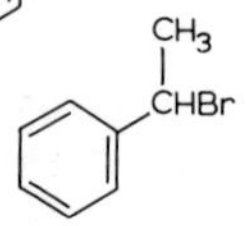 CH_3 / $CHBr$	$C_8H_9^+ \rightarrow C_6H_7^+ + C_2H_2$	D	X			79,107
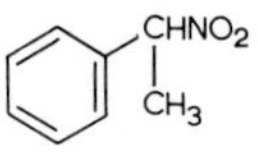 $CHNO_2$ / CH_3						
Triptycene	$M^{+\cdot} \rightarrow C_{20}H_{13}^+ + H^\cdot$	D	X			97
$(C_6H_5)_3CH$	$M^{+\cdot} \rightarrow C_{19}H_{15}^+ + H^\cdot$	D	X			97
	$\rightarrow C_{18}H_{13}^+ + CH_3^\cdot$	D	X			97
CH_2	$\rightarrow C_{18}H_{12}^{+\cdot} + CH_4$	D	X			97
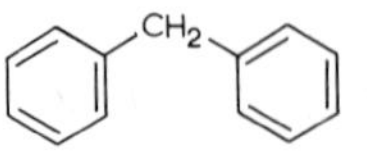	$M^{+\cdot} \rightarrow C_{13}H_{11}^+ + H^\cdot$	D	X			97
	$\rightarrow C_{12}H_9^+ + CH_3^\cdot$	D	X			97
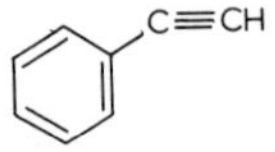 $C{\equiv}CH$	$M^{+\cdot} \rightarrow C_6H_4^{+\cdot} + C_2H_2$	D	X		X	98
	$C_8H_5^+ \rightarrow C_6H_3^+ + C_2H_2$	D	X			98
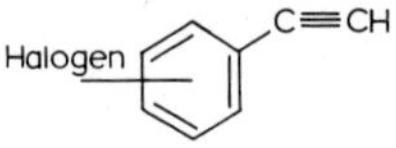 Halogen, $C{\equiv}CH$						
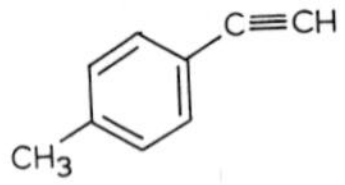 $C{\equiv}CH$, CH_3	$M^{+\cdot} \rightarrow C_9H_7^+ + H^\cdot$	D	X			98
	$C_9H_7^+ \rightarrow C_7H_5^+ + C_2H_2$	D	X			98
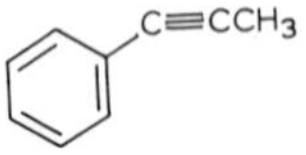 $C{\equiv}CCH_3$	$M^{+\cdot} \rightarrow C_9H_7^+ + H^\cdot$	D	X			99
	$C_9H_7^+ \rightarrow C_7H_5^+ + C_2H_2$	D	X			99
$(C_6H_5)_4C_4X$	$(C_6H_5)_4C_4^+ \rightarrow (C_6H_5)C_2^+$	D		X		100
	$+ (C_6H_5)_2C_2$					250
$C_{10}H_{11}X$	$C_{10}H_{11}^+ \rightarrow C_9H_8^{+\cdot} + CH_3^\cdot$	D, ^{13}C			X	101
	$\rightarrow C_9H_7^+ + CH_4$	D, ^{13}C			X	101
	$\rightarrow C_7H_7^+ + C_3H_4$	D, ^{13}C			X	101

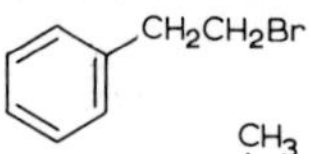

TABLE 3 (continued)

DISSOCIATIONS OF AROMATIC IONS FOR WHICH COMPLETE ATOM RANDOMISATION HAS BEEN IDENTIFIED

Precursor molecule	Fragmentation	Label	Energy region			Ref.
			s	$^1m^*$	$^2m^*$	
$C_8H_7(CH_2)_nX$ (cyclooctatetraenyl–$(CH_2)_nX$)	$C_9H_9^+ \rightarrow C_9H_7^+ + H_2$	D			X	102
	$\rightarrow C_7H_7^+ + C_2H_2$	D			X	102
	$C_{11}H_{13}^+ \rightarrow C_7H_7^+ + C_4H_6$	D, ^{13}C			X	34, 102
	$\rightarrow C_9H_9^+ + C_2H_4$	D, ^{13}C			X	102
	$C_{12}H_{15}^+ \rightarrow C_7H_7^+ + C_5H_8$	D, ^{13}C			X	102
$(C_6H_5)_2CHX$	$C_{13}H_{11}^+ \rightarrow C_{12}H_8^{+\cdot} + CH_3^{\cdot}$	D, ^{13}C			X	103
	$C_{13}H_{12}^{+\cdot} \rightarrow C_{12}H_9^+ + CH_3^{\cdot}$ (X=H)	D, ^{13}C		X	X	103
$C_9H_{11}X$ [d]	$C_9H_{11}^+ \rightarrow C_7H_7^+ + C_2H_4$	D		X		87
	$\rightarrow C_6H_7^+ + C_3H_4$	D		X		87
	$\rightarrow C_3H_5^+ + C_6H_6$	D		X		87
$C_6H_5CH{=}CHCH_2CH_2Br$	$C_{10}H_9^+ \rightarrow C_7H_7^+ + C_3H_2$	D		X		34

[a] Scrambling only ca. 50%.
[b] It is possible that these $C_{13}H_9^+$ daughter ions are structurally different.
[c] These results are at varience with refs. 104 and 105. The mechanisms for methyl loss are not fully understood and have recently been re-evaluated [106].
[d] See p. 80 for discussion.

Table 3 lists the many aromatic ions (not discussed above) for which complete or very extensive loss of positional identity of hydrogen and/or carbon atoms has been identified. In the majority of cases, the randomisation has been observed prior to fragmentation in the ion source and it can therefore be assumed that ions fragmenting more slowly (metastable ion peak generating species) will behave similarly.

The early work of McCollum and Meyerson [73] and McLafferty [74] on deuterium-labelled alkylbenzenes was repeated and extended in 1971 by Lightner et al. [75]. They showed that the predominant hydrogen transfer from a γ-carbon atom in the formation of $C_7H_8^{+\cdot}$ remained, regardless of whether that hydrogen be primary, secondary or tertiary.

$$[C_6H_5CH_2(\alpha)CH_2(\beta)CHR_1R_2]^{+\cdot} \longrightarrow C_7H_8^{+\cdot} + CH_2{=}CR_1R_2$$

As with analogous work [76] on the McLafferty rearrangement of ketones, the importance of such a hydrogen transfer increased with increasing substitution at the γ-carbon atom. An additional complication for the above fragmentation was the inferred presence of some exchange between side chain and *ortho*-ring hydrogen atoms. It was also concluded that the majority of $C_7H_7^+$ ions in the normal mass spectrum originated from the molecular ion by direct C—C cleavage rather than from fragmentation of $C_7H_8^{+\cdot}$.

The mass spectra of a series of deuterium and carbon-13 labelled isomeric 1-phenylheptenes have been described [77]. The major daughter ions in the normal mass spectra, $C_9H_9^+$ and $C_8H_8^{+\cdot}$, were found to arise from complex fragmentation pathways.

C. Other Aromatic Molecules

The mass spectra of deuterium- and carbon-13-labelled 1,3-diphenylpropenes were described by Johnstone and Millard [78]. Loss of a methyl radical from the molecular ion involves only the C-2 atom in the propene chain. Loss of vinyl and ethyl radicals were described but it was assumed that these occurred solely from the molecular ions and not via consecutive processes such as $(M - H - C_2H_2)$ and $(M - H - C_2H_4)$.

The structures of the ions $C_8H_9^+$ and $C_9H_{11}^+$ (produced from molecules of type $C_6H_5C_2H_4X$ and $C_6H_5C_3H_6X$ respectively) have been discussed by Nibbering and de Boer [79]. The $C_8H_9^+$ ion is of interest because earlier work [50] on labelled xylenes (C_8H_{10}) indicated considerable hydrogen randomisation prior to its formation as did observations on its production in the mass spectra of ethyl benzenes [50]. The production of $C_8H_9^+$ from *p*-methylethylbenzene [68] was shown to involve partial loss of both the ring and β-methyl groups. From the above results it was concluded that the $C_8H_9^+$ ion is best represented as a methyltropyl cation. Nibbering and de Boer observed complete randomisation of hydrogen atoms for acetylene loss from $C_8H_9^+$ and proposed that this involved equilibration between methyltropylium and an eight membered ring. Later experiments [80] with carbon-13 showed a clear distinction between the rapidly dissociating, source-generated ions which lost predominantly *side chain* carbon and the slower, lower energy, metastable ion peak generating ions which showed increasing carbon randomisation with increasing lifetime. Similarly for the analogous process in styrene

$$C_8H_8^{+\cdot} \rightarrow C_6H_6^{+\cdot} + C_2H_2$$

hydrogen atom randomisation is complete at all observed lifetimes whereas carbon atom scrambling tends to completion only in second field-free region metastable ion peaks. The above experiments provide a particularly good example of independent atom scrambling. In the former paper [79] it was reported that the $C_9H_{11}^+$ ion (formed by loss of nitrogen dioxide from 1-phenyl-1-nitropropane or 1-phenyl-2-nitropropane) dissociated by loss of ethylene in which only the hydrogen atoms from the propane chain were randomly selected

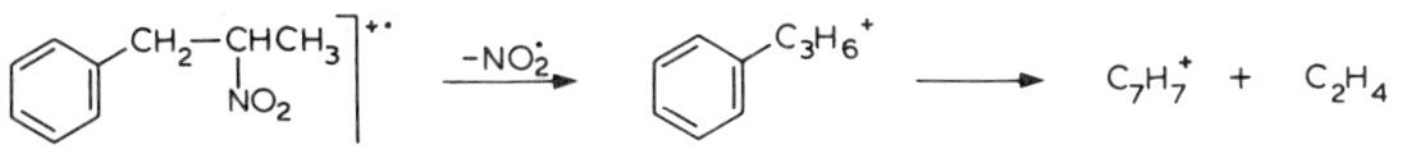

$$C_6H_5CH_2\text{—}CH(NO_2)CH_3^{+\cdot} \xrightarrow{-NO_2^{\cdot}} C_6H_5C_3H_6^{+} \longrightarrow C_7H_7^{+} + C_2H_4$$

and these observations support the phenylated cyclopropane structure for $C_9H_{11}^{+}$, suggested much earlier in a study of carbon-13 labelled *tert*-butylbenzene [50].

The *p*-xylene experiments [50] were extended by Meyerson and Fields to polymethylbenzenes [81] again with particular regard to loss of methyl from the molecular ions. For *p*-xylene the deuterium-labelling experiments were explained by assuming that the original methyl groups are lost following a single 1:1 exchange of methyl hydrogen with an *ortho*-ring hydrogen, i.e.

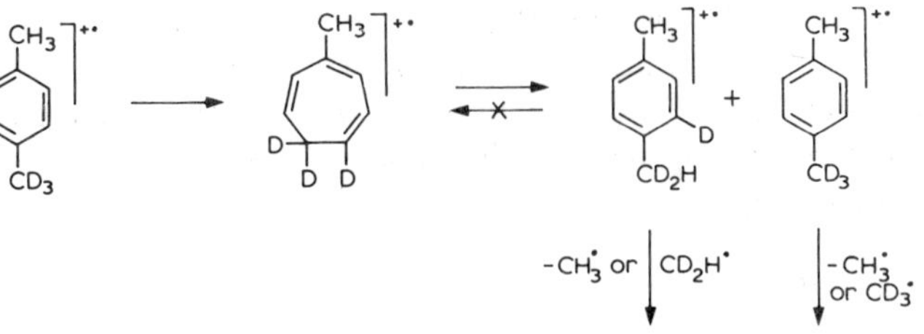

CH_3, CD_3 → CH_3, D, D, D → CH_3, D, CD_2H + CH_3, CD_3

$-CH_3^{\cdot}$ or $CD_2H^{\cdot}$; $-CH_3^{\cdot}$ or $CD_3^{\cdot}$

However, the carbon-13 labelling experiments showed that 16% of the ejected methyl carbons originated from the ring. This ring expansion–contraction mechanism preceding methyl loss received support from some observations of Venema et al. [82] who found that for 7-^{13}C-methylcycloheptatriene, ring carbon atoms were lost about 25% of the time in the normal mass spectrum (70 eV), (35% at 15 eV and with equal probability for second field-free region dissociations). 7-d_3-Methylcycloheptatriene showed mainly CD_3 loss (∼ 80% in the normal mass spectrum) with increasing hydrogen/deuterium atom mixing at lower electron energies and longer ion lifetimes. For the deuterium-labelled polymethylbenzenes, the above explanation proved satisfactory with one additional requirement, namely that a methyl group flanked by two methyl groups undergoes no atom exchange. That this explanation may be an oversimplification is indicated by some recent experiments by Dawson and Gillis on ring deuterium-labelled tri- and tetramethylbenzenes [83]. They proposed that more than one such expansion–contraction cycle is required to give quantitative agreement with their results.

Safe [84] reported that only partial hydrogen/deuterium atom scrambling was observed in the second fragmentation shown below.

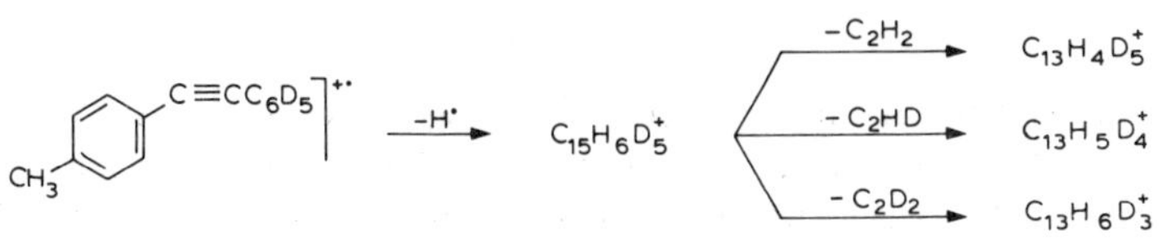

$$CH_3C_6H_4C{\equiv}CC_6D_5^{+\cdot} \xrightarrow{-H^{\cdot}} C_{15}H_6D_5^{+}$$

$-C_2H_2$ → $C_{13}H_4D_5^{+}$

$-C_2HD$ → $C_{13}H_5D_4^{+}$

$-C_2D_2$ → $C_{13}H_6D_3^{+}$

A similar result was found [85] for

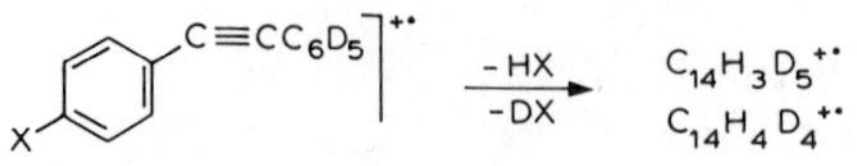

$$X\text{-}C_6H_4\text{-}C{\equiv}CC_6D_5^{+\cdot} \xrightarrow[-DX]{-HX} C_{14}H_3D_5^{+\cdot},\ C_{14}H_4D_4^{+\cdot}$$

X = F, Cl, Br

indicating that some hydrogen/deuterium equilibration may take place across the intervening acetylene bridge. Extensive, but incomplete, hydrogen/deuterium atom randomisation also takes place prior to fragmentation of substituted biphenyls [86].

Uccella and Williams [87] have described the fragmentation behaviour of deuterium-labelled $C_9H_{11}^+$ ions as function of their precursor molecules' structure. Those which are generated as $C_6H_5C_3H_6^+$ units display only localised atom scrambling within the C_3 group prior to dissociation by loss of C_2H_4, C_3H_4 or C_6H_6 (see above and ref. 79). Those generated from di- or trisubstituted benzene ring compounds having side chains containing only one or two carbon atoms undergo the same fragmentations but with complete hydrogen atom scrambling involving the whole ion (see Table 3). It is apparent that $C_9H_{11}^+$ ions formally represented as

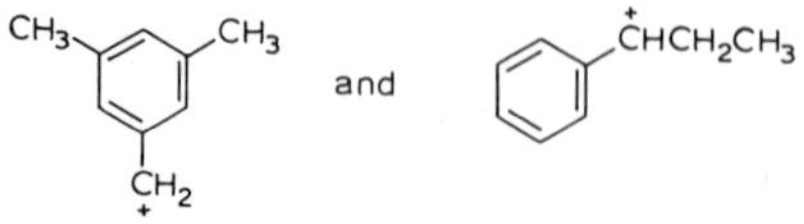

CH_3, CH_3, $\overset{+}{C}H_2$ (3,5-dimethylbenzyl) and $C_6H_5\overset{+}{C}HCH_2CH_3$

neither interconvert nor rearrange to a common intermediate prior to unimolecular dissociation.

IV. KETONES

A. Simple Aliphatic and Aromatic Ketones

The recent literature presents an abundance of information concerning the behaviour of the carbonyl group in a myriad of different environments. In this section only the more important results derived from labelling experiments will be described.

In general, the mass spectra of carbonyl compounds have the McLafferty rearrangement (shown below) as an important fragmentation route.

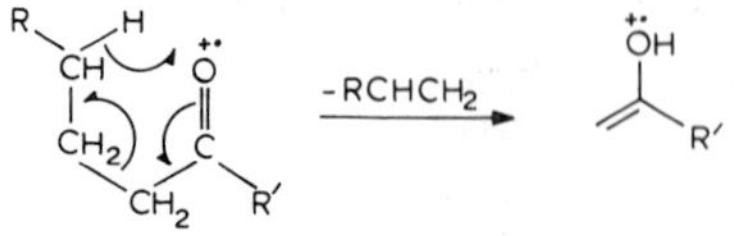

$$R\text{-}CH(H)\text{-}CH_2\text{-}CH_2\text{-}C(=\overset{+\cdot}{O})R' \xrightarrow{-RCHCH_2} CH_2{=}C(\overset{+\cdot}{O}H)R'$$

The site-specific nature of this reaction for ketones, appropriately deuterium labelled, was reported in 1963 by Seibl and Gäumann [111] and is discussed in detail in a book by Budzikiewicz et al. [112]. Further information was obtained by substituting methyl groups at the γ-position [113] and by placing a double bond in the β- and γ-positions [114,115]. In addition, it has been shown that a γ tertiary hydrogen atom is more readily transferred than a secondary hydrogen atom which in turn is preferred over a primary hydrogen atom [76,116]. This rearrangement is often compared with the analogous photochemical process; however, experiments designed to compare the details of the electron impact-induced reaction with photochemical counterparts often show that the analogy is only superficial [117].

Isotope effects are readily identifiable in this rearrangement because of its specific nature. For example, MacLeod and Djerassi [118] studied a series of molecules in which one of the two γ-hydrogen atoms available for participation in a McLafferty rearrangement was replaced by deuterium. The isotope effect (I.E.) was equated to "the number of atoms of deuterium per atom of hydrogen transferred for the hypothetical case in which equal numbers of deuterium and hydrogen are available." For γ-d_1 and γ-d_2 methyl butyrates a value of 0.88 had been observed [119]; in the aliphatic ketones 3-heptanone-6-d_1, 3-octanone-6-d_2 and 5-nonanone-2,8-d_2, there was no primary isotope effect, I.E. = 1.0. The cyclic ketone 2-propyl-2$'$-d_1-cyclohexanone, however, showed a small isotope effect, I.E. = 0.87. This apparent discrimination against deuterium was explained as possibly arising from some hydrogen atom transfer other than from the γ-carbon atom. While there is no doubt that the simple site-specific mechanism described above predominates in the conventional 70 eV mass spectrum, the behaviour of low energy ions may be much less clear cut. This is well exemplified by the observations of Carpenter et al. [120] on the (nominally) 10 eV mass spectra of labelled 3-octanones, where there is apparently considerable intramolecular scrambling of hydrogen and deuterium atoms prior to both acylic loss of ethyl and the McLafferty rearrangement. It is noteworthy that the published low voltage spectra show prominent $(M + 1)^+$ and $(M + 2)^+$ peaks, indicative of ion–molecule reactions. Generally, conclusions drawn from results of this kind should be regarded with reservations. However, in this case it was reported that, when recorded on other instruments, the mass spectra were free from these peaks, but otherwise were similar to those shown. The above observations were repeated and extended by Yeo and Williams [121] who studied the extent of internal hydrogen rearrangement as a function of ion lifetime (and hence internal energy content) by examining normal spectra at 70 and 10 eV and first and second field-free region metastable ion peaks. Hexan-, heptan- and octan-3-ones and their 2,2,4,4-d_4 analogues were selected for these experiments. Their results can be summarised as follows.

(i) The 70 eV spectra showed little ($<$ 10%) or no evidence of hydrogen/deuterium atom mixing prior to cleavage by ethyl loss and by olefin elimination resulting from a McLafferty rearrangement.

(ii) The low voltage (10 eV) spectra showed that considerable loss of positional identity of hydrogen and deuterium atoms took place before the above fragmentations.

(iii) For ions of increasing lifetime (first ($^1m^*$) and second ($^2m^*$) field-free region metastable ion peaks observed at 70 eV) hydrogen/deuterium atom mixing also increased. For the McLafferty rearrangement processes the $^2m^*$ relative abundances approached those for the random participation of *all* hydrogen/deuterium atoms. It is noteworthy that for these long-lived ions, ethyl loss never involved more than two deuterium atoms, perhaps indicative of a localised hydrogen/deuterium scrambling process. Localised scrambling has for example been reported in ethyl acetate [122].

It seemed probable, therefore, that although low-energy spectra and metastable ion studies of ketones would be difficult to interpret owing to atom randomisation, 70 eV spectra would result from rapid reactions, (relatively) free of mechanistic complexities. However, Yeo showed soon afterwards [123], that extensive hydrogen randomisation took place even prior to α-cleavage (methyl loss) in a series of aliphatic methyl ketones. The degree of scrambling increased with aliphatic chain length and was cited as evidence for a degrees-of-freedom effect – an increase in the number of vibrational modes among which the internal energy of the molecular ion can be distributed, which leads to a longer-lived and hence more "scrambled" species.

A great deal of discussion has centred about properly identifying and distinguishing between the three isomeric $C_3H_6O^{+\cdot}$ ions produced from different precursors by a single or double McLafferty rearrangement. The three ions and their potential origins are shown below. McLafferty and Pike [124] showed that the $^2m^*$ abundance ratios

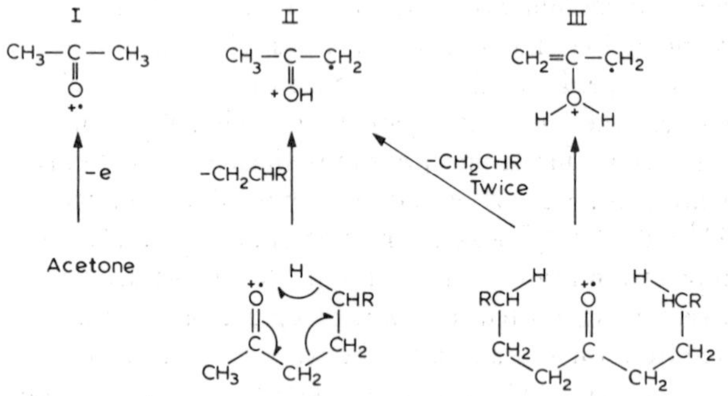

I: $CH_3-C(=O^{+\cdot})-CH_3$ ← −e ← Acetone

II: $CH_3-C(=\overset{+}{O}H)-\dot{C}H_2$ ← $-CH_2CHR$ (from 2-alkanone); ← $-CH_2CHR$ Twice (from di-*n*-alkanone)

III: $CH_2=C(-\overset{+}{O}H_2)-\dot{C}H_2$

for the dissociation of I, II generated from 2-alkanones and II or III generated from seven di-*n*-alkanones, were 23:280:110. For this comparison the competing processes were

$$C_3H_6O^{+\cdot} \rightarrow CH_3^{\cdot} + C_2H_3O^+$$

$$C_3H_6O^{+\cdot} \rightarrow CH_3^+ + C_2H_3O^{\cdot}$$

It was felt that the latter two ratios were sufficiently different to provide evidence in favour of (*but not proof for*) the participation of structure III. However, in 1968, a series of elegant experiments by Diekman et al. [125] showed that the double McLafferty rearrangement generated the enolic form II, rather than the oxonium ion III. In this work the reactivity of $C_3H_6O^{+\cdot}$ ions, produced from a variety of sources, was studied using i.c.r. spectroscopy. No fewer than seven ion–molecule reactions were found which were capable of distinguishing between keto and enol $C_3H_6O^{+\cdot}$ ions. It is important to remember that labelled ions in which atom scrambling has taken place (vide infra) will represent only a small fraction of those reacting in the i.c.r. experiments. Thus virtually all $C_3H_6O^{+\cdot}$ ions with insufficient energy to react further will be sampled; in a conventional mass spectrometer the (scrambled) metastable ions represent only a small range of the precursor ion, and hence daughter ion, energies. Daughter ions generated in the source are much more abundant and these rapidly generated ions are largely unscrambled.

Further evidence against the presence of form III in the mass spectra of di-*n*-alkanones was provided by Eadon et al. [26] from an i.c.r. study of 4-nonanone-1,1,1-d_3. A lengthier description of this work was published a year later [127]. The aim of these experiments was to ascertain whether the enol ion is formed directly or by isomerisation of the oxonium form, *or* by ketonisation of the intermediate enol ion followed by hydrogen rearrangement, i.e.

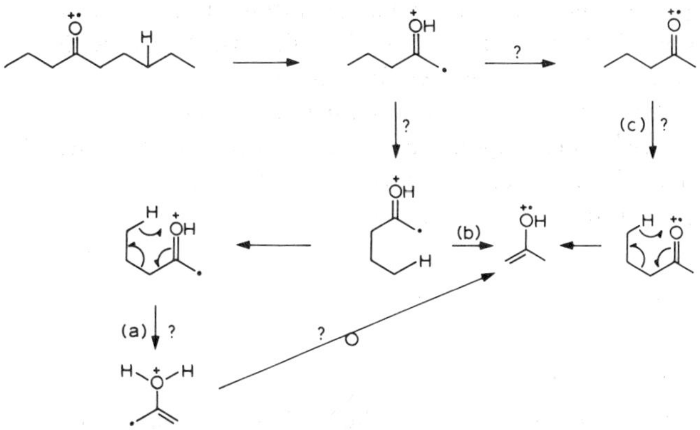

This experiment is a particularly good example of the power of i.c.r. spectroscopy and so the results will be described in some detail. 4-Nonanone-1,1,1-d_3 was chosen as the reactant molecule because its fragmentation would lead to a uniquely labelled $C_3H_6O^{+\cdot}$ ion after the double McLafferty rearrangement. This is because in the first dissociation, a secondary hydrogen atom is preferentially involved by a factor of ∿ 50 over *primary* deuterium.

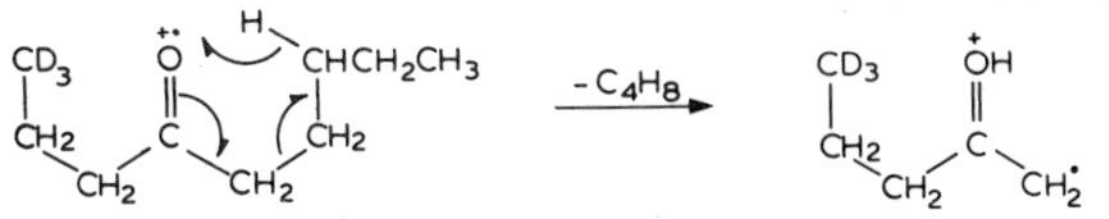

The ions resulting from the second McLafferty rearrangement will be one or more of those shown below.

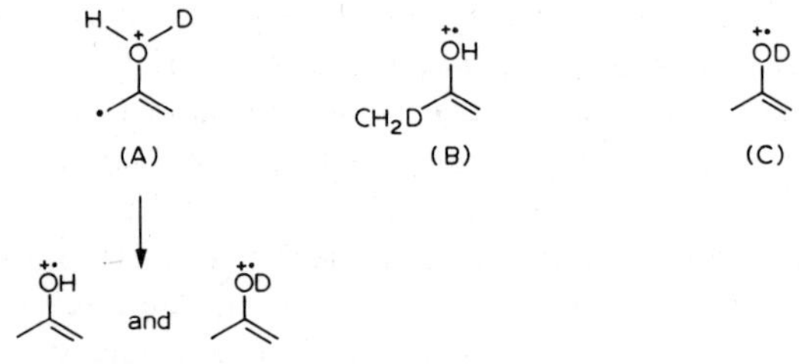

(A) (B) (C)

and

(in approx. equal amounts)

Now it was known that neutral ketones react specifically with hydrogen attached to oxygen in ions such as (A) and (B). Thus using the pulsed double resonance technique for the (M + 1) and (M + 2) ions of 4-nonanone-1,1,1-d_3 and of 4-nonanone-7,7-d_2 it was shown that m/e 59 contributed to (M + 1) over (M + 2) in a ratio > 5:1 for the former and that m/e 59 contributed to (M + 2) over (M + 1) in a ratio > 6:1 for the latter. These results show that the double McLafferty ion from the d_3 ketone has *hydrogen* attached to oxygen and that the corresponding ion from the d_2 ketone has *deuterium* attached to oxygen. Thus reaction pathway (b) is that which operates in the double rearrangement and the enolic ion is therefore produced.

An independent verification of the above results appeared shortly afterwards in a paper by McLafferty et al. [128]. These experiments were designed to discover whether decomposition of the enolic $C_3H_6O^{+\cdot}$ ion by methyl loss involved direct cleavage or cleavage following rearrangement to the keto form, e.g.

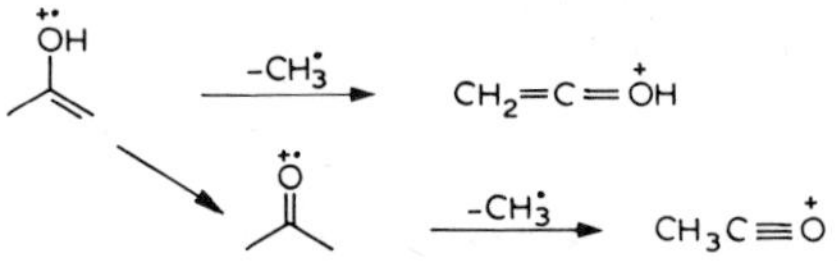

$$\xrightarrow{-CH_3^{\cdot}} CH_2{=}C{=}\overset{+}{O}H$$

$$\xrightarrow{-CH_3^{\cdot}} CH_3C{\equiv}\overset{+}{O}$$

By a suitable choice of precursor molecules the nine labelled $C_3H_6O^{+\cdot}$ ions shown below were generated from single and/or double McLafferty rearrangements.

$CD_3C(OH) = CH_2]^{+\cdot}$ $CH_3C(OD) = CH_2]^{+\cdot}$

$CD_3C(OD) = CH_2]^{+\cdot}$ $CH_3C(OH) = CD_2]^{+\cdot}$

$CD_3C(OH) = CD_2]^{+\cdot}$ $CH_3C(OD) = CD_2]^{+\cdot}$

$CD_2HC(OH) = CD_2]^{+\cdot}$ $CH_3C(OH) = CHD]^{+\cdot}$

$CH_2DC(OH) = CD_2]^{+\cdot}$

Metastable ion peak abundances were compared for the various methyl radical losses from this wide range of labelled ions. The results clearly showed that rearrangement to an excited acetone molecular ion took place prior to dissociation, but the observations also revealed the presence of substantial primary isotope effects. Furthermore, an additional isomerisation reaction was proposed in which a methyl hydrogen is transferred to the methylene group before ketonisation; this too had an appreciable primary isotope effect. After analysis of the results for isotope effects a discrepancy remained indicating that the methyl group, formed by accepting the enolic hydrogen atom, was lost more readily than the *in situ* methyl group. This interesting result was interpreted as arising from incomplete internal energy randomisation in the keto form. Additional work in this area (e.g. the use of carbon-13 to distinguish between the methyl and enol–methylene groups) is required to substantiate this proposal.

The reketonisation of an ion produced by a McLafferty rearrangement has been illustrated by i.c.r. spectrometry [320] for the ion $C_5H_8O^{+\cdot}$, *m/e* 84. This ion is prominent in the mass spectra of 2-propylcyclopentanone and 2-ethyl-5-propylcyclopentanone (where it results from a *double* McLafferty rearrangement). At residence times of from about 4×10^{-4} to 1×10^{-3} sec the ion behaves as expected for an enol, donating a hydrogen atom to other ketones on collision. However, at longer lifetimes, 5×10^{-3} to 2×10^{-1} sec, the ion behaves like a ketonic species.

The above work on $C_3H_6O^{+\cdot}$ was extended in the following year to the homologous $C_4H_8O^{+\cdot}$ ions [129] of which deuterium-labelled analogues of the isomeric species

$C_2H_5C(OH)CH_2]^{+\cdot}$ (D)

and

$CH_3CHC(OH)CH_3]^{+\cdot}$ (E)

were studied. Again the bulk of the evidence was gathered from metastable ion peak abundances. It was found that dissociation took place only from the ketonic ion $C_2H_5COCH_3]^{+\cdot}$ to yield $C_2H_5CO^+$ and CH_3CO^+. It was further shown that in the case of (D) two 1,4 hydrogen shift rearrangements predominated in the slow, metastable ion peak generating fragmentations, whereas for (E) the favoured low energy process is a 1,2 followed by a 1,4 hydrogen shift, i.e.

$CH_3CH_2C(=\overset{+}{O}H)\dot{C}H_2 \rightarrow \dot{C}H_2CH_2C(=\overset{+}{O}H)CH_3 \rightarrow CH_3CH_2C(=O^{+\cdot})CH_3$

$CH_3\dot{C}HC(=\overset{+}{O}H)CH_3 \rightarrow \dot{C}H_2CH_2C(=\overset{+}{O}H)CH_3 \nearrow$

Fragmentation

For the source-generated fragmentations it was seen (from normal ion abundances) that 1,3 hydrogen rearrangements compete effectively with 1,4 shifts, producing the uncommon situation that hydrogen scrambling is greater in ions of higher internal energy than in those observed as metastable ion peaks. This latter result appears to be in conflict with the conclusions of Yeo and Williams [121], but they did not, however, examine the fate of isotopic labels in $C_4H_8O^{+\cdot}$ ions generated in the ion source.

The ion kinetic energy spectrum of nonan-4-one and its 7,7-d_2 and 1,1,1-d_3 analogues were studied by Eadon et al. [130]. Although some new minor fragmentation modes were discovered, the use of this technique added little to previous knowledge. A related experiment was reported by Beynon et al. [131] in 1971. Here, the average energy released in the fragmentation

$$C_6H_5COCH_3^{+\cdot} \rightarrow C_6H_5CO^+ + CH_3^{\cdot}$$

was measured (from the half-height width of the metastable ion peak for this dissociation) as a function of precursor molecule. For acetophenone the $T_{0.5}$ value was small, 7.2 meV. For a series of compounds $C_6H_5COCH_2CH_2CH_2R$ (R = alkyl or aryl) $T_{0.5}$ was larger, 50 ± 5 meV and essentially was independent of R. This result was presented as proof that for the latter series of compounds the McLafferty rearrangement produces the enolic ion shown below which must then rearrange prior to dissociation by methyl loss.

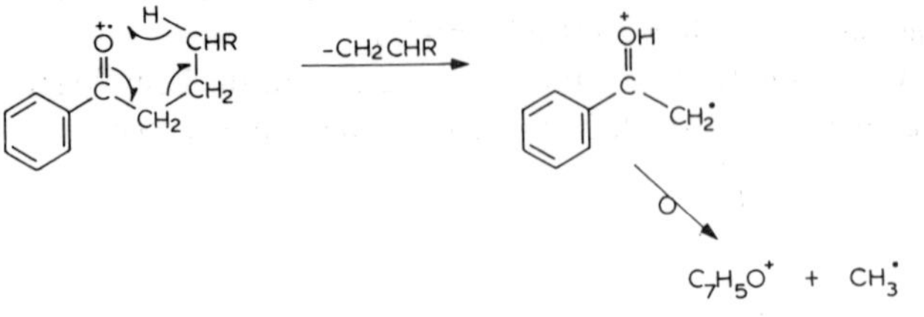

$-CH_2CHR$

$C_7H_5O^+ + CH_3^{\cdot}$

However, this explanation, although attractive, was shown to be incorrect by Tomer and Djerassi [132]. They considered that the enolic ion could behave similarly to the $(M - C_2H_4)^{+\cdot}$ ion produced in the mass spectrum of ethyl benzoate.

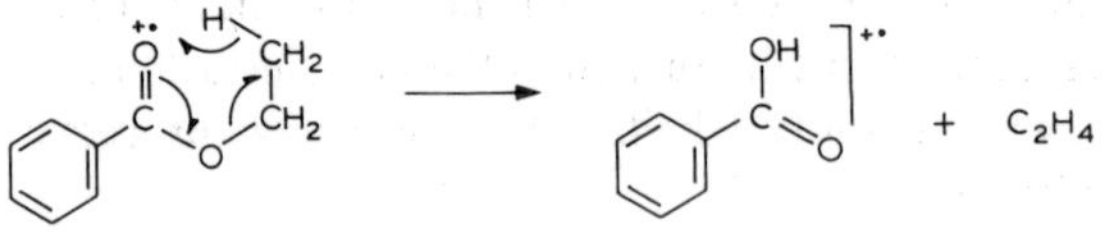

$+ \ C_2H_4$

This ion dissociates by loss of hydroxyl which involves not only the hydroxyl hydrogen (originally in the ester group) but also the two *ortho* ring hydrogen atoms (see pp. 92–94). Thus for the ketone a similar process is feasible, the enolic ion dissociating as shown

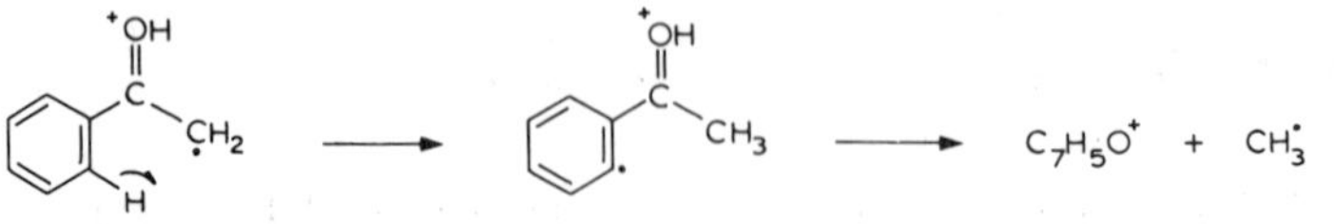

$C_7H_5O^+ + CH_3^{\cdot}$

Tomer and Djerassi found that butyrophenone-ring-d_5 lost mostly $CH_2D^{\cdot}$ (< 10% $CH_3^{\cdot}$) from the $(M - C_2H_4)^+$ ion and that butyrophenone-d_7 lost essentially CD_3 (< 10% CD_2H) (see below). These observations were reported for second field-free region metastable ion peaks only. Thus this latter work shows that ketonisation *does not occur in this system.* The energy release experiments did not prove the absence of

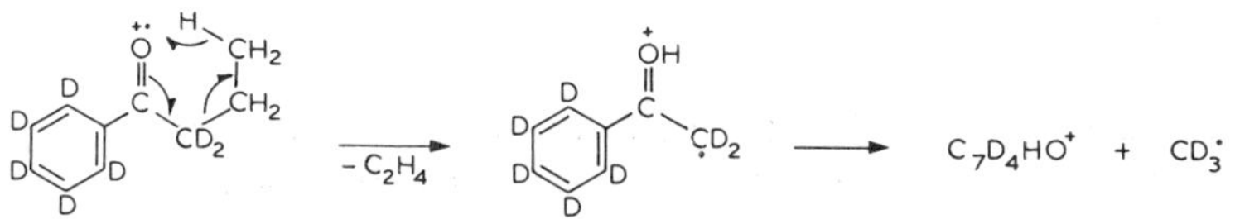

$$\text{C}_7\text{D}_4\text{HO}^+ + \text{CD}_3^{\cdot}$$

ketonisation but only provided additional evidence for the enolic form of the ion. Further studies of the physico-chemical aspects of this problem would certainly be worthwhile.

The ketones discussed above were not branched; Eadon and Djerassi [133] drew particular attention to the behaviour of branched ketones and esters. It was evident from the normal (70 eV) mass spectra of deuterium-labelled 7-methyl-4-octanone and several other γ-and δ-branched compounds that extensive hydrogen/deuterium atom randomisation was taking place prior to fragmentation of the molecular ion. The effect was markedly greater than that reported by Yeo and Williams [121] ; in addition the α-cleaved molecular ions underwent an important fragmentation by water loss giving rise to an abundant peak. This dissociation was considered to have diagnostic utility as an aid to distinguishing branched from straight-chain ketones, for in the latter this fragmentation is a minor process.

The effect of α, β-unsaturation on the mass spectra of aliphatic ketones was studied by Sheikh et al. [134] . Both product ions resulting from α-cleavage were present. The most unexpected fragmentation was the loss of acetone from the molecular ions of a series of ketones $CH_3COCH{=}CH{-}R$ ($R \geqslant C_4$), a process which must involve a double hydrogen transfer. Deuterium labelling allowed partial elucidation of the mechanism, which, for example, does not involve the first CH_2 group in R. The ion $C_3H_3O^+$, *m/e* 55, is also prominent; it, too, results from an unexpected fission through or adjacent to the vinylic bond.

The effect of non-conjugated double bonds on the McLafferty rearrangement was examined by Dias et al. [135] . The most obvious changes in the normal (70 eV) mass spectra are the appearance of relatively abundant ($M - CH_3COCH_3$) and ($M - H_2O$) hydrocarbon cation peaks.

Labelling experiments showed that the double bond was mobile, rearrangement to a δ,ϵ position taking place provided that the double bond initially was not too remote from the carbonyl group, e.g.

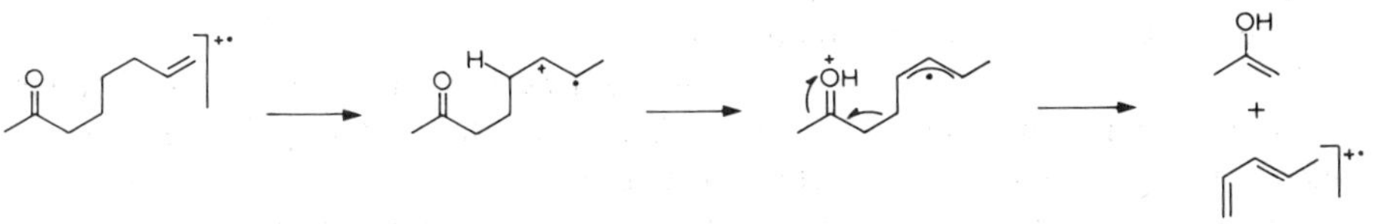

Similar effects were observed in the mass spectra of related esters and acids. These experiments were quickly followed by a detailed examination of the behaviour of phenyl substituted α, β-unsaturated ketones [136] by both conventional and i.c.r. mass spectrometric techniques.

B. Miscellaneous Ketones

Thomas and Willhalm [137] examined the possibility of McLafferty rearrangements taking place in the fragmentation of the molecular ions of *exo*- and *endo*-acetyl norbornanes and some related compounds. Deuterium-labelled camphor (and isoborneol) have been examined by Dimmel and Wolinsky [138].

An ion whose possible structures have given rise to controversy is $C_4H_4O^{+\cdot}$, which is readily produced by loss of carbon monoxide from the molecular ion of 2-pyrone. Early labelling experiments by Pirkle [139] indicated that the furan cation was an inappropriate structure.

3 6 $\xrightarrow{-CO}$

However, Brown and Green [140] disagreed with Pirkle's conclusions; the chief points being as follows.

The mass spectrum of 2-pyrone below *m/e* 68 is very similar to that of furan. Deuterium placed at C-3 (in the 2-pyrone) is largely retained in the $C_3H_3^+$ fragment ion, while a label placed at position 6 is not retained. The origin of $C_3H_3^+$, however, is not solely from $C_4H_4O^{+\cdot}$ but comes, at least in part from $(M-H)^+$, wherein the C-6 hydrogen atom has been lost. Pirkle and Dines resumed their investigation and reported a much more detailed series of experiments on 2-pyrones, singly labelled with deuterium at each ring position [141].

The origins of the $C_3H_3^+$ and $C_3H_2D^+$ ions were discussed by considering their relative abundances only in the normal mass spectrum. The results, although inconclusive, were hard to interpret in terms of a symmetrical furan-like $C_4H_3DO^{+\cdot}$ precursor. More striking evidence came from metastable ion peak abundance ratios for loss of C_2HD and acetylene from $C_4H_3DO^{+\cdot}$. For the 3-,4-,5- and 6-d_1 labelled compounds the ratios were 1.3, > 5, 1.2 and 0 respectively. This fragmentation also seems unlikely to proceed from a furan precursor. Various ion structures were proposed, but since none were substantiated (e.g. by a carbon-13 labelling experiment) they will not be presented here. The conclusions drawn from these labelling experiments provide additional evidence in support of those derived from other data [142–144].

The properties of the $(M-H_2O)^{+\cdot}$ ion in the mass spectrum of *o*-methoxybenzaldehyde are similar to those of benzofuran [145].

The mass spectra of a series of non-enolised (at least in the liquid phase) α-diketones have been examined, with some deuterium labelling, by Bowie et al. [146]. Simple bond cleavages predominated in aliphatic analogues while double loss of carbon monoxide was observed in quinonoid species. No evidence was found for participation of a McLafferty rearrangement.

A few β-diketones have been studied with the aid of deuterium labelling [147].

2-Phenoxy-4,5-benztropone displays an unusual fragmentation; Kinstle et al. [148] by oxygen-18 and deuterium labelling, showed that the loss of hydroxyl from the molecular ion involved both oxygen atoms with about equal probability, the hydrogen atom coming from an *ortho*-site of the phenoxy group.

The proposed migration of the phenyl group from one oxygen to the other was given additional support from a carbon-13 labelling experiment. 2-Phenoxy-4,5-benztropone-1-^{13}C loses 20% of its label in the first loss of carbon monoxide from the molecular ion. This agrees well with the observation that the ether oxygen is lost preferentially (by a factor of 4:1) over the carbonyl oxygen in the same fragmentation of the oxygen-18 labelled pair of molecular ions.

Although the fragmentation behaviour of cyclohexanone has been extensively examined [149], and is reasonably well understood, the same cannot be said for its isomers cyclohexene oxide and 2-cyclohexene-1-ol. The latter pair were studied by Strong et al. [150] (see also ref. 243). Examination of the mass spectra of analogues of cyclohexene oxide labelled at each ring position gave some insight into the mechanism but allowed no firm conclusions to be drawn. The principal dissociation is by loss of methyl to give the base peak; several possible pathways for this fragmentation were proposed but the experimental results could not be fitted to any simple combination thereof.

The mass spectra and fragmentation mechanisms of a series of deuterium-labelled 4-substituted cyclohexanones have been reported by Gray et al. [151].

Spiro-ketones (and olefins) of varying ring size were studied by Christiansen and Lightner [152]. The most noteworthy process involved formation of an intense hydrocarbon radical cation rather than oxygen containing species; $C_5H_7^+$, m/e 67, was base peak or a prominent ion in the mass spectra of the following spiro-compounds.

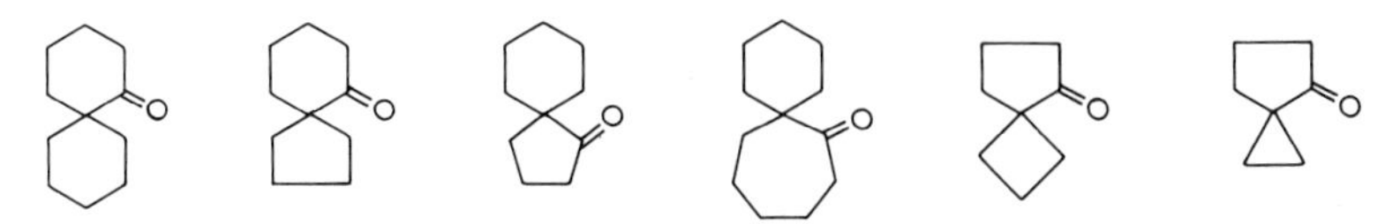

Similar results were reported by Weringa [153] who studied five- and six-membered ring spiro-alkanones. In general, agreement between the two sets of data is good; however there are a few discrepancies which indicate some incorrect m/e assignments in the former work.

Eadon and Djerassi [154] showed, by appropriate deuterium labelling, that in the reciprocal hydrogen transfer shown below the transition state ring sizes were not unusually large, being five-, six- and seven-membered.

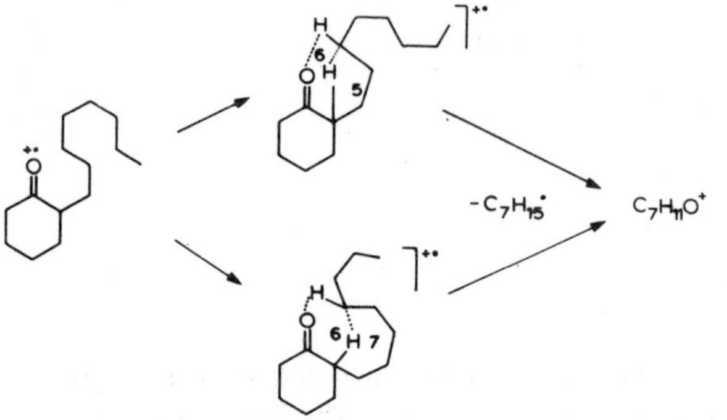

$-C_7H_{15}^{\cdot}$ $C_7H_{11}O^+$

Fenselau et al. [155] examined the effect of methylene chain length between benzoyl and styryl entities.

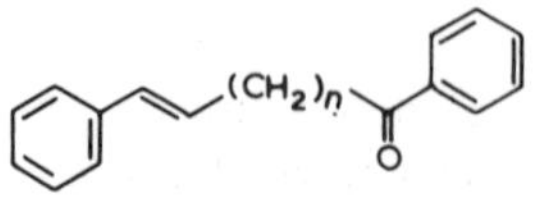

$(CH_2)_n$

For n = 1, 2 and 3 the fragmentations were unremarkable but for n = 4 (1,7-diphenylhept-6-ene-1-one) loss of C_7H_7 from the molecular ion was a prominent process. Deuterium, inserted at C-2, indicated that it was at least partially involved in the rearrangement, necessarily preceding loss of a benzyl radical.

An interesting hydrogen migration was shown to take place in a primary fragmentation of *ortho*-substituted benzophenones by Ballantine and Pillinger [156]. In this process a hydrogen atom from the *ortho* substituent was shown, by appropriate deuterium labelling, to be transferred to the other aromatic ring yielding the ions indicated below.

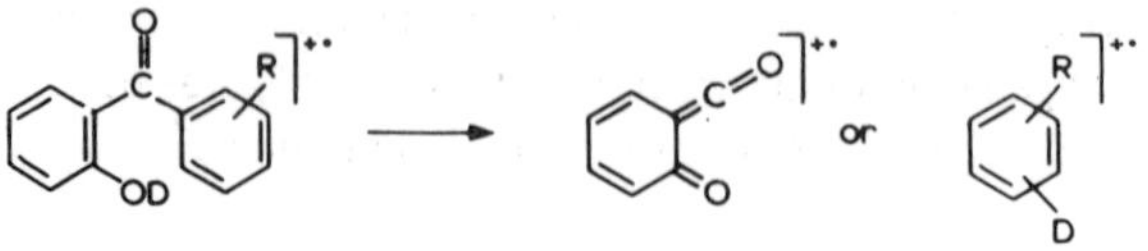

They proposed the transition state

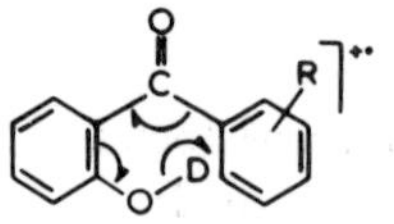

It also seems reasonable that the process involves formation of the enolic ion via a six-membered intermediate, followed by hydrogen transfer yielding the keto form.

V. ALDEHYDES

Investigations of the behaviour of the carbonyl function in the mass spectra of aldehydes are conspicuous by their absence. Derivatives of aldehydes have been studied in moderate detail (see below). Liedtke and Djerassi [157] performed the first "modern" study of the fragmentation behaviour of aliphatic aldehydes in 1969, using deuterium labelling and metastable ion peaks. *n*-Hexanal and *n*-heptanal were chosen for this study and almost every position in the molecules was labelled in turn. In spite of this exhaustive approach, even the mechanism of water loss from the molecular ions (which is absent in ketones and thus serves as a structure-identifying feature) could not be thoroughly understood. (It is worth noting that the hexanal mass spectrum has features in common with that of cyclohexanol [158], another molecule whose fragmentation mechanism is fraught with difficulties.) Similarly, loss of ethylene involves C-2 and C-3, C-5 and C-6 as units, but *not* C-4 in the case of hexanal, whereas in heptanal C-2 and C-3 are predominantly lost. The above results refer only to normal ions; relative abundances of metastable ion peaks were not reported. Fenselau et al. [159] concurrently obtained similar results, as did Harrison [160]. They reported further results soon afterwards [161] on deuterium-labelled hexanal, heptanal and nonanal. All the above experiments confirmed that γ-hydrogen migration followed by β-cleavage, as shown below, are responsible for the genesis of the prominent $C_2H_4O^{+\cdot}$ and $(M{-}C_2H_4O)^{+\cdot}$ species. However, the latter paper raises an interesting and apparently paradoxical observation. Whereas the $C_2H_4O^{+\cdot}$ ion generated in the mass spectra of the labelled compounds retains its isotopic integrity, the complementary ions $C_4H_8^{+\cdot}$ and $C_5H_{10}^{+\cdot}$ (for hexanal and heptanal respectively) display non-specific retention of label. Thus the process which generates the charge-retaining hydrocarbon ion involves hydrogen/deuterium mixing. This may result from the appropriate hydrocarbon having a lower ionisation potential than C_2H_4O and its being produced from ions having lower internal energy. From the few observations on metastable ion peaks, in hexanal and heptanal, those corresponding to $M^{+\cdot} \rightarrow C_2H_4O^{+\cdot} + RH$ are much weaker than those for $M^{+\cdot} \rightarrow C_2H_4O + RH^{+\cdot}$; this too is in keeping with the energy difference implied above.

$$\mathrm{HC(=\overset{+\cdot}{O})CH_2CH_2CHHC_2H_5} \longrightarrow \mathrm{HC(=\overset{+}{O}H)CH_2CH_2\dot{C}HC_2H_5} \longrightarrow C_2H_4O^{+\cdot} + C_4H_8 \text{ and } C_4H_8^{+\cdot} + C_2H_4O$$

Venema et al. [162] showed by deuterium labelling that for the major fragmentations of the molecular ion of 3-phenyl propionaldehyde (hydrocinnamaldehyde), namely loss of C_3H_4O and C_2H_2O, both pathways involve transfer of an α-hydrogen atom to the ring but with extensive randomisation of all hydrogen atoms before dissociation. This is a surprising result because the reactions could reasonably have been expected to proceed via a six-centred intermediate *specifically* involving the aldehydic hydrogen atom, i.e.

$$C_6H_5CH_2CH_2CHO \longrightarrow C_6H_6^{+\cdot} + C_3H_4O$$

$$\longrightarrow C_7H_8^{+\cdot} + C_2H_2O$$

Schwarz and Bohlmann [163] have recently described observations on carbon-13 and deuterium-labelled trimethylbenzaldehyde. They found that the $(M - HCO)^+$ ion underwent complete hydrogen randomisation but only partial carbon atom mixing prior to its dissociation by ethylene loss.

VI. KETONE AND ALDEHYDE DERIVATIVES

Some common derivatives of carbonyl compounds have received attention. *N,N*-Dimethyl hydrazones and dinitrophenyl hydrazones of butyraldehyde, valeraldehyde, di-*n*-propyl and di-*n*-butyl ketones, cyclopentanone and cyclohexanone, and deuterated analogues were studied by Goldsmith and Djerassi [164]. Results showed that specific hydrogen atoms in the alkyl groups were involved in the major fragmentations. Holmes and Benoit [165] and Seibl [166] examined the role of the *o*-nitro group in dissociations of molecular ions of some simple aldehydes and ketones. The results were interpreted as involving transfer of an *o*-nitro oxygen atom plus a hydrazine hydrogen atom to the hydrocarbon portion of the molecule yielding a protonated aldehyde or ketone ion. Unfortunately, little correlation could be seen between fragmentation and the structure of the original carbonyl compound. Thus such derivitisation is of little help in structure elucidation by mass spectrometry.

VII. CARBOXYLIC ACIDS AND ESTERS

In the early period of the expansion of mass spectrometry as a tool for the organic chemist, the problem of sample volatility frequently presented itself. If a compound was believed to contain an acid function, it was normally converted to the methyl ester. This expedient certainly raised the vapour pressure of the substance but, in general, the differences in the mass spectra of isomeric acids are sharply reduced by esterification. Nowadays, with the development of versatile sample probes, problems of low volatility are rare.

A. Benzoic Acid

This compound was one of the first carboxylic acids to be studied with deuterium labelling. In retrospect it is surprising that it was labelled at all because the mass spectrum of the unlabelled compound shows no major fragments which cannot be accounted for by direct bond rupture, i.e. the only major peaks are $M^{+\cdot}$, $C_6H_5CO^+$, $C_6H_5^+$ and $C_4H_3^+$, *m/e* 122, 105, 77 and 51, respectively. The unexpected feature in the mass spectrum of C_6H_5COOD is the retention of label shown by the partial displacement of *m/e* 105 to 106. This was first reported by Beynon et al. [167] and very shortly afterwards by Meyerson and Corbin [168]. Beynon found that the metastable ion peaks for the generation of $(M - OH)^+$ and $(M - OD)^+$ were in the ratio 2:1 for the carboxyl-labelled acid; the daughter ions are not, *m/e* 105 being substantially larger than *m/e* 106. It was proposed that in the slowly dissociating, metastable ion peak generating molecular ions, the carboxyl deuterium was equilibrating with the two *ortho* hydrogen atoms (thus yielding the observed metastable ion peak ratio). That the *ortho* hydrogen atoms were involved was proved by Meyerson and Corbin [168] who labelled in turn the 2, 3 and 4 positions with deuterium. Only the former lost OD from its molecular ion and it was estimated that 36% of the molecular ions had rearranged. Labelling experiments were repeated by Holmes and Benoit [169] in 1970; these authors showed that the equilibrated daughter ion which retained the label did not dissociate further and must be generated from only a small range of energies above its threshold.

They also discussed possible structures for the daughter ions, using observations on the mass spectra of phthalaldehydic acid and its carboxyl-d_1 analogue to support their proposals. It remains questionable, however, as to whether a discussion of daughter ion structures can be separated from that of energy distribution in the precursor. This point was emphasized by Howe and McLafferty [64] who showed that the ratio $(M - OD)^+$:$(M - OH)^+$ in carboxyl-d_1-benzoic acid varied strongly with energy in the range 0–4 V above the threshold for the dissociation. Beynon and co-workers [170] extended these observations using metastable ion studies and by generating benzoic acid molecular ions from ethyl benzoate, a fragmentation of known mechanism (see below). The ester was labelled with oxygen-18 (carbonyl oxygen) and the ethyl group

$$C_6H_5C(={}^{*}O^{+\cdot})OCH_2CH_2H \longrightarrow C_6H_5C({}^{*}OH)=O^{+\cdot} + C_2H_4$$

$$\downarrow$$

$$C_6H_5CO^+ + OH^\cdot$$

was C_2D_5. Their results showed that the deuterium transfer to the other oxygen atom (the one not marked above) can only take place via exchange reactions involving the *ortho* hydrogen atoms. A carboxyl group rotation mechanism was described in detail and it was concluded that transfer of deuterium from acid oxygen to carbonyl oxygen required at least two complete rotations of the substituent. Shortly afterwards [171] the work was extended to include the ring deuterium-labelled ethyl esters; the results proved again that in the benzoic acid molecular ions generated by ethylene loss, only *ortho* hydrogen atoms are involved in the equilibration preceding hydroxyl loss. These latter experiments were probably stimulated by the observations of Shapiro and Tomer [172,173] on the behaviour of β-bromoethyl benzoate. They showed that the ion m/e 122 dissociated by loss of hydroxyl in a process where the oxygen atoms became equivalent; the metastable ion peaks m/e 124 $\rightarrow$ m/e 107 and $\rightarrow$ m/e 105 were of equal abundance.

$C_6H_5\overset{*}{C}O^+$ and $C_6H_5CO^+$

The mass spectrum of thiobenzoic acid is discussed elsewhere (p. 123). The mass spectrum of carboxyl deuterium-labelled phthalic, iso- and terephthalic acids were described by Beynon et al. [167]. The latter pair behaved similarly to benzoic acid with equilibration preceding the fragmentation of the low-energy molecular ions. The mass spectrum of phthalic acid is complicated by both thermal decarboxylation and dehydration [167,174]; a spectrum free from these interferences was reported [175] in 1969. The carboxyl deuterated acid, however, loses neither hydroxyl nor HDO from its molecular ion [169] showing that *ortho* hydrogen atom exchange is prevented, probably as a result of the strong intramolecular association of the adjacent carboxyl groups. Loss of carbon dioxide from the molecular ion produced two isomers of the benzoic acid molecular ion; one dissociated by water (> 90% D_2O) loss only, while the other behaved like benzoic acid itself.

$C_6H_4O^{+\cdot}$ $C_6H_4DO^+$, $C_6H_3D_2O^+$

Note that equilibration between the isomeric species cannot have taken place.

The mass spectra of deuterium-labelled *o*-toluic acids have been discussed by Shannon et al. [176]. Water loss from the molecular ion had been considered to proceed via a six-membered transition state [174]; i.e.

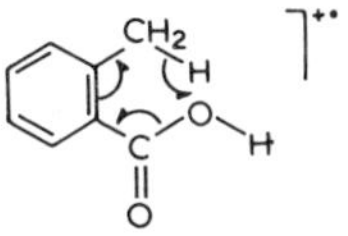

The labelling experiments showed, however, that there was exchange between the methyl and hydroxyl hydrogens prior to water loss but *no* involvement of the *ortho* hydrogen atom; the equilibration was not complete and an isotope effect operating against deuterium in the water loss was also evident. The atom mixing was seen to be greatest in ions which had the longest lifetime – those observed fragmenting in the second field-free region. The loss of water from the molecular ion of salicylic acid was considered to be another process in which aromatic ring hydrogen atoms could be equilibrated with carboxyl; labelling experiments by Benezra and Bursey [177] (salicylic acid-O,O'-d_2) showed that no such hydrogen exchange occurred and that at least 97% of the water lost from the molecular ion was deuterium oxide. This result is also readily explained by invoking the strong intramolecular attraction between the functional groups.

The behaviour of the carboxyl group in nicotinic and iso-nicotinic acids [178] is very similar to that of benzoic acid, except that the molecular ions of the pyridine derivatives eliminate both hydroxyl and water after exchange of acid and *ortho* ring hydrogen atoms. For iso-nicotinic acid both β-hydrogens are involved, for nicotinic acid only the α-hydrogen atom is exchanged. As in the case of benzoic acid, the amount of exchange varies inversely with the ions' energy content. This work was extended to the methyl ester of iso-nicotinic acid [179] where the surprising observation was made that the molecular ion successively eliminates water and carbon monoxide; the water contains β and methyl hydrogen atoms. It was proposed that the ester rearranges (as shown below) to an acid with accompanying hydrogen exchange.

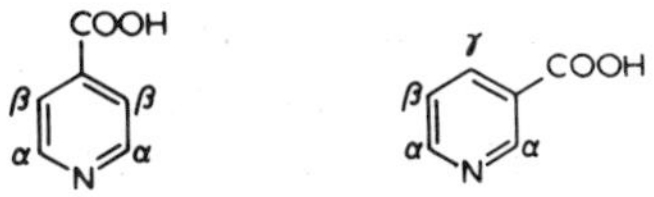

These processes were studied by examination of the appropriate first and second field-free region metastable ion peaks; the mechanism below is certainly an oversimplification because some participation of α-hydrogen atoms was observed. The notion that an acid intermediate is involved was also supported by the observation of a metastable ion peak corresponding to –COOH loss from the molecular ion; the latter also eliminates carbon dioxide generating a flat-topped metastable ion peak.

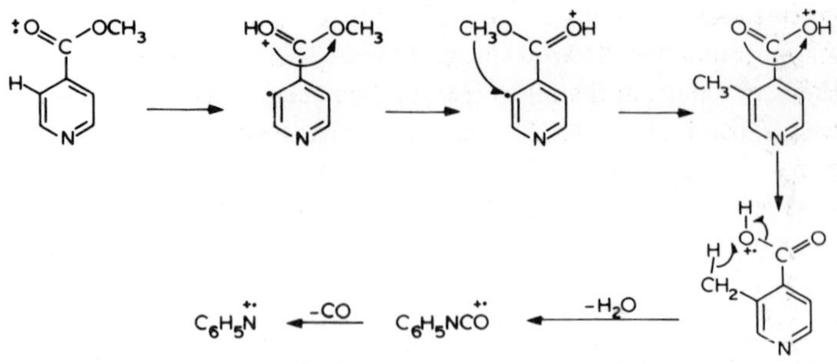

$C_6H_5N^{+\cdot}$ ← −CO — $C_6H_5NCO^{+\cdot}$ ← $-H_2O$ —

B. Other Mono-carboxylic Acids

The formation of the $(M - C_2H_4)^{+\cdot}$ ion in the mass spectrum of butyric acid was discussed by Fairweather and McLafferty [180] with regard to the detailed mechanism of the reaction. They observed that in O-d_1-butyric acid the label was completely retained; thus in the accepted mechanism for this fragmentation there cannot be any atom exchange which could have resulted from rotation of the carboxyl group in the

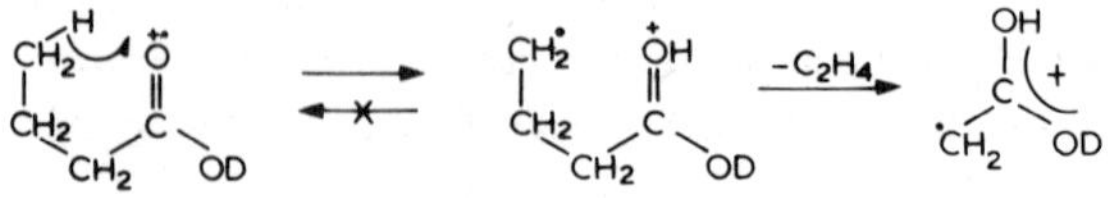

$-C_2H_4$

intermediate species. These observations were made on peaks in the normal mass spectrum; they were extended to the slower processes by Smith and McLafferty [181] who studied the metastable ion peaks. The lack of atom exchange in the earlier observations allowed the process to be either concerted or stepwise. However, for the O-d_1-acid, metastable ion peaks for loss of ethylene *and* C_2H_3D were observed; for the 4,4,4-d_3-acid C_2H_3D, $C_2H_2D_2$ and C_2HD_3 losses were all observed and for the 4,4,4-d_3, O-d_1-acid $C_2H_2D_2$ and C_2HD_3 losses were seen. These results show not only that olefin loss from the molecular ion is a stepwise process but that it also involves exchange with β-hydrogen atoms. 5-d_2-5-Phenylpentanoic acid loses some of the label when it dissociates by loss of water [182]. The probable ring intermediate is seven-membered and is thus analogous to that partially responsible for loss of water from the molecular ion of adipic acid [183].

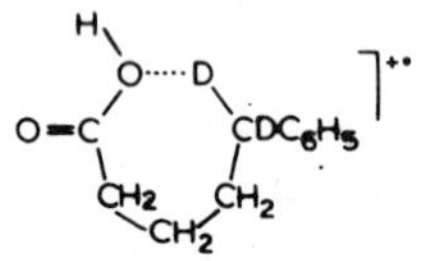

C. Dicarboxylic Acids

The mass spectra of a large number and variety of dicarboxylic acids have been studied by Holmes and his co-workers [183–186,169,175]. The aim of this work was to study the effect of molecular structure on the interaction of a pair of carboxyl groups. Where the groups can readily interact, this manifests itself by hydrogen atom transfer from one carboxyl to the other followed by loss of water and/or carbon dioxide. This is most clearly seen in the mass spectra of fumaric and maleic acids and those of mesaconic, citraconic, itaconic and phthalic acids [175]. Thus *cis* acids all displayed loss of carbon dioxide followed by water loss involving both carboxyl hydrogen atoms. The *trans* acids lost water (HDO only in the carboxyl-d_2 compounds) and unexpectedly, carbon monoxide. Loss of the latter was proposed to follow hydroxyl migration to doubly bonded carbon, e.g.

The same reaction is observed as a minor process in benzoic acid [169].

Carboxyl–carboxyl interaction was shown to be governed by ring size in the manner expected from classical organic chemistry, in the study of the $HOOC(CH_2)_nCOOH$ homologues [183].

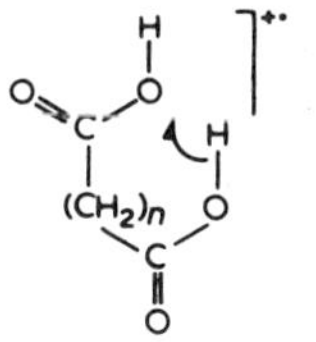

For $n = 0, 1, 2$ (ring intermediate 5-, 6- or 7-membered) loss of carbon dioxide and D_2O from carboxyl-d_2 acids was observed as an important fragmentation. It was absent for $n = 3–8$ and reappeared for $n = 9–12$ (ring 14–17-membered). Loss of water from molecular ions also took place mainly via interaction with the CH_2, γ to the carboxyl, with small contributions from the β and δ methylenes. Ring size of the intermediate for γ interaction is six-membered and this behaviour parallels that of the alkanols. Detailed fragmentation mechanisms for the lower homologues, oxalic, malonic, succinic and glutaric acids were proposed. The prominent ions $(M - CO_2)^{+\cdot}$ in oxalic and malonic acids may be isomeric with the molecular ions of formic and acetic acids respectively. This possibility has been studied with regard to the metastable ion phenomena accompanying their further dissociation [186] and it was concluded that isomeric species different with respect to structure and/or energy are involved.

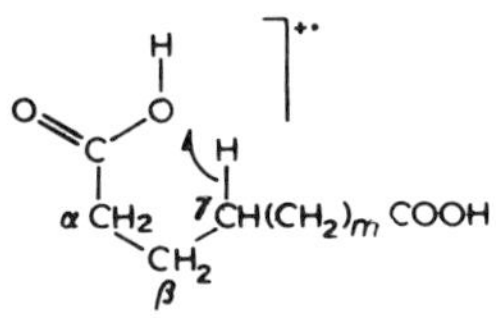

The effect of molecular geometry on carboxyl—carboxyl group interactions was extended to flexible ring systems [184,185]. The groups interact as expected in the fragmentation of *cis*-cyclobutane-1,2-dicarboxylic acid [184], $(M - CO_2)^{+\cdot}$ being a prominent peak in its mass spectrum. This is a minor process for the *trans* isomer. It is, however, *trans*-cyclohexane-1,2-dicarboxylic acid in which carboxyl—carboxyl interaction is most important. For the *cis* isomer, in the axial—equatorial arrangement, deuterium-labelling experiments showed that the carboxyl groups interact with the ring methylenes in preference to each other, giving rise to a prominent $(M - H_2O)^{+\cdot}$ peak. For the *trans* acid the di-equatorial orientation of carboxyl groups is favoured over the di-axial arrangement and in the former configuration the carboxyl groups interact without competition from ring methylenes and thus $(M - CO_2)^{+\cdot}$ is a prominent daughter ion. The fragmentation behaviour of six isomeric cyclohexene-1,2-dicarboxylic acids were found to be very similar [185], only the 1-ene acid being clearly distinguishable from the rest; this compound behaved similarly to maleic or phthalic acids with $(M - H_2O)^{+\cdot}$ (involving only the carboxyl hydrogen atoms) being followed by loss of carbon dioxide. 1-Cyclopentene- and 1-cyclobutene-1,2-dicarboxylic acids also behaved predictably, showing strong functional group interaction.

D. Aliphatic Esters

The mass spectra of simple aliphatic esters have received some careful attention. McFadden et al. [187] studied deuterium-labelled butyl hexanoates and similar esters having at least two carbons in the alcohol portion and at least four carbons in the acid chain. The common feature in all these esters are prominent ions at *m/e* 60 and *m/e* 61. The labelling experiments showed that the primary fragmentation of the molecular ion involved ester oxygen interaction with the γ-methylene group in the acid chain followed by carbonyl oxygen interaction with the alcohol chain.

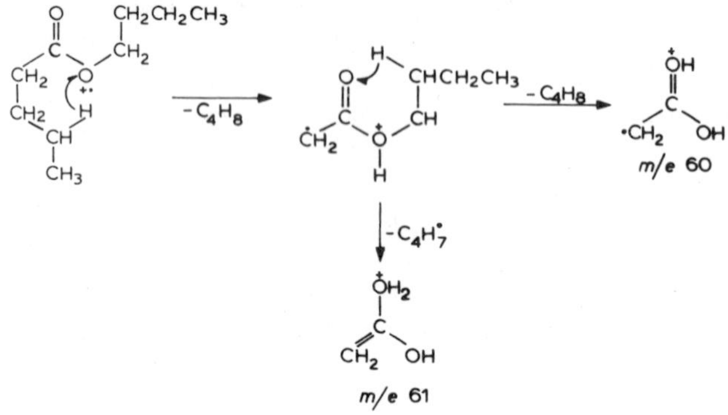

$-C_4H_8$ → $-C_4H_8$ → *m/e* 60

$-C_4H_7^{\cdot}$ → *m/e* 61

The field-ionisation mass spectra of labelled *n*-butyl acetates show similar specific involvement of a six-centred transition state [188].

A similar site specific hydrogen transfer process has been identified in the fragmentation of phenyl valerate and butanoate [189].

$-COCHR$

Aliphatic dimethyl esters have been shown to undergo methoxy group migration in the cases where favourable five- or six-membered cyclic transition states are possible; for example in dimethyl glutarate [190]

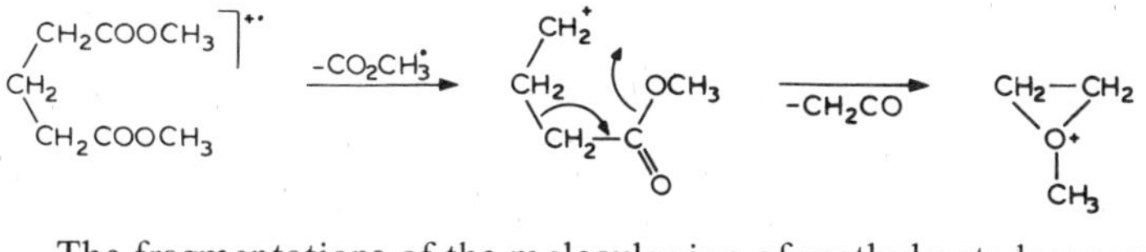

$-CO_2CH_3^{\bullet}$ $-CH_2CO$

The fragmentations of the molecular ion of methyl octadecanoate to yield *m/e* 143 and *m/e* 199, ($CH_3OOC(CH_2)_n$, $n = 6$ and 10, respectively), was investigated by means of carbon-13 labelling in the 8 and 14 positions. The former ion was produced essentially by simple bond cleavage while that at *m/e* 199 arose in almost equal parts from simple bond cleavage and by elimination of the alkyl chain from C-2 to C-9. The mechanism of the latter process was not described.

Some stereospecific methoxy and methanol eliminations were observed [191] in the mass spectra of dimethyl esters of the four acids shown below

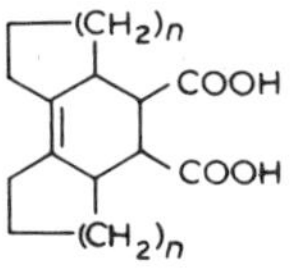

$n = 1,2,3,4$

Endo dimethyl esters eliminate methanol through a seven-centred intermediate while the *exo* diesters lose methoxy instead. In contrast the *trans* diesters lose methanol via a five-centred transition state.

The question of fragmentation mechanisms in esters of low molecular weight, where the well established six-centred olefin elimination (see p. 98 [118,119,192]) cannot take place, has been discussed recently by Yeo [122]. He showed that water loss from the molecular ion of ethyl acetate involved at least two distinct mechanisms, one in which only ethyl hydrogens are scrambled and one in which all hydrogens are scrambled. In $CD_3COOC_2H_5$ the $(M - H_2O)^{+\cdot}$ daughter ion showed complete retention of label while the first and second field-free region metastable ion peaks showed increasing participation of label. For $CH_3COOCD_2CH_3$ the daughter ion showed considerable loss of label approaching the random statistical values. Metastable ion peaks also indicated that scrambling was extensive.

Ketene elimination from *O*-acetyl methyl salicylate was studied by Nakata and

Tatematsu [193]. They wished to establish whether the $(M - 42)^{+\cdot}$ peak in this compound, which fragmented in the same manner as the molecular ion of methyl salicylate, could indeed be represented by the latter structure and by what mechanism it was produced. If the ketene elimination from the labelled acetate involved a six-membered cyclic transition state then the subsequent losses of methanol and CH_3OD would be expected to occur in approximately equal amounts.

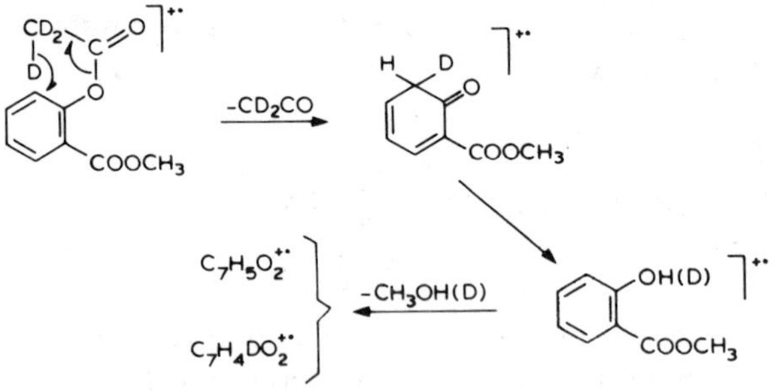

However, only CH_3OD was lost in the second step indicating that a four-membered ring intermediate is more attractive.

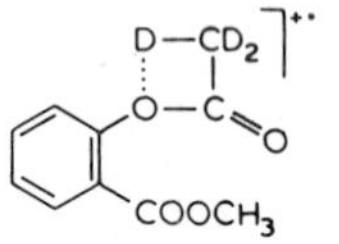

A synchronous elimination of ketene and methanol is also possible via an eight-membered ring species.

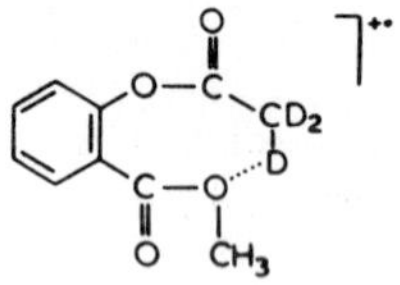

Kossanyi et al. [194] have proposed similar four-centre mechanisms for ketene elimination from the molecular ions of isomeric bicyclo[2.2.1]-heptyl acetates. The mass spectra of a series of labelled dialkyl malonates have been studied by Wilson and McCloskey [195]. The purpose of their experiments was to discover the mechanism of intermolecular hydrogen atom transfer which gives rise to prominent $(M + 1)^{+}$ peaks in the low-pressure mass spectra of such compounds.

VIII. ETHERS

The first detailed studies on the mass spectra of deuterium-labelled aliphatic ethers were those of Djerassi and Fenselau [196]. The most characteristic dissociations of a simple aliphatic ether, namely α-cleavage of the molecular ion, e.g. for isopropyl *n*-butyl ether

$$\begin{matrix} CH_3 \\ CH_3 \end{matrix} \!\!>\! CHOCH_2CH_2CH_2CH_3]^{+\cdot} \xrightarrow{-CH_3} CH_3CH = \overset{+}{O}CH_2CH_2CH_2CH_3$$

followed by elimination of the long alkyl chain accompanied by hydrogen migration

$$CH_3CH = \overset{+}{O}\overset{\alpha}{C}H_2\overset{\beta}{C}H_2\overset{\gamma}{C}H_2\overset{\delta}{C}H_3 \xrightarrow{-C_4H_8} C_2H_5O^+$$

had been established in 1957 by McLafferty [197]. Djerassi and Fenselau [196] showed that the first step is the specific process shown, but in the second step hydrogen atoms from all four alkane positions, α, β, γ and δ, were involved. Although these observations, which were for normal daughter ion peaks only (generated by 70 and 15 eV electrons), could be interpreted as resulting from hydrogen transfer to oxygen via three-, four-, five- and six-membered cyclic transition states, other mechanisms are possible. Smith and Williams [198] studied the low energy, metastably dissociating $(M - CH_3)^+$ ions of deuterium-labelled isopropyl *n*-butyl ethers, in order to extend the observed range of lifetimes of precursor ions. Their observations and conclusions are listed below.

(1) Even in the lowest energy ions examined (those dissociating in the second field-free region) there was no hydrogen/deuterium mixing between the isopropyl and *n*-butyl groups prior to fragmentation of the molecular ion by methyl loss.

(2) The participation of α- and δ-hydrogen atoms each fell from about 15% in normal ion peaks to only a few percent in $^2m^*$ processes; β- and γ-hydrogen atom involvement correspondingly increased. Now direct transfer of an α-hydrogen atom would not only involve a geometrically unfavourable three-membered ring transition state but would also involve the breaking of two bonds to the same carbon atom. It seems unlikely that such a process could have an appreciable rate (and hence be observable taking place in the field-free regions) comparable with that of (say) γ-hydrogen transfer involving a five-membered ring. It was therefore concluded from the experimental results and the above argument that there is a rapid 1,2 exchange of β- and γ-hydrogen atoms and slower exchange reactions involving the α and δ positions. Alternatively, there is no β–γ randomisation but the four- and five-membered transition state reactions have similar rate constants throughout the ion internal energy range studied.

(3) There was an energy dependent isotope effect operating with a small discrimination against deuterium in 70 eV spectra reversing to a small discrimination in favour of deuterium in the lowest energy ions.

(4) There was no firm evidence to suggest that the carbon skeleton of the *n*-butyl chain had undergone rearrangement before loss of C_4H_8 (or H_2O) from the $(M - CH_3)^+$ ions.

Tsang and Harrison [199] examined the behaviour of $[CH_3\overset{+}{O}{=}CHCH_3]$ ions generated (by simple bond cleavage) from ethers of formula $CH_3OCH(CH_3)R$, (R = H or alkyl), with respect to their fragmentation to form (a) $C_2H_5^+$ and (b) $CH_2\overset{+}{O}H$. Both reactions involve rearrangements; the study included comparison of metastable ion peak abundances for the competing dissociations and the use of deuterium-labelled precursor molecules. It was noteworthy that the metastable abundance ratio for processes (a) and (b) was independent of R, having a value close to 0.25; however, for ions generated from $C_2H_5OCH_2R$ process (a) was minimal and $m_b{}^*/m_a{}^* > 1000$.

These and other results were taken as indicating that the $CH_3O{=}CHCH_3$ ion did not rearrange to $CH_2{=}\overset{+}{O}CH_2CH_3$ prior to dissociation to $C_2H_5^+ + CH_2O$ and that the reaction probably proceeded via a four-centred transition state, i.e.

$$\begin{matrix} H_2C - H \\ O - \underset{+}{C}HCH_3 \end{matrix} \longrightarrow C_2H_5^+ \; + \; CH_2O$$

Carpenter et al. [200] examined the origin of the ion $R\overset{+}{O}H_2$ resulting from a double hydrogen transfer in the mass spectra of aliphatic ethers ROR′(R′> *n*-C_5). Ethyl *n*-hexyl ethers, successively deuterium-labelled in the hexyl chain, were studied; it was found that hydrogen atoms from all positions participated in the formation of $C_2H_5\overset{+}{O}H_2$. In the 5,5-d_2 compound an appreciable fraction (45–50%) of $C_2H_5\overset{+}{O}D_2$ was observed. The observations were not extended to studies of metastable ion peaks but the results, unlike those of Smith and Williams [198] discussed above, indicate extensive hydrogen randomisation in the molecular ion prior to dissociation by this particular pathway.

The relatively minor β-cleavage reaction in *n*-butyl ethyl ether was studied by Bernasek and Cooks [201]. This dissociation was shown to involve neither simple bond fission nor a mechanism analogous to the γ-cleavage found in carbonyl compounds; instead, skeletal rearrangement within the butyl group was evident.

Smakman and de Boer [202] repeated and extended the earlier experiments by Duffield et al. [203] on the mass spectrum of deuterium-labelled tetrahydrofuran and they also studied the fragmentation behaviour of tetrahydropyran and hexamethylene oxide. The five- and six-membered cyclic ethers show a prominent $(M - H)^+$ ion. Deuterium labelling showed that predominantly an α hydrogen is lost in its formation. In hexamethylene oxide the $(M - H)^+$ peak is very small. The latter molecule is the first homologue to show a $(M - H_2O)^{+\cdot}$ peak; the labelling experiments indicated that the two hydrogen atoms came from both the β and γ positions. It should be noted that this molecule is an isomer of cyclohexanol (q.v.) and hexanal (q.v.) both of which yield $(M - H_2O)^{+\cdot}$ ions.

The participation of an unusually large, eight-membered transition state in the fragmentation of vinyl ethers was proven with the aid of deuterium labelling by Djerassi and Katoh [204,205]. For *n*-butyl vinyl ether the chief source of the intense $(M - CH_3)^+$ peak involves hydrogen from C-3 of the butyl chain and the vinyl methylene group, implicating a seven-membered transition state for this molecule.

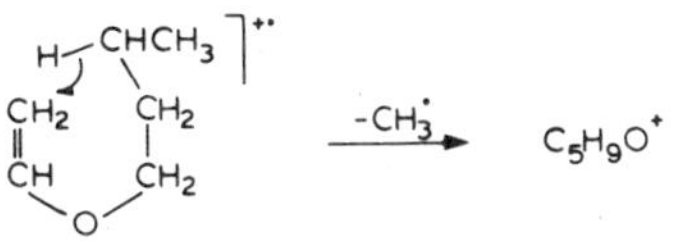

$$\xrightarrow{-CH_3^{\bullet}} C_5H_9O^+$$

For deuterium-labelled *n*-heptyl vinyl ethers, selected for study because of the greater number of secondary hydrogen atoms available for abstraction, over 80% of the heptyl hydrogen atom lost originated at position C-4. This result indicates that an eight-membered cyclic intermediate is involved.

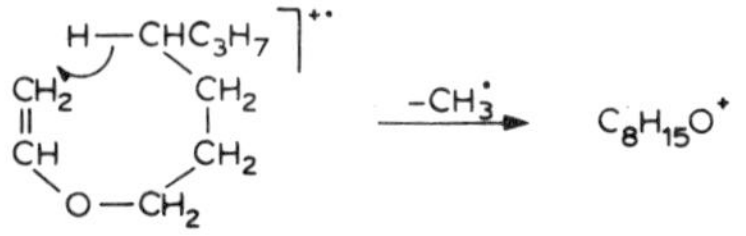

$$\xrightarrow{-CH_3^{\bullet}} C_8H_{15}O^+$$

The observed loss of methyl from the molecular ions of the tetrahydrofuran and tetrahydropyran (shown below) was presented as evidence for the intermediate participation of such species in the respective methyl radical losses discussed above.

In contrast to these dissociations, generation of the $C_2H_4O^{+\cdot}$ ion, *m/e* 44, involves extensive hydrogen randomisation prior to the postulated rearrangement and cleavage shown below.

$$\longrightarrow C_4H_8 + C_2H_4O^{+\cdot}$$

Another noteworthy feature of the fragmentation behaviour of these ethers is the triple hydrogen migration proposed by Katoh and Djerassi [204,205] to explain the loss of C_2H_6O from the molecular ion of vinyl-R (alkyl $> C_5$) ethers. The genesis of $(M - C_2H_6O)^{+\cdot}$ from the molecular ions was confirmed by the observation of the appropriate metastable ion peak in each case, but it should be stressed that such an observation does not unequivocally prove that it is a one-step process. Deuterium labelling showed that the vinyl group was lost in this fragmentation and therefore three hydrogen atoms from the alkyl chain must migrate there. The bulk of the transferred hydrogen originates from positions 3,4 and 5. Similar fragmentations were observed in the mass spectra of some thio analogues [205].

The fragmentation mechanism of methyl cyclohexyl ether has been reported by Klein and Smith [206] who studied six deuterium-labelled analogues. Major dissociations are believed to originate from a β-cleaved ion (i.e. very similar to cyclohexanol, q.v.)

OCH_3]+· → $^+OCH_3$ —$-C_3H_7^{\cdot}$→ $C_4H_7O^+$
m/e 71 base peak

The similarity to cyclohexanol is also apparent in the loss of methanol from the molecular ion. The gross process is complex, hydrogens from all carbon atoms except C-1 being involved (mainly C-3, C-4 and C-5) but it is probable that the analogous

OCH_3]+· → + CH_3OH
H

cis-1,4-elimination takes place. A relatively small matter, but one which has aroused some controversy, concerns the nature of the ion $C_6H_6O^{+\cdot}$ produced from phenyl alkyl ethers. MacLeod and Djerassi [207] showed by deuterium labelling that hydrogen transfer from the alkyl group was non-specific and proposed that it led to a phenol-like $C_6H_6O^{+\cdot}$ ion. McLafferty et al. [208] arguing from evidence obtained by studying metastable ion peak abundances and substituent effects in phenol and phenetole, claimed that loss of ethylene from the latter involved loss of identity of ring position perhaps via ring expansion. Woodgate and Djerassi [209], however, labelled phenyl *n*-butyl ether with carbon-13 and observed *no* retention of label in the $C_5H_6^{+\cdot}$ ion. This effectively confirms that the $C_6H_6O^{+\cdot}$ ion should properly be represented in phenolic form.

O C_4H_9]+· —$-C_4H_8$→ $^{13}CC_5H_6O^{+\cdot}$ —$-^{13}CO$→ $C_5H_6^{+\cdot}$

The dissociation of phenyl *n*-butyl ether was further examined in 1972 by Yeo and Djerassi [210]. They reinvestigated the hydrogen transfer from the 1,2,3 and 4 positions of the *n*-butyl group by using the appropriately deuterium-labelled analogues. The transfer previously [207] had been reported as non-specific and is thus very different from the analogous but *specific* γ-hydrogen transfer in ketones. In this work, the effect of internal energy was investigated by examining normal low electron energy spectra and $^2m^*$. For the latter, specificities for the alkyl positions, 1,2,3 and 4 were 12%, 23%, 55%, 12%, respectively, in marked contrast to the 70 eV daughter ions, 18%, 29%, 34% and 19%, respectively. Ions with even higher internal energy – which can dissociate by loss of carbon monoxide within the ion source – show a further loss of specificity, namely, 20%, 33%, 25% and 22% for the respective alkyl positions. These results were interpreted not in terms of hydrogen randomisation because a low energy, low frequency factor scrambling process would predominate at low internal energies, but rather as representing competitive hydrogen transfer processes involving transition states of different ring sizes.

Functional group interactions in a series of bi-functional benzyloxy ethers has been investigated by Sheehan et al. [211] with particular regard to C—O cleavage at the ether function.

Some cyclic diethers, based on catechol, were examined by Vouros and Biemann [212]. These were of general formula

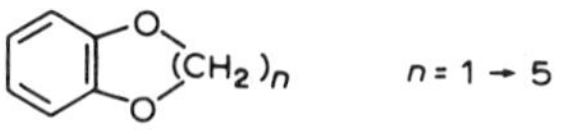

$(CH_2)_n$ $\quad n = 1 \rightarrow 5$

where $n = 1$–5. All produced $C_7H_5O_2^+$, m/e 121, as a common fragment ion by loss of the appropriate hydrocarbon $(CH_2)_{n\text{-}1}$ and a hydrogen atom. This ion, represented as

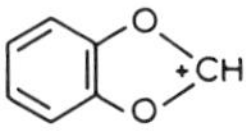

$\overset{+}{C}H$

was shown, by deuterium labelling of the tetramethylene compound, to retain an α-carbon and one of its hydrogen pair. Hydrogen scrambling was not in evidence for this process but the loss of $C_nH_{2n\text{-}2}$, to yield a catechol fragment $C_6H_6O_2^{+\cdot}$, involved extensive loss of positional identity among the methylene hydrogen atoms.

IX. ALCOHOLS

A. Alkanols

As long ago as 1951 it had been shown [213], by deuterium labelling, that the prominent (M – H) peak in the mass spectra of simple alkanols arose from specific loss of the hydrogen atom attached to the carbinol carbon.

$$RCH_2OH^+ \rightarrow RCH = \overset{+}{O}H + H^{\cdot}$$

Loss of water [214] was shown to take place preferentially by a 1,4 elimination via a six-membered intermediate.

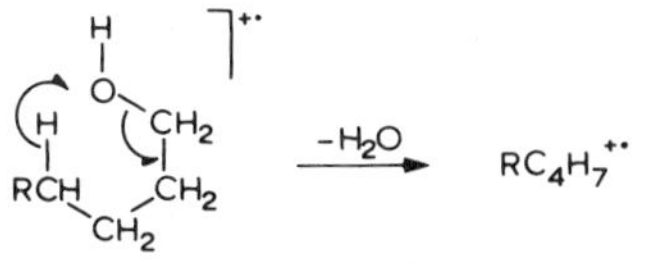

$$\xrightarrow{-H_2O} RC_4H_7^{+\cdot}$$

The mass spectra of long-chain alcohols display both prominent hydrocarbon and oxygen containing peaks, and labelling experiments [215,216] suggest that complex rearrangements occur even in the molecular ion prior to fragmentation.

In surprising contrast to the behaviour of the simple alkanols, Kurland and Lutz [217] showed that it is not exclusively loss of the α-hydrogen atom that produces the base peak, m/e 57, $(M - H)^+$, in the mass spectrum of allyl alcohol. Their deuterium-labelling study showed that *all* hydrogen atoms, except the hydroxyl hydrogen, can be lost in this primary fragmentation; some 40% comes from C-1, 20% from C-2 and 40% from C-3, indicative of complete randomisation of these hydrogen atoms. This possibly results from 1,2 hydrogen shifts in a symmetrical molecular ion.

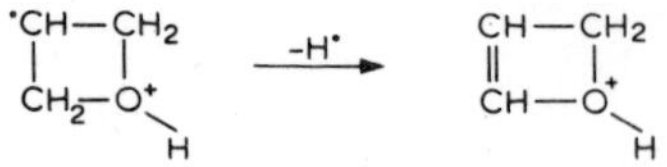

Scrambling of deuterium appears to be incomplete, and fragmentation mechanisms more complex, in the mass spectrum of pent-3-en-2-ol [218].

A number of phenyl-substituted alkanols have been studied by Nibbering and de Boer. In the fragmentation of γ-phenyl propanol [219], successive deuterium substitution along the alkyl chain and in the *ortho* and *para* aromatic positions was employed. The results showed that elimination of water from the molecular ion is indeed a complex process; although the *para*, α and β hydrogen atoms are clearly not involved, the hydroxyl hydrogen is not always lost. The γ and *ortho* hydrogen atoms are partially involved and it was proposed that there was partial exchange (not complete scrambling) between these atoms and the hydroxyl hydrogen prior to water loss. In a related publication [220], 1-phenylpropanol-2 was found not to lose water from its molecular ion as a major process. The chief fragmentations of the latter were by loss of methyl radical and acetaldehyde; the former involved specifically the β-methyl group and the latter was proposed to occur via a McLafferty rearrangement in which the daughter ion retains the hydroxyl hydrogen atom.

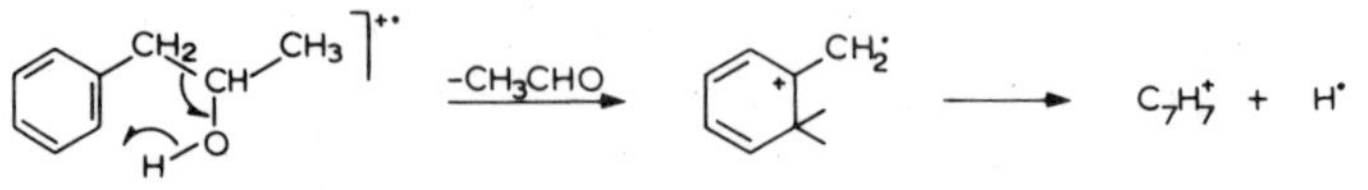

$$\xrightarrow{-CH_3CHO} \longrightarrow C_7H_7^+ + H^\bullet$$

The mass spectra of 1- and 2-phenylethanols were described in the same paper. For the former compound, the specific loss of the methyl group was considered to yield the same ion as that produced by the loss of a hydrogen atom from the molecular ion of benzyl alcohol; this latter compound completely scrambles all hydrogen atoms before fragmentation and its ring expansion to hydroxytropylium was proposed [221]. It was uncertain in the substituted ethanol as to whether ring expansion prior to fragmentation should be invoked; if it were so, then scrambling must involve all but the methyl hydrogen atoms. The mass spectrum of 2-phenylethanol is dominated by intense peaks at *m/e* 91 and *m/e* 92 (C_7H_8). The latter is generated by a McLafferty rearrangement.

$$\xrightarrow{-CH_2O}$$

m/e 92

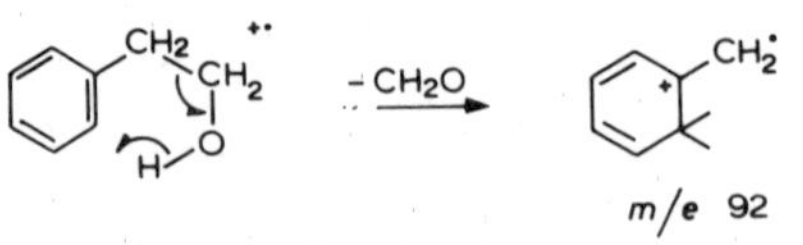

Loss of C_2H_4O from the molecular ion was preceded by exchange of aromatic and hydroxylic hydrogen atoms.

The mass spectra of *m*-hydroxybenzyl alcohols labelled with oxygen-18 (benzylic hydroxyl group) and with deuterium (CH_2 and both OH groups) were studied in order to determine whether the molecular ion underwent ring expansion prior to fragmentation [222]. In this case dihydroxycycloheptatriene and dihydroxytropylium ions would represent the $M^{+\cdot}$ and $(M - 1)^+$ species. The formation of $(M - 1)^+$ was found to involve both benzylic and ring hydrogens but not the hydroxyl hydrogen atoms. Although the above apparently requires ring expansion as a satisfactory explanation for the scrambling, loss of carbon monoxide from $(M - H)^+$ involved *only* the benzylic oxygen, a result which is incompatible with the loss of identity of the hydroxyl groups that would accompany ring expansion.

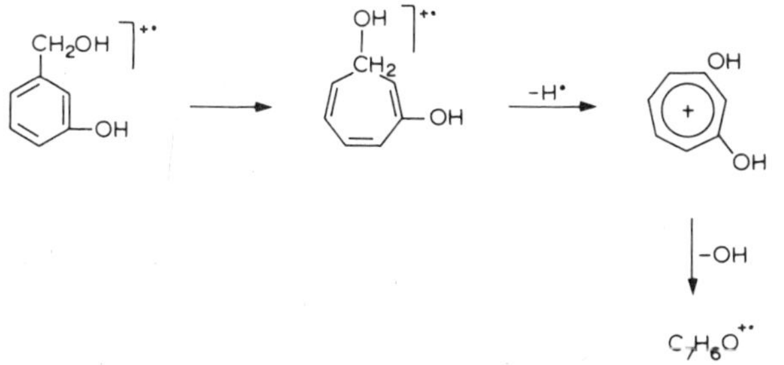

These results, and the further observations described by Molenaar-Langeveld and Nibbering [222], prove that the scrambling takes place in a molecular ion which retains its six-membered ring structure.

The possibility of scrambling between phenyl rings separated by a methylene bridge was investigated by Williams et al. [223]. Here again, the attractive explanation for scrambling is via ring expansion.

For X = OH, the primary fragmentation is the production of the benzoyl cation, $C_6H_5CO^+$; from a study of four deuterium-labelled diphenyl methanols it was concluded that some exchange occurs in the low-energy ions (nominal 12 eV spectrum) between phenyl and benzylic hydrogens but that the hydroxyl hydrogen does not do so. The exchange is not so extensive as to cause complete randomisation of hydrogen and deuterium atoms. In contrast, for X = Cl, the major fragment ion $(M - Cl)^+$, $C_{13}H_{11}^+$ has undergone total hydrogen scrambling prior to its dissociation by loss of methyl (see Table 3). It is noteworthy that this is in disagreement with an earlier observation [224] of the behaviour of this ion, where it was concluded, also from deuterium-labelling experiments, that for the methyl loss the central CH unit was lost together with hydrogen atoms from the *ortho* ring positions. However, the bulk of evidence indicates that this was an erroneous conclusion (see Table 3).

The principal fragmentation of phenol is by loss of carbon monoxide from the molecular ion; Dickinson and Williams [57] showed that in this dissociation 1-^{13}C-phenol loses the entire label showing that no carbon atom scrambling has taken place. This is, of course, in marked contrast to the behavior of benzene (see pp. 71–72). Hydrogen scrambling does, however, take place prior to the production of the $(M - HCO)^+$ ion in the mass spectrum of 1-naphthol [225], but only to the extent of about 25%. Dissociations of this daughter ion are accompanied by complete randomisation of hydrogen atoms.

B. *The Structures of* $C_2H_5O^+$, $C_3H_7O^+$ *and* $C_4H_9O^+$

The structures of the above ions have long been a matter of considerable interest and it is only comparatively recently that the means of providing proof has become available. Van Raalte and Harrison [226] proposed structure I for $C_2H_5O^+$ ions (which fragmented by loss of H_3O^+) on the basis of appearance potential data and deuterium-labelling experiments with 2-alkanols.

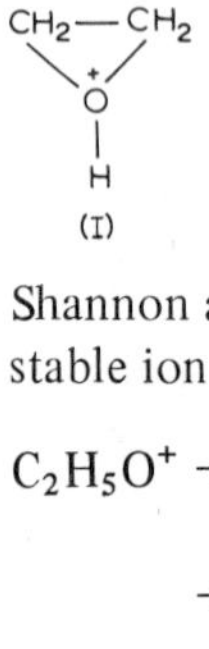

CH_2—CH_2

O^+

H

(I)

Shannon and McLafferty [109] proposed the same structure; they examined the metastable ion peak abundance ratios for the competing reactions

$$C_2H_5O^+ \rightarrow H_3O^+ + C_2H_2$$

$$\rightarrow HCO^+ + CH_4$$

observing $C_2H_5O^+$ ions produced from a wide variety of 2-alkanols, ethoxy compounds (CH_3CH_2O–R) and β-substituted ethanols. The metastable abundance ratios proved to be constant and, therefore, independent of the precursor molecule which generated $C_2H_5O^+$, indicating that all had rearranged to a common species. Thus structures II, III and IV were considered less likely than I. Harrison and Keyes [227] prepared 2-^{13}C-2-propanol in an attempt to obtain definitive evidence

$$CH_3CH=\overset{+}{O}H \qquad CH_3CH_2O^+ \qquad HOCH_2CH_2^+$$

$$\text{II} \qquad\qquad \text{III} \qquad\qquad \text{IV}$$

concerning the $C_2H_5O^+$ ion structure. The first field-free region metastable ion peaks were found to be of equal abundance for the fragmentation

$$^{13}CCH_5O^+ \rightarrow HCO^+ + {}^{13}CH_4$$

$$\rightarrow H^{13}CO^+ + CH_4$$

indicating an equal probability for loss and retention of label; this is best explained on the basis of structure I. Daughter HCO^+ ions in the normal mass spectrum (70 eV and 12 eV) displayed 70% retention and this was interpreted as resulting from the metastable generating fragmentation plus a fast, specific, label retaining (non-metastable generating) dissociation.

It should be emphasised at this point that appearance potential data for the $C_2H_5O^+$ ion generated from 2-alkanols show that *at threshold* the ion must have structure II and that it must therefore assume structure I only in species with sufficient energy to decompose. Tsang and Harrison [228] extended their observations to ^{13}C-2-labelled 2-methyl-2-propanol; these molecular ions generate $C_3H_7O^+$ by loss of methyl, which at threshold has the structure $(CH_3)_2C{=}\overset{+}{O}H$ [229]. A major dissociation of these ions is

$$C_3H_7O^+ \rightarrow CH_2OH^+ + C_2H_4$$

Siegel [230] had previously observed that for the 1-^{13}C-alcohol, loss of methyl therefrom generated $C_3H_7O^+$ containing two-thirds of the label. In the subsequent dissociation (shown above) it was found that one-third of the remaining label was retained in the CH_2OH^+ ion. Similarly, for the 2-^{13}C-alcohol, no label was lost in the expelled methyl group and about 40% of the total label was found in the CH_2OH^+ species. These results were taken as evidence favouring a symmetrical intermediate such as

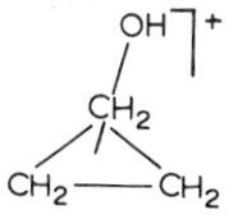

OH]+
CH2
CH2 —— CH2

a hydroxylated cyclopropane, for $C_3H_7O^+$.

Tsang and Harrison's observations were, however, quite at variance with those of Siegel. Their CH_2OH^+ daughter ion retained 90% of the label for those ions produced in the source and 100% for those ions generating first field-free region metastable ion peaks; dissociation of the symmetrical intermediate proposed by Siegel would have resulted in loss of two-thirds of the label. According to Tsang and Harrison, therefore, the C-2—O bond is *not* broken in processes leading to CH_2OH^+. The loss of carbon-13 from 2-propanol-2-^{13}C in the generation of CH_2OH^+ was also examined; retention was a constant 70% in the electron energy range 30—70 eV, but near the threshold for CH_2OH^+, it increased to ∿ 85%. First field-free region metastable ion peaks confirmed that the label was not being lost. Finally, a study of the mass spectrum of $CD_3(CH_3)_2COH$ indicated that hydrogen atoms were extensively (though not completely) scrambled prior to dissociation of the $(M - CH_3)^+$ fragment. Possible reasons for the disparity between these results were discussed at length by the later authors [228], but no satisfactory conclusion was reached.

The third homologue, $C_4H_9O^+$, has been studied by Mead and Williams [231], who generated isomeric ions from a variety of alcohols, ethers, acetals and ketals. They used the test of competing metastable abundance ratios (see pp. 68–69 for a brief explanation thereof) in order to classify the ions into five structures (or mixture of structures). In this work, the use of deuterium labelling permitted the fragmentation behaviour of each of the proposed structures to be studied. The five species were

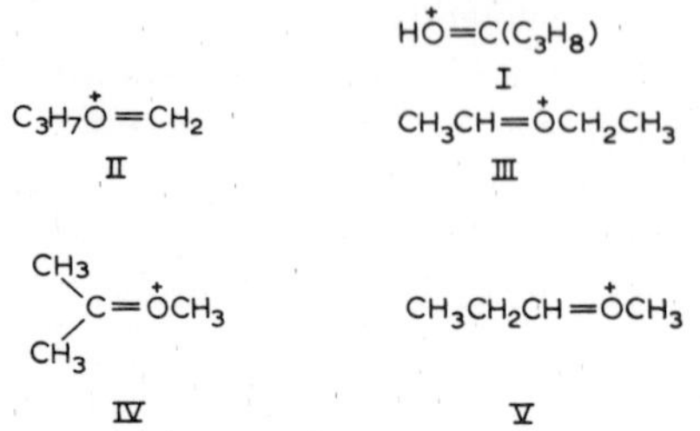

$$H\overset{+}{O}{=}C(C_3H_8) \quad \text{I}$$

$$C_3H_7\overset{+}{O}{=}CH_2 \quad \text{II} \qquad CH_3CH{=}\overset{+}{O}CH_2CH_3 \quad \text{III}$$

$$(CH_3)_2C{=}\overset{+}{O}CH_3 \quad \text{IV} \qquad CH_3CH_2CH{=}\overset{+}{O}CH_3 \quad \text{V}$$

For ions of type I, which dissociate mainly by water loss, the O–H bond remains largely intact during that dissociation and there is little atom scrambling. The hydroxyl group interacts most favourably with hydrogen atoms on a δ-carbon. Type II ions fragment by loss of water and formaldehyde; the latter is lost by direct cleavage (no label mixing) but water loss is a complex process. Species III dissociates mainly by loss of ethylene, from only the intact ethyl group. Ions IV and V appeared to undergo some reactions via a common intermediate. Loss of ethylene from V involves only the propyl group (with hydrogen scrambling therein) and loss of formaldehyde involves only the lone methyl group. For IV, formaldehyde loss predominates.

The mechanisms of some elimination reactions in the mass spectrum of hexane-2,5-diol have recently been described [232]. It is noteworthy that whereas the elimination of water from the molecular ion of monohydric alcohols proceeds most favourably via a six-centred intermediate, the corresponding interaction (shown below) is absent in this molecule. The labelling experiments showed that hydroxyl–hydroxyl interaction was responsible for water elimination (a seven-membered transition state).

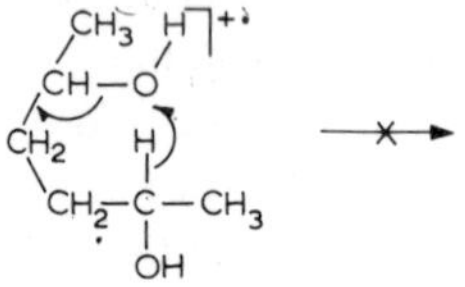

$$\left[CH_3\text{–}CH(\text{–}O\text{–}H)\text{–}CH_2\text{–}CH_2\text{–}C(H)(OH)\text{–}CH_3\right]^{+\cdot} \not\longrightarrow$$

C. Cyclic Alcohols

The mass spectrum of cyclohexanol has been studied, with the aid of deuterium labelling, by many workers. A summary of the position obtaining in 1967 may be found in Budzikiewicz et al. [233]. Fresh summaries by Duffield and co-workers [234] and by Bowie [235] appeared in 1971 and the most recent data are those of Holmes et al. [158]. There is no doubt that the fragmentation behaviour of this molecule is complex and the mechanism of some dissociations of the molecular ion remain unelucidated. The chief sources of controversy to date have been the mechanisms of formation of the base peak, $C_3H_5O^+(M - C_3H_7)^+$, *m/e* 57; of water loss from the molecular ion and of methyl loss from $(M - H_2O)^{+\cdot}$.

It seems clear now that formation of the base peak

(i) does *not* involve scrambling of the hydroxyl hydrogen [233,158] (in contrast with a recent proposal [234]) and

(ii) can be represented by the simple process shown below.

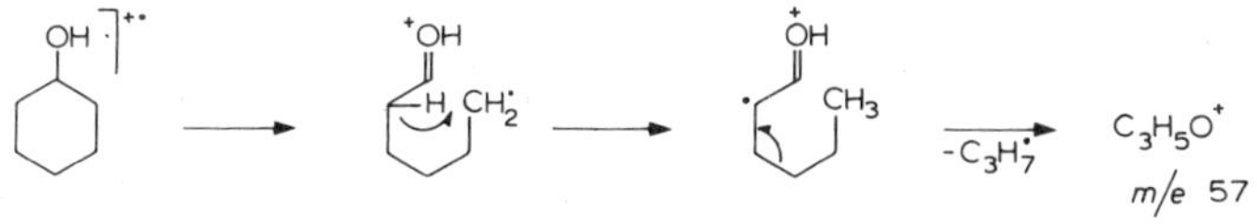

Water loss from the molecular ion is complex; however, it can be mechanistically subdivided as follows.

(i) Water is eliminated stereospecifically via a *cis*-1,4 interaction involving no hydrogen scrambling [233,158,236].

(ii) Water is lost from the α-cleaved ion. There is disagreement as to the importance of the hydrogen scrambling preceding this mode of water loss, but it has been found that all hydrogen atoms are involved [234,237–240]. In the most recent publication [158], in which eleven deuterated cyclohexanols were studied, it was concluded from observations of daughter and metastable ion peak abundances that the only atom mixing needed to explain the observations was that of the hydroxyl hydrogen with those at C-2 and C-6. The mechanisms proposed are shown below and Table 4 indicates the change of importance of the various mechanistic pathways with ion lifetime.

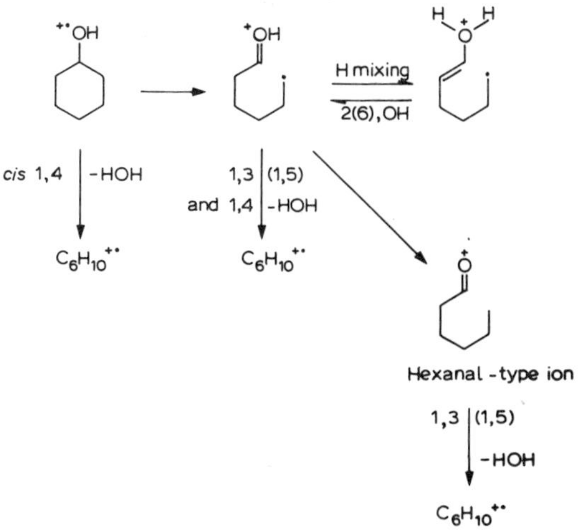

TABLE 4

CONTRIBUTION OF VARIOUS PROCESSES TO WATER LOSS FROM THE MOLECULAR ION OF CYCLOHEXANOL [158]

Process	% of total loss in	
	daughter ions	metastable ions ($^1m^*$)
Cis-1,4 elimination from intact ring	42	6
1,4 elimination (α-cleaved ion)	10	26
1,3 (1,5) elimination (α-cleaved ion)	33	48
Loss from "hexanal" ion	15	20

The $(M - H_2O)^{+\cdot}$ (m/e 82) ion might be expected to behave in a manner similar to cyclohexene. Methyl loss therefrom was believed to involve complete atom scrambling [238] but the possibility of an additional specific methyl loss has been proposed [158]. Loss of ethylene and ethyl from m/e 82 has yet to be satisfactorily explained [158].

The stereospecific *cis*-1,4 water elimination is also observed in substituted cyclohexanols [239–242]. The mass spectrum of cyclohex-2-ene-1-ol and related molecules [243] display an intense $(M - CH_3)^+$ peak. This was shown by deuterium labelling to result from the specific process shown below.

OH +· → +OH H · —$-CH_3^{\cdot}$→ +OH

The question of the interaction of hydroxyl groups as a function of their separation in the ground state geometry of a molecule has been considered by Fenselau and Robinson [244] who used the steroid carbon skeleton as a rigid template. Their results were not complicated by hydrogen scrambling and they concluded that if hydroxyl groups lie within the distance normally expected for hydrogen bonding, e.g. < 2.5 Å, then the molecular ion can be expected to display fragmentations resulting from hydroxyl interaction, i.e. the OD,OD compound will have a $(M - D_2O)^{+\cdot}$ peak in its mass spectrum.

However, this may not necessarily be the case where other fragmentations involving a hydroxyl group may proceed more rapidly. Fragmentation mechanisms for cyclohexane-1,3- and 1,4-diols were proposed by Grützmacher et al. [245] in 1966 and the 1,2-diol was examined, with deuterium labelling, by Buchs [246] in 1968 and by Strong and Djerassi [247] in 1969. The loss of water from the molecular ion of the latter compound was obviously complex. Buchs observed negligible difference in the spectra of the *cis*- and *trans*-1,2-diols (confirmed by Strong and Djerassi) but from his series of thirteen deuterium-labelled compounds he was unable to propose any definite ion structures. He concluded, however, that water loss from the molecular ion involved three mechanisms

(i) hydroxyl–hydroxyl interaction;
(ii) interaction between hydroxyl and C-4 and C-5 hydrogens;
(iii) without participation of the hydroxyl hydrogen atoms.

Strong and Djerassi proposed similar mechanisms. Holmes and Benoit [248] reduced the number of mechanisms to two; namely (i) above and a scrambling mechanism involving the 1,2 cleaved ion and proceeding via a symmetrical intermediate possibly such

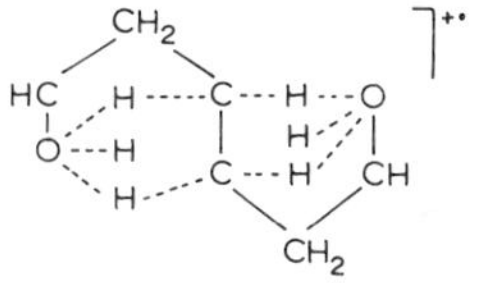

as that shown above, in which the hydroxyl and 4 and 5 methylene hydrogen atoms are randomised. This conclusion was drawn from a study of separated metastable ion peaks; the direct hydroxyl–hydroxyl interaction accounted for only 10% of the total water loss from the molecular ion. This serves as a useful *caveat* against making premature predictions on the basis of ground state molecular structure. Further fragmentions of this molecule were discussed by Buchs and by Strong and Djerassi but firm conclusions were not reached.

The mass spectral behaviour of *trans*-1,2-cyclopentanediol was reported by Singy and Buchs [249]. From results accumulated from a study of daughter and metastable ion peaks in five deuterium-labelled species they drew the following conclusions. Water loss from the molecular ion involved the hydroxyl hydrogens but *not* those on C-1, C-2, C-3 or C-5 because the OD,OD analogue lost mostly HDO (plus D_2O and a trace of H_2O). As with their earlier study of cyclohexanediol, mechanisms for this process (other than hydroxyl–hydroxyl interaction) were not detailed; hydroxyl hydrogen scrambling with C-4 hydrogen atoms does not lead to a satisfactory explanation of all the observations and the mechanism remains uncertain. The formation of *m/e* 57, $C_3H_5O^+$, was proposed to take place as shown below.

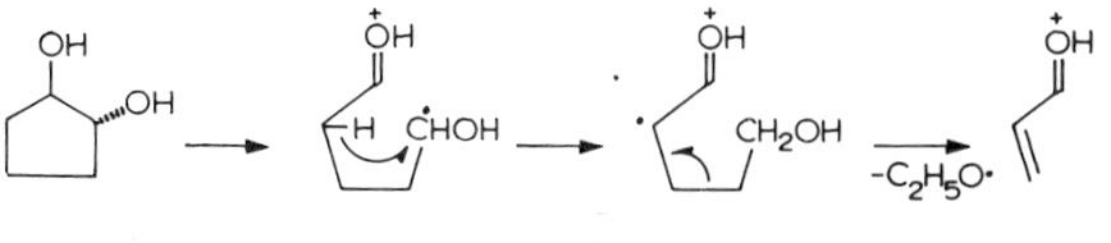

Bursey and Elwood [250] observed that the ion $C_4(C_6H_5)_4^+$ produced in the mass spectrum of pentaphenylcyclopentadienol was unexpectedly prominent. From a variety of labelling experiments they argued in favour of a tetrahedral structure for the ion

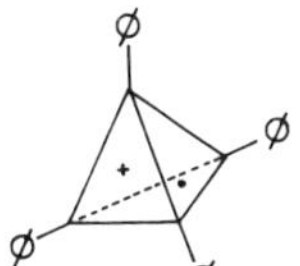

The mass spectral fragmentation mechanism of *exo*- and *endo*-norborneols has been studied in detail by several groups of workers [251–254].

The base peak, *m/e* 94, is produced by loss of water from the molecular ion. It has recently been shown [254], in contradiction to earlier proposals [252], that this ion does not behave similarly to the molecular ion of norbornene [43]. Its chief mode of fragmentation is by methyl loss, involving two mechanisms, one specific and the other involving complete hydrogen scrambling. Norbornene on the other hand decomposed predominantly by a retro-Diels–Alder reaction [43]. The mechanism of the water loss is complex, as is shown by water loss from the –OD alcohol, and certainly follows more than one pathway. Hydrogen atoms at C-2 and C-3 are not involved [252,253, 255], but those at positions 5, 6 and 7 all participate [254].

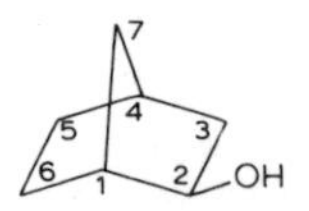

In the recent work [254], it was proposed that two mechanisms operated, one involving a C-2, C-3 cleaved molecular ion and the other a C-1, C-2 cleaved species.

The related bicyclo[2.2.1]heptanediols have been studied by Grützmacher and Fechner [256]. It was concluded that hydroxyl–hydroxyl interactions were involved in fragmentation reactions in spite of rapid ring cleavage, particularly when their spatial juxtaposition was favourable. It is worth noting that the mass spectra of the corresponding dimethyl ethers were also sufficiently different from each other to allow the isomers to be identified.

X. NITROGEN-CONTAINING COMPOUNDS

Again, for this series of molecules, chief attention will be given to the simpler compounds for which extensive labelling experiments have been performed.

A. Aniline and Derivatives

The electron-impact induced fragmentation of this compound contains processes analogous to those observed in the isoelectronic molecules toluene and phenol.

Early studies [257] proved that the major fragmentation, loss of hydrogen cyanide, involved mainly amino but also some ring hydrogen atoms. The origin of the expelled carbon atom was investigated by Rinehart et al. [258] who particularly wished to establish the possibility of the formation of the azepinium (azatropylium) ion.

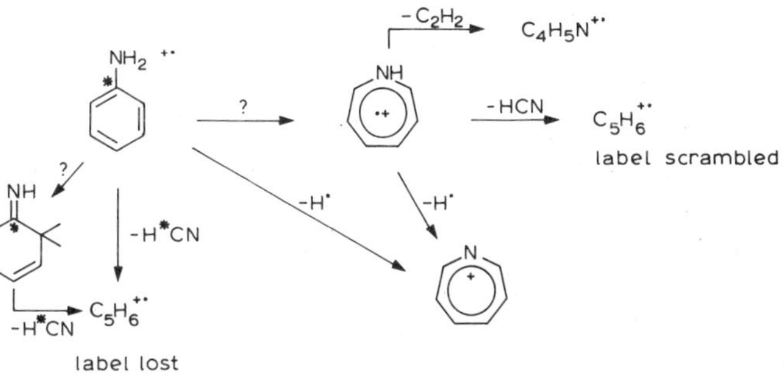

The above scheme illustrates their experiment; the label was mostly lost, but not entirely, among the ions *m/e* 65–68. For the ions $C_5H_6^{+\cdot}$, $C_4H_5N^{+\cdot}$ and $C_4H_3N^{+\cdot}$ (which arise from acetylene loss from the molecular ion) at least 90% of them come from an unrearranged molecular ion. However, careful high-resolution studies of the *m/e* 65–68 group of daughter ions indicated that for those arising from the $(M-H)^+$ species the precursor must have undergone rearrangement. The authors therefore concluded that, whereas the $C_6H_7N^{+\cdot}$ ion retains its structure prior to fragmentation, the $C_6H_6N^+$ ion can reasonably be assigned the ring-expanded azepinium form. That the loss of hydrogen cyanide from the molecular ion does not involve carbon atoms other than C-1 was concurrently confirmed by Djerassi and co-workers [259,260] from examination of the appropriate metastable ion peaks. It is worth noting here that in the latter paper it was also shown that in the carbon-13 labelled $C_6H_6N^+$ ion (produced in the fragmentation of 1-^{13}C-sulphanilamide) appreciable loss of positional identity of the label took place prior to hydrogen cyanide loss. These observations serve as a warning that postulation of ring-expanded structures without support from appropriate labelling experiments may be unjustified.

Uccella et al. [261] have used the competitive deuterium isotope effect [see p. 125 for discussion] to show that the loss of ketene from acetanilides proceeds via a four-membered intermediate (a) and not via the geometrically more attractive *ortho*-interaction (b).

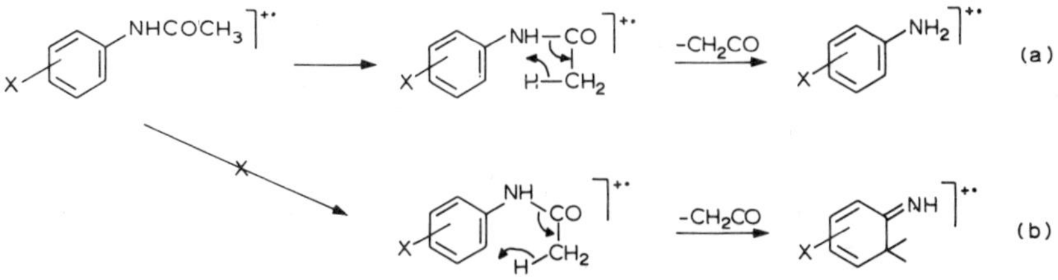

The relative abundances of metastable ion peaks for the loss of X (=Cl), hydrogen cyanide and DCN from the $(M-CH_2CO)^{+\cdot}$ ion were measured. The deuterium label was incorporated in the methyl group or on the nitrogen atom. No isotope effect on the ratio $(m^*_{HCN} + m^*_{DCN}):m^*_{Cl}$ would indicate that (b) was generated directly by the ketene loss; (b) is the species believed to be involved in hydrogen cyanide loss (see above). An isotope effect was found however, 1.4 for the first field-free region metastable ion peaks and 2.2 for the second field-free region peaks. This indicated, therefore, that hydrogen cyanide and DCN losses were competing with chlorine loss and that a hydrogen transfer process was involved in the former. These results were confirmed by Hammerum and Tomer who extended the observations to ring labelled anilines [262].

The problem of the azatropylium ion has been discussed by Nibbering and coworkers [263] with respect to its possible participation in the fragmentation of γ-picoline. The azatropylium cation has also been postulated as being produced by the loss of nitrogen from the molecular ion of phenyl azide. Kingston and Henion [264] showed that the $C_6H_5N^{+\cdot}$ ion loses hydrogen cyanide with complete hydrogen scrambling when it is generated from 2,4,6-d_3-phenyl azide. Woodgate and Djerassi [265] prepared 1-^{13}C-phenyl azide; unfortunately, the high-resolution instrument, necessary for unequivocal separation of ions arising from $H^{12}CN$ and $^{13}CCH_2$ elimination, could not be used (because of sample inlet problems) and so calculation of peak height contributions had to replace their direct experimental measurement. Nevertheless it was concluded that 50–80% carbon scrambling had preceded fragmentation, depending upon the chosen mechanism for ring expansion. Although intermediate formation of the azatropylium cation is an attractive device for explaining mass spectral observations, its existence in such systems remains unproven.

B. Other Amines

Siegel [230] examined the mass spectra of carbon-13 labelled 2-methyl-2-aminopropanes; the results indicated that complex and extensive rearrangements took place and ion structure assignments were not presented. The loss of ammonia from phenyl alkylamines was studied by Lightner et al. [266]. Deuterium labelling was used to prove that C-3 was the major source of hydrogen abstracted when the molecular ion of 3-phenylpropylamine fragments by loss of ammonia. Competing metastable abundance ratios and extensive deuterium and carbon-13 labelling were used by Uccella et al. [267] in their experiments on the structures of $C_3H_8N^+$ ions produced from a wide variety of precursor molecules.

C. Pyridine and Related Molecules

The first labelling experiment on pyridine was reported in 1967 in which the fragmentation behaviours of 2-d_1- and 2,6-d_2-pyridines were described [268]. The principal dissociation of the pyridine molecular ion, loss of hydrogen cyanide, was shown to proceed with complete hydrogen atom scrambling from a study of the appropriate metastable ion peaks ($^2m^*$) in the labelled compounds. That carbon atom scrambling is both complete and independent of hydrogen atom scrambling prior to the above fragmentation was proven in 1972 by Dickinson and Williams [269] using 2,5-d_2-2-^{13}C-pyridine. The methylpyridines were shown by deuterium labelling to behave in a manner similar to that of toluene insofar as hydrogen atom scrambling among low energy (metastable peak generating) ions was complete before expulsion of hydrogen cyanide; the amount of scrambling was appreciably less among the high energy ions [270].

The mass spectra of quinoline [270] and alkyl quinolines [271] have been investigated by deuterium labelling; the behaviour of the methyl, dimethyl, ethyl and propyl derivatives were found to be very similar to those of their aromatic hydrocarbon counterparts. Tetrahydro- and decahydroquinolines have also been studied by MacLean and his co-workers [272,273].

The pyridine derivatives, nicotinic and isonicotinic acids, are discussed on p. 95 with respect to their behavioural relationship with benzoic acid.

Although ring expansion (most commonly a six- to a seven-membered ring, by analogy with the toluene–tropyl systems) is commonly cited to explain the loss of identity among isotopic labels, there are examples where ring contraction is inferred. Burlingame and co-workers [274] concluded from deuterium labelling studies that *ring* carbon atoms 2 and 5 were involved in the loss of methyl from the molecular ion of *N*-acetyl piperidine. This they confirmed by showing that loss of CD_3 placed in the acetyl methyl group, was a minor portion of total methyl loss. Their ring contraction mechanism is shown below.

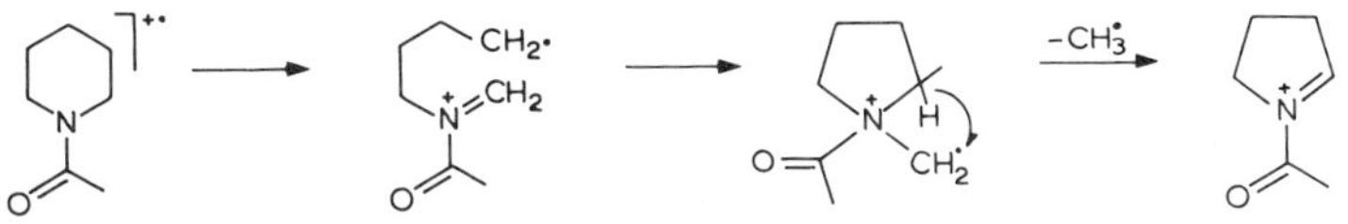

$CH_2\cdot$ $\overset{+}{N}=CH_2$ $-CH_3^{\bullet}$ $CH_2^{\bullet}$ H

D. Nitro Compounds

Ortho substituted nitro-arenes generally display fragmentations indicative of interaction between the substituent and the nitro group. The primary loss of nitric oxide from such compounds has often been cited as being aided by the *ortho* substituent [275] and is generally regarded as a nitro–nitrite conversion [276]. Many of the mechanistic details of the fragmentations of *ortho*-nitro-arenes have been worked out by the use of deuterium and oxygen-18 labelling [275]. However, nitric oxide loss also is a major decomposition route in the fragmentation of nitrobenzene and Holmes and Benoit [277] utilised carbon-13 labelling to examine the nitro–nitrite process. Two possible mechanisms for nitric oxide loss are in situ isomerisation or *ortho*-hydrogen aided isomerisation; as can be seen below a ring carbon labelling experiment would allow a clear distinction to be made.

$$C_6H_5NO_2^{+\bullet} \longrightarrow C_6H_5ONO^{+\bullet} \xrightarrow{-NO^{\bullet}} C_6H_5O^{+} \xrightarrow{-{}^{*}CO} C_5H_5^{+}$$

$$[NO, OH]^{+\bullet} \xrightarrow{-NO^{\bullet}} [\ ]^{+} \xrightarrow{-CO} {}^{*}C_5H_5^{+}$$

It was found [277] that 1-^{13}C was completely lost in the secondary carbon monoxide fragmentation indicating that nitro–oxygen became attached to the original ring site before or during nitric oxide expulsion. One reason for the attractiveness of the incorrect mechanism was its similarity to the *ortho*-hydrogen interaction observed in the primary fragmentation of benzoic acid (see pp. 92–94). The other commonly observed result of an *ortho* effect in substituted nitro-arenes is the loss of hydroxyl from the molecular ion; that this is the result of an *ortho*-interaction has been readily proven by labelling experiments [275] but the mechanism for loss of hydroxyl can in some instances be unexpectedly complex [278], and a “double” *ortho*-effect has been identified in 4-methyl-5-nitroimidazole [279]. An interesting sidelight on the nitro–nitrite rearrangement and the loss of nitric oxide from nitro compounds has been provided by the study which Nibbering and de Boer [280] made of the comparative behaviour of 3-phenylnitropropane and 3-phenylpropyl nitrite. Although both molecular ions lose nitric oxide, the behaviour of the daughter ions are significantly different. Deuterium labelling of these compounds showed that water loss from these two molecules involved either both α-hydrogen atoms or a γ- and an *ortho*-hydrogen atom.

E. Nitriles and Isonitriles

Djerassi and co-workers [281] extended their earlier studies [282] of labelled alkyl nitriles with a detailed examination of the behaviour of *n*-hexyl cyanide labelled with deuterium at each carbon atom in turn. The major fragmentations of this molecule involved loss of ethylene and ethyl radical. The labelling experiments showed that carbon atoms 2,3 and 6,7 were chiefly (but not solely) involved in these fragmentations and the authors postulated suitable cyclic intermediates to explain these observations. At the same time, Rol [283] reported on the behaviour of labelled isopentyl cyanide and in 1969 Heerma et al. [284] described the mass spectra of the labelled *n*-alkyl nitriles from methyl to pentyl. It was shown that the major alkyl fragments arise from direct C–C scission without prior atom scrambling. Nitrogen-retaining fragments in the *n*-propyl, *n*-butyl and *n*-pentyl homologues arise from alkene eliminations probably via cyclic intermediates but the presence of all except α-hydrogen atoms in these ions shows the complexity of these dissociations.

The mass spectrum of benzyl cyanide, and the possibility of its rearrangement to the cyanotropylium ion by loss of a hydrogen atom, has been thoroughly investigated by Nibbering and his co-workers [285–287] with the help of deuterium, carbon-13 and nitrogen-15 labelling.

The first experiments [285] arose from the observation that 2-phenyl-1-nitroethane loses successively a water molecule and oxygen atom to yield C_8H_7N which resembles benzyl cyanide in its ability to fragment by elimination of hydrogen cyanide, perhaps with the hydrogen atom coming from the *ortho* ring position. However, the mass spectrum of o-d_2-2-phenyl-1-nitroethane revealed that in the hydrogen cyanide elimination hydrogen/deuterium was lost in statistical ratio. The mass spectra of α-d_2-, *ortho*-d_2- and *para*-d_1-benzyl cyanides on the other hand show that prior to losing a hydrogen or deuterium atom the scrambling is incomplete – a situation similar to that found for toluene (see pp. 72–74). The toluene analogy was pursued in the second study [286] in which benzylic and cyano carbon-13 and nitrogen-15 labelling was added. Second field-free region metastable ion peaks for hydrogen cyanide loss from the molecular ion were examined and for these low energy ions both benzylic and cyano carbon atoms were found to be involved (22% benzylic, 78% cyano) after *complete randomisation of all* the hydrogen atoms. For the high energy ions (those dissociating in the ion source) the original cyano group is exclusively involved but again random selection of hydrogen and deuterium are observed. To test the hypothesis that a seven-membered ring intermediate is involved in these fragmentations, the mass spectra of appropriately labelled *o*-, *m*- and *p*-cyanobenzyl cyanides were examined; here it was found that for both high- and low-energy molecular ions, hydrogen cyanide loss predominantly involved the side chain cyano group and that extensive hydrogen randomisation preceded this fragmentation. These results, therefore, do not support the proposal that a cyanotropylium intermediate is important in the benzyl cyanide fragmentation. This idea received additional support in a third publication [287] in which the fragmentation behaviour of labelled 7-cyanocycloheptatrienes was reported. Indeed, the results could well be explained by proposing a ring *contraction,* because in the cyano-^{13}C compound's spectrum, $H^{12}CN$ as well as $H^{13}CN$ was eliminated from the low energy ions. The structure of the six-membered ring species remains speculative but the suggestion of ring contraction is important.

Phenyl cyanide [288] and isocyanide [289,290] behave identically with respect to hydrogen cyanide loss from the molecular ion in which both high and low energy dissociations are preceded by complete hydrogen scrambling.

F. Amides and Imides

Little recent work has been reported on these groups of compounds, the main fragmentation features of which are reviewed elsewhere [291]. A series of substituted *N*-methylbenzamides was studied by Prox and Schmid [292]; Biellmann and Hirth [293] showed that the major $(M - 1)^+$ peak in *N,N*-dimethylbenzamide originated almost entirely from loss of a ring hydrogen atom. Benzamide itself shows few features which cannot be explained by direct bond cleavages; some minor fragmentations observed in deuterium-labelled benzamide and thiobenzamide require the participation of the imide form of the molecular ion [294]. The behaviour of some substituted imides has been described by Bowie and co-workers [295]. The fragmentations of fumaric and maleic diamides and related compounds [296] have been examined in order to compare amide–amide hydrogen atom transfer with carboxyl–carboxyl interaction in the corresponding acids.

XI. SULPHUR-CONTAINING COMPOUNDS

Since the publication in 1967 of the mass spectrometrists vade mecum, "Mass Spectrometry of Organic Compounds", relatively few labelling experiments involving thio-compounds have been reported. In general, sulphur-containing compounds behave similarly to their oxygen counterparts and it is the minor differences which have received most attention.

A. Thiophene

This molecule has been extensively studied by both deuterium and carbon-13 labelling. Williams et al. [297] compared the behaviours of furan and thiophene by deuterium labelling and concluded that, prior to dissociation, hydrogen scrambling does not occur in the molecular ion of furan, but is virtually complete in thiophene before acetylene loss ($^2m^*$ and "normal" spectrum at low eV) and extensive before the CHS^+ fragment ion is formed. The authors also pointed out that photo-induced positional interchange of carbon atoms is observed in aryl thiophenes but is absent in corresponding furans [298,299]. The deuterium-labelling results were confirmed later by Meyerson and Fields [300] who showed that for ion source generated fragmentations, 40% of the CHS^+ species had arisen from completely scrambled precursors; (44% [297]). In addition Meyerson and Fields showed that $(M - CH_3)^+$ is generated following complete hydrogen scrambling. The point of interest remained as to whether, like benzene, both hydrogen and carbon atom scrambling were complete.

Carbon-13 experiments have been performed by de Jong et al. [301] and by Siegel [302]. The former authors prepared 2-^{13}C- and 2,5-$^{13}C_2$-thiophenes of partial isotopic purity. For the CHS^+ fragment ions they found, in agreement with Williams et al. [297], that like the hydrogen scrambling process, carbon scrambling increased with decrease of ionizing electron energy showing that the lower energy ions were the more extensively rearranged. For the $C_3H_3^+$ ion, observations were difficult to interpret owing to uncertainty as to its origin (i.e. the number of steps by which it is generated from the molecular ion). However, for the acetylene loss fragmentation, whereas hydrogen scrambling is complete prior to dissociation, de Jong et al. [301] found that this was not accompanied by total carbon scrambling. The appropriate daughter ion and first and second field-free region metastable ion peak ratios for 2-^{13}C-thiophene were essentially the same (see Table 5), indicating that rearrangement and fragmentation do not differ appreciably in activation energies and frequency factors. However, results from the 2,5-$^{13}C_2$ compound (see Table 5) allowed the authors to suggest a

TABLE 5

RELATIVE DAUGHTER AND METASTABLE ION PEAK ABUNDANCES IN ^{13}C LABELLED THIOPHENES FOR THE ACETYLENE LOSS FRAGMENTATION [301]

Compound	*m/e* 58	*m/e* 59	*m/e* 60
2-^{13}C-thiophene			
Daughter ions	0.73 ± 0.02	1.0	
$^{1}m^*$	0.77 ± 0.04	1.0	
$^{2}m^*$	0.85 ± 0.05	1.0	
Random	1.0	1.0	
2,5-$^{13}C_2$-thiophene			
Daughter ions	0.38 ± 0.02	1.88 ± 0.03	1
$^{1}m^*$	0.60 ± 0.03	1.90 ± 0.10	1
$^{2}m^*$	~0.9	2.12 ± 0.10	1
Random	1	4	1

plausible mechanism for the skeletal rearrangement. In the ions of lowest internal energy ($^{2}m^*$ generating) the relative abundances are considerably removed from those predicted from complete scrambling but they can be explained by postulating that the molecular ions have the "Ladenburg"-type structure shown below; alternatively this could be visualised as the intermediate in an equilibration represented thus

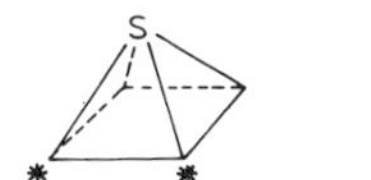

Finally, for the $(M-CH_3)^+$ ion, carbon scrambling is incomplete and the α-carbon atom is most frequently lost. Shortly afterwards, the results of de Jong et al. were semi-quantitatively confirmed by Siegel [302].

B. Substituted Thiophenes

The mass spectra of 2- and 3-phenylthiophenes and their C_6D_5 analogues were studied by Meyerson and Fields [303] in 1968. Their most important observation was that the process leading to loss of C_2H_2S involves only the thienyl ring atoms whereas in all other fragmentations some atom mixing had occurred prior to dissociation. No fragmentations involved completely statistical distribution of the label. The presence of $C_6H_5CS^+$ in the mass spectrum of 3-phenylthiophene indicated that phenyl group migration took place in the molecular ion. A related publication in 1971 by Weringa et al. [304] extended the above observations to ^{13}C-labelled 2-phenyl- and 2,5-diphenylthiophenes. The aim of this study was to distinguish between mechanism (a) migration and (b) isomerisation for the phenyl shift in 2-phenylthiophene. Such a rearrangement was considered necessary to explain the presence of a $(M-HCS)^+$ peak in the mass spectrum of 2,5-diphenylthiophene; furthermore, mechanism (b) has a photochemical analogy [298].

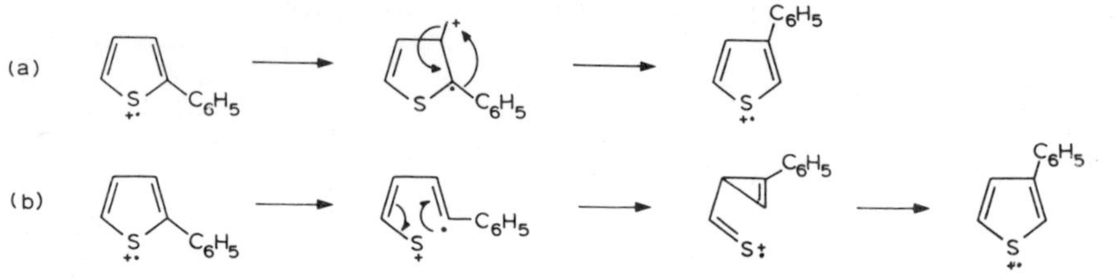

(a)

(b)

The mechanism operating remains uncertain, but the formation of $(H^{13}CS)^+$ in 2,5-$^{13}C_2$-2,5-diphenylthiophene led the authors to claim confirmation of phenyl migration in the molecular ion prior to dissociation, it being a slow process of high activation energy relative to carbon skeletal rearrangement in the thiophene ring. Carbon scrambling between the thiophene and phenyl rings was completely ruled out and the results were generally accounted for by participation of a Ladenburg intermediate. More recently, Cooks and co-workers [305] have re-examined in detail the behaviour of 3-phenylthiophene and its 2-^{13}C analogue. Daughter and metastable ion peak abundances were measured for the loss of C_2H_2, C_2H_3 and C_2H_2S from the molecular ion. Similar experiments were performed with 2-bromo-3-phenylthiophene and its 2-^{13}C analogue and 2-bromo-4-phenylthiophene and its 5-^{13}C analogue. In spite of the wealth of accumulated material, the complex reactions in these systems remain incompletely understood. Nevertheless, it is certain that substituent migration is important; the thiophene molecular ion fragmentations are of particular interest with respect to their differences from the benzene system where extensive atom scrambling is found to take place irrespective of ion energy or lifetime.

Deuterium-labelled 3-methylthiophene has recently been examined by Wakkers et al. [306]. Interest here centered on the possible participation of and scrambling in the more stable protonated thiopyran structure, a system somewhat analogous to toluene–cycloheptatriene–tropylium.

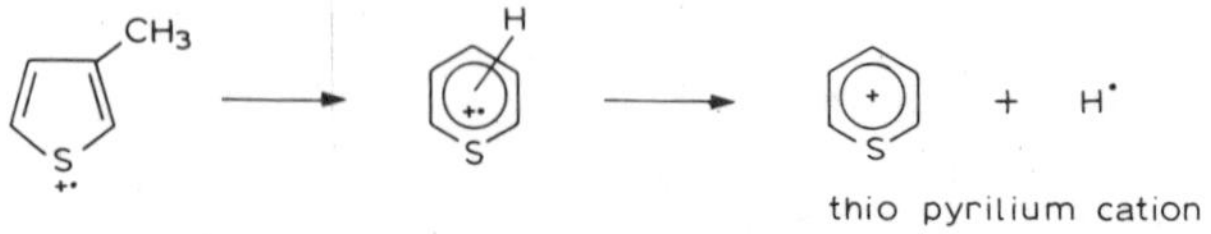

thio pyrilium cation

It was concluded from examination of daughter ion abundances that extensive but incomplete scrambling preceded hydrogen atom loss. In competition with scrambling modes, specific hydrogen atom loss from the α- and 2-positions was proposed. Dissociations of the $C_5H_5S^+$ ions also involved some label scrambling. It is perhaps unfortunate that observations of scrambling among the lower energy ions were not pursued into the metastable regions of the mass spectrum.

The major fragmentations of benzothiophene have been shown to be preceded by both hydrogen [88] and carbon atom [307] scrambling.

C. Thiols, Thio ethers and Thio acids

The mass spectral behaviour of these compounds bears close resemblance to their oxygen counterparts; perhaps the most noteworthy general difference is that in the thio-compounds molecular ions are more abundant than in their oxy-analogues. Elimination of hydrogen sulphide from primary mercaptan molecular ions was shown by deuterium labelling [308] to be a mixed 1,4 (60%) and 1,3 (40%) process in contrast to the $>$ 90% 1,4 elimination found in the corresponding alcohols.

Diphenyl disulphide-1-^{13}C was shown to undergo skeletal rearrangement on electron impact by Henion and Kingston [309]. In particular they showed that loss of CS from the molecular ion exceeded that predicted for complete scrambling. Label mixing was proposed to take place via expanded-ring intermediates by analogy with the behaviour of toluene and aniline.

The mass spectrum of labelled thiobenzoic acid has been reported twice; in the first report [169] the mass spectrum of C_6H_5COSD showed no mixing of *ortho*-hydrogen atoms with the acidic hydrogen, an important difference from the behaviour of benzoic acid [169]. The molecular ion dissociated chiefly via simple bond cleavages; nevertheless there were minor fragmentation routes in which group migrations took place. These were the losses of CSH and COH from the molecular ion. Tomer and Djerassi [310] produced the molecular ion of thiobenzoic acid, labelled with carbon-13 at the substituted ring position, by the electron impact induced ethylene elimination from *S*-ethyl thiobenzoate (a fragmentation analogous to that of ethyl benzoate, see p. 93). Now in the ions generated by loss of CHO from $C_6H_5COSH^{+\cdot}$, sulphur must have become directly attached to the ring; the most probable position is *ortho* to the original –COSH group. It was shown that in the secondary fragmentation the carbon-13 label was fully retained, ruling out the possibility that sulphur migrated to the 1-position. This is shown in the scheme

$C_6H_5C(=O)SC_2H_5^{+\cdot}$ —($-C_2H_4$)→ $C_6H_5C(OH^{\dagger})=S^{+\cdot}$ → ortho-thioquinonoid $C=\overset{+}{O}H^{\dagger}$ ion

|$-CH^{\dagger}O$ ↓

$^{13}CC_4H_5^{+}$ ←($-CS$, no label loss)— $^{13}CC_5H_5S^{+}$

The deuterium-labelling results [169] showed that only the acidic hydrogen atom ($H^{\dagger}$) was lost in the above fragmentation.

A similar carbon-13 labelling study, to try to identify the ring position to which a substituent had migrated, was performed by Siegel [311]. The problem was to identify the carbon atom involved in the loss of carbon monoxide from the molecular ion of thionylaniline. The 1-^{13}C compound was synthesised and it was found that the label was fully retained in the fragmentation. The simplest conclusion (i.e. involving the least drastic rearrangements) is that the oxygen atom is transferred to the *ortho*-position.

XII. HALOGEN-CONTAINING COMPOUNDS

Although many halogenated compounds have been isotopically labelled, the intent was rarely to study the fate of the isotope in halogen-containing fragment ions. This is because for most of such molecules the primary fragmentation of the molecular ion is to lose the halogen, it being the most weakly bonded atom. Thus a popular method of generating labelled ionised hydrocarbon radicals is via the appropriate bromo compound.

$$R\,Br^{+\cdot} \rightarrow R^{+} + Br^{\cdot}$$

Dehydrohalogenation processes have, however, received attention and McFadden and Lounsbury [312] and Djerassi and co-workers [313,314] showed, by appropriate deuterium labelling, that the loss of halogen acid in primary alkyl halides proceeds preferentially via a 1,3 elimination for chlorides and bromides and via a 1,5 elimination for fluoro compounds. Thus the preferred transition state ring sizes appear to be 5 for bromine and chlorine, 6 for oxygen and sulphur and 7 for fluorine.

Green [315,239] has presented evidence in support of a stereoselective halogen acid elimination from two diastereomers which reflected their structural differences. The pair of compounds synthesised are shown below.

(A) (B)

Elimination of DCl from A was relatively slightly more facile than from B and this was considered as reflecting the 5-membered transition state for A being weakly thermochemically favoured over that for B. i.e.

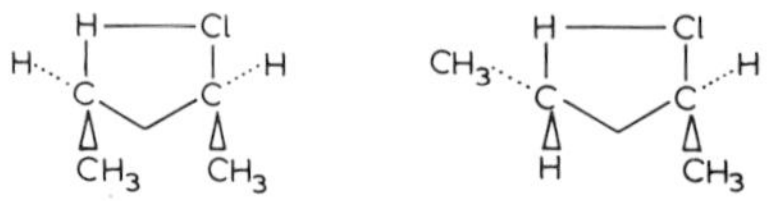

(A, 55%) (B, 45%)

Cyclohexyl chloride was labelled with deuterium and it was observed that (similar to cyclohexanol, q.v.) loss of hydrogen chloride proceeded by a stereospecific *cis*-1,4 elimination and (unlike cyclohexanol) by a largely stereo-specific *cis*-1,3 elimination. Thus for the latter process, α-cleavage in the molecular ion prior to hydrogen chloride loss is essentially absent whereas in cyclohexanol there is no stereospecificity associated with the 1,3-H_2O elimination.

The eliminations of ethylene and bromoethylene from the molecular ion of γ-phenylpropyl bromide were studied by Nibbering and de Boer [316] using analogues specifically deuterated in the ring and side chain. The results showed that exchange of α-methylene and *ortho*-ring hydrogens took place in the molecular ion and that a bromine atom or an α-hydrogen atom were transferred to the *ortho*-ring position in the elimination of ethylene and vinyl bromide, respectively, i.e.

$$\xrightarrow{-C_2H_4} C_7H_7Br^{+\bullet}$$

$$\xrightarrow{-C_2H_3Br} C_7H_8^{+\bullet}$$

The molecular ion of fluorobenzene decomposes by acetylene loss to yield the base peak in its mass spectrum. In 1-^{13}C-fluorobenzene, the metastable ion peaks ($^1m^*$ and $^2m^*$) for this process (loss of acetylene and $^{13}CCH_2$) have relative abundances indicative of the complete randomisation of fluorine and hydrogen atoms prior to fragmentation [269].

Hoffman and Amos [317] have recently shown that the ions formed by loss of a hydrogen atom from 3,4-dichlorocycloheptatriene and from 2,6-dichlorotoluene behaved identically insofar as their subsequent competing fragmentations by loss of acetylene, ethynyl chloride and dichloroethyne were concerned. This result suggests that complete equilibration of structure has taken place in the $C_7H_5Cl_2^+$ ions prior to their dissociation.

The direct use of a deuterium isotope effect in determining a fragmentation mechanism was reported by Howe and Williams [318] in 1971. The loss of carbon monoxide from the molecular ion of phenol must involve a hydrogen transfer, perhaps to yield a cyclohexadienone-like transition state. That this transfer has an isotope effect was illustrated by comparing metastable abundance ratios ($^1m^*$) for loss of bromine and of carbon monoxide from the molecular ions of *p*-BrC_6H_4OH, 4-Br-2,6-d_2-phenol and *p*-BrC_6H_4OD; they were 20, 22 and 60:1 respectively. Thus in these low-energy ions (assuming no deuterium isotope effect on the C–Br cleavage) the hydrogen transfer is clearly faster than deuterium transfer. The elimination of ethylene from the molecular ion of substituted phenetoles may yield (A) via a four-centred transition state or (B) via a six-membered transition state.

$$\longrightarrow + C_2H_4 \quad (A)$$

$$\longrightarrow + C_2H_4 \quad (B)$$

The corresponding bromine/carbon monoxide eliminations from the $C_6H_5OBr^+$ and $C_6H_4DOBr^+$ ions (generated from p-$BrC_6H_4OC_2H_5$ and p-$BrC_6H_4OC_2D_5$ respectively) yielded $^1m^*$ abundance ratios of 19 and 50, respectively. This observation clearly supports (A) rather than sequence (B) for which no isotope effect would be expected for carbon monoxide loss.

An unusual stereoselective hydrogen atom abstraction was observed in the fragmentation of *exo*-2-norbornyl chloride by Holmes and McGillivray [319]. They showed, by appropriate deuterium-labelling experiments and metastable ion studies, that the loss of a chloroethyl radical from the molecular ion of *exo*-2-norbornyl chloride involves C-2, C-3, the atoms attached thereto and the 6-*endo* hydrogen atom viz.

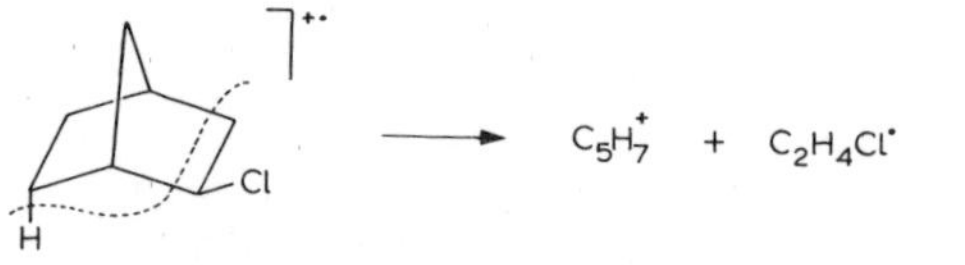

$$\longrightarrow C_5H_7^+ + C_2H_4Cl^{\cdot}$$

REFERENCES

1 J.L. Holmes and F. Benoit, in A. Maccoll (Ed.), Mass Spectrometry, M.T.P. International Review of Science, Vol. 5, Butterworths, London, 1972, pp. 273–274.
2 N.J. Turro, D.C. Neckers, P.A. Leermakers, D. Seldner and P. D'Angelo, J. Amer. Chem. Soc., 87 (1965) 4097.
3 E. Caspi, J. Wicha and A. Mandelbaum, Chem. Commun., (1967) 1161.
4 S.D. Sample and C. Djerassi, Nature (London), 208 (1965) 1314.
5 See for example, R.G. Cooks, J.H. Beynon, R.M. Caprioli and G.R. Lester, Metastable Ions, Elsevier, Amsterdam, 1973, Ch. 3, or ref. 1, pp. 274–277.
6 M. Barber and R.M. Elliott, 12th Annual Conference on Mass Spectrometry and Allied Topics, Montreal, 1964, A.S.T.M. Committee E14.
7 V.H. Dibeler and H.M. Rosenstock, J. Chem. Phys., 39 (1963) 1326.
8 Ch. Ottinger, Z. Naturforsch., 20a (1965) 1232.
9 J.H. Beynon, R.M. Caprioli, W.E. Baitinger and J.W. Amy, Org. Mass Spectrom., 3 (1970) 479.
10 Y. Amenomiya and R.F. Pottie, Can. J. Chem., 46 (1968) 1735.
11 U. Löhle and Ch. Ottinger, J. Chem. Phys., 51 (1969) 3097.
12 C. Lifshitz and R. Sternberg, Int. J. Mass Spectrom. Ion Phys., 2 (1969) 303.
13 J.L. Holmes and F. Benoit, in A. Maccoll (Ed.), Mass Spectrometry, M.T.P. International Review of Science, Vol. 4, Butterworths, London, 1972, pp. 282–283, 291.
14 H.H. Jaffé and S. Billets, J. Amer. Chem. Soc., 94, (1972) 674.
15 P. Ausloos, R.E. Rebbert, L.W. Sieck and T.O. Tiernan, J. Amer. Chem. Soc., 94 (1972) 8939.
16 C. Lifshitz and M. Shapiro, J. Chem. Phys., 45 (1966) 4242.
17 C. Lifshitz and M. Shapiro, J. Chem. Phys., 46 (1967) 4912.
18 Ch. Ottinger, J. Chem. Phys., 47 (1967) 1452.
19 F.P. Lossing and G.P. Semeluk, Can. J. Chem., 48 (1970) 955.
20 M. Vestal and J.H. Futrell, J. Chem. Phys., 52 (1970) 978.
21 D.J. McAdoo, F.W. McLafferty and P.F. Bente, J. Amer. Chem. Soc., 94 (1972) 2027.
22 R. Liardon and T. Gäumann, Helv. Chim. Acta, 52 (1969) 1042.
23 W.A. Bryce and P. Kebarle, Can. J. Chem., 34 (1956) 1249.

24 B.J. Millard and D.F. Shaw, J. Chem. Soc., B (1966) 664.
25 G.G. Meisels, J.Y. Park and B.G. Giessner, J. Amer. Chem. Soc., 91 (1969) 1555.
26 F.P. Lossing, Can. J. Chem., 50 (1972) 3973.
27 S.G. Lias and P. Ausloos, J. Amer. Chem. Soc., 92 (1970) 1840.
28 B. Davis, D.H. Williams and A.N.H. Yeo, J. Chem. Soc., B (1970) 81.
29 A.N.H. Yeo and D.H. Williams, Chem. Commun., (1970) 737.
30 J.L. Holmes, unpublished results.
31 R. Liardon and T. Gäumann, Helv. Chim. Acta, 54 (1971) 1968.
32 H.M. Grubb and S. Meyerson in F.W. McLafferty (Ed.), Mass Spectrometry of Organic Ions, Academic Press, New York, 1963, pp. 516–519.
33 M.S. Shaw, R. Westwood and D.H. Williams, J. Chem. Soc., B (1970) 1773.
34 K.B. Tomer, J. Turk and R.H. Shapiro, Org. Mass Spectrom., 6 (1972) 235.
35 J.L. Holmes, D.C.M. Tong and R.T.B. Rye, Org. Mass Spectrom., 6 (1972) 897.
36 R. Liardon and T. Gäumann, Helv. Chim. Acta, 52 (1969) 528.
37 D.S. Weinberg and M.W. Scoggins, Org. Mass Spectrom., 2 (1969) 553.
38 I.V. Goldenfeld and I.Z. Korostyshevsky, Int. J. Mass Spectrom. Ion Phys., 3 (1969) 404.
39 M.V. Guriev and M.V. Tikhomirov, Zh. Fiz. Khim., 32 (1958) 2731.
40 M.V. Guriev, M.V. Tikhomirov and N.N. Tunitsky, Dokl. Akad. Nauk SSSR, 123 (1958) 1.
41 T.H. Kinstle and R.E. Stark, J. Org. Chem., 32 (1967) 1318.
42 K.K. Mayer and C. Djerassi, Org. Mass Spectrom., 5 (1971) 817.
43 J.L. Holmes and D. McGillivray, Org. Mass Spectrom., 5 (1971) 1349.
44 M. Kraft and G. Spiteller, Org. Mass Spectrom., 2 (1969) 865.
45 P.J. Derrick, A.M. Falick and A.L. Burlingame, J. Amer. Chem. Soc., 94 (1972) 6794.
46 M.L. Gross, C.L. Wilkins and T.G. Regulski, Org. Mass Spectrom., 5 (1971) 99.
47 P.D. Woodgate, K.K. Mayer and C. Djerassi, J. Amer. Chem. Soc., 94 (1972) 3115.
48 F.P. Lossing and J. Collin, J. Amer. Chem. Soc., 80 (1958) 1568.
49 H. Luftmann and G. Spiteller, Org. Mass Spectrom., 5 (1971) 1073.
50 H.M. Grubb and S. Meyerson in F.W. McLafferty (Ed.), Mass Spectrometry of Organic Ions, Academic Press, New York, 1963, Ch. 10.
51 C.G. McDonald and J. Shannon, Aust. J. Chem., 15 (1962) 771.
52 K.R. Jennings, Z. Naturforsch., 22a (1967) 454.
53 K.E. Wilzbach, A.L. Harkness and L. Kaplan, J. Amer. Chem. Soc., 90 (1968) 1116.
54 I. Horman, A.N.H. Yeo and D.H. Williams, J. Amer. Chem. Soc., 92 (1970) 2131.
55 W.O. Perry, J.H. Beynon, W.E. Baitinger, J.W. Amy, R.M. Caprioli, R.N. Renaud, L.C. Leitch and S. Meyerson, J. Amer. Chem. Soc., 92 (1970) 7236.
56 J.H. Beynon, R.M. Caprioli, W.O. Perry and W.E. Baitinger, J. Amer. Chem. Soc., 94 (1972) 6828.
57 R.J. Dickinson and D.H. Williams, J. Chem. Soc., B (1971) 249.
58 D.H. Williams, S.W. Tam and R.G. Cooks, J. Amer. Chem. Soc., 90 (1968) 2150.
59 P.N. Rylander, S. Meyerson and H.M. Grubb, J. Amer. Chem. Soc., 79 (1957) 842.
60 S. Meyerson, H. Hart and L.C. Leitch, J. Amer. Chem. Soc., 90 (1968) 3419.
61 S. Meyerson, P.N. Rylander, F.L. Eliel and J.D. McCollum, J. Amer. Chem. Soc., 81 (1959) 2606.
62 K.L. Rinehart, A.C. Buchholz, G.E. Van Lear and H.L. Cantrill, J. Amer. Chem. Soc., 90 (1968) 2983.
63 A.S. Siegel, J. Amer. Chem. Soc., 92 (1970) 5277.
64 I. Howe and F.W. McLafferty, J. Amer. Chem. Soc., 92 (1970) 3797.
65 I. Howe and F.W. McLafferty, J. Amer. Chem. Soc., 93 (1971) 99.
66 J.H. Beynon, J.E. Corn, W.E. Baitinger, R.M. Caprioli and R.A. Benkeser, Org. Mass Spectrom., 3 (1970) 1371.

67 M. Bertrand, J.H. Beynon and R.G. Cooks, Org. Mass Spectrom., 7 (1973) 193.
68 F. Meyer and A.G. Harrison, J. Amer. Chem. Soc., 86 (1964) 4757.
69 T. Ast, J.H. Beynon and R.G. Cooks, J. Amer. Chem. Soc., 94 (1972) 1834.
70 T. Ast, J.H. Beynon and R.G. Cooks, Org. Mass Spectrom., 6 (1972) 741.
71 Y. Yamamoto, S. Takamuku and H. Sakurai, J. Amer. Chem. Soc., 94 (1972) 661.
72 A.P. Bruins, N.M.M. Nibbering and Th. J. de Boer, Tetrahedron Lett., (1972) 1109.
73 J.D. McCollum and S. Meyerson, J. Amer. Chem. Soc., 81 (1959) 4116.
74 F.W. McLafferty in F.W. McLafferty (Ed.), Mass Spectrometry of Organic Ions, Academic Press, New York, 1963, p. 337.
75 D.A. Lightner, G.B. Quistad and E. Irwin, Appl. Spectrosc., 25 (1971) 253.
76 H. Budzikiewicz, C. Fenselau and C. Djerassi, Tetrahedron, 22 (1966) 1391.
77 A.F. Gerrard and C. Djerassi, J. Amer. Chem. Soc., 91 (1969) 6808.
78 R.A.W. Johnstone and B.J. Millard, J. Chem. Soc., C (1966) 1955.
79 N.M.M. Nibbering and Th.J. de Boer, Org. Mass Spectrom., 2 (1969) 157.
80 A. Venema, N.M.M. Nibbering and Th.J. de Boer, Org. Mass Spectrom., 3 (1970) 1589.
81 S. Meyerson and E.K. Fields, Org. Mass Spectrom., 2 (1969) 1309.
82 A. Venema, N.M.M. Nibbering and Th.J. de Boer. Tetrahedron Lett., (1971) 2141.
83 R.M. Dawson and R.G. Gillis, Org. Mass Spectrom., 6 (1972) 1003.
84 S. Safe, Chem. Commun., (1969) 534.
85 S. Safe, Org. Mass Spectrom., 3 (1970) 239.
86 S. Safe, Org. Mass Spectrom., 5 (1971) 1221.
87 N.A. Uccella and D.H. Williams, J. Amer. Chem. Soc., 94 (1972) 8778.
88 R.G. Cooks, I. Howe, S.W. Tam and D.H. Williams, J. Amer. Chem. Soc., 90 (1968) 4064.
89 R.T. Aplin and S. Safe, Can. J. Chem., 47 (1969) 1599.
90 T.K. Bradshaw, J.H. Bowie and P.Y. White, Chem. Commun., (1970) 537.
91 J.H. Bowie and T.K. Bradshaw, Aust. J. Chem., 23 (1970) 1431.
92 P. Nounou, J. Chim. Phys., 65 (1968) 700.
93 P.F. Donaghue, P.Y. White, J.H. Bowie, B.D. Roney and H.J. Rodda, Org. Mass Spectrom., 2 (1969) 1061.
94 J.H. Bowie and P.Y. White, Aust. J. Chem., 24 (1971) 205.
95 J.H. Bowie, Aust. J. Chem., 25 (1972) 903.
96 T. Blumenthal and J.H. Bowie, Org. Mass Spectrom., 6 (1972) 1083.
97 S. Meyerson, Org. Mass Spectrom., 3 (1970) 119.
98 S. Safe, J. Chem. Soc., B (1971) 962.
99 S. Safe, W.D. Jamieson and S. Wolfe, Can. J. Chem., 48 (1970) 1171.
100 M.K. Hoffmann, T.A. Elwood, P.F. Rogerson, J.M. Tesarek, M.M. Bursey and D. Rosenthal, Org. Mass Spectrom., 3 (1970) 891.
101 J.H. Bowie and P.Y. White, Aust. J. Chem., 25 (1972) 439.
102 J.H. Bowie, G.E. Gream and M. Mular, Aust. J. Chem., 25 (1972) 1107.
103 J.H. Bowie, P.Y. White and T.K. Bradshaw, J. Chem. Soc. Perkin II, (1972) 1966.
104 R.A.W. Johnstone and S.D. Ward, J. Chem. Soc., C (1968) 2540.
105 H. Gusten, L. Klasinc, J. Marsel and D. Milivojevic, Org. Mass Spectrom., 5 (1971) 357.
106 J.H. Bowie and P.Y. White, Org. Mass Spectrom., 6 (1972) 135.
107 H-Fr. Grützmacher, Org. Mass Spectrom., 3 (1970) 131.
108 H.M. Rosenstock, V.H. Dibeler and F.N. Harllee, J. Chem. Phys., 40 (1964) 591.
109 T.S. Shannon and F.W. McLafferty, J. Amer. Chem. Soc., 88 (1966) 5021.
110 A.N.H. Yeo and D.H. Williams, J. Amer. Chem. Soc., 93 (1971) 395.
111 J. Seibl and T. Gäumann, Z. Anal. Chem., 197 (1963) 33; Helv. Chim. Acta, 46 (1963) 2857.
112 H. Budzikiewicz, C. Djerassi and D.H. Williams, Mass Spectrometry of Organic Compounds, Holden-Day, San Francisco, 1967, Ch. 3.

113 R.R. Arndt and C. Djerassi, Chem. Commun., (1965) 578.
114 N.C. Yang and D. Thap, Tetrahedron Lett., (1966) 3671.
115 L. Ahlquist, R. Ryhage, E. Stenhagen and E. Von Sydow, Ark. Kemi, 14 (1959) 211.
116 H. Fritz, H. Budzikiewicz and C. Djerassi, Chem. Ber., 99 (1966) 35.
117 A.F. Gerrard, R.L. Hale, R. Liedtke, W.H. Faul and C.A. Brown, Org. Mass Spectrom., 3 (1970) 683.
118 J.K. MacLeod and C. Djerassi, J. Amer. Chem. Soc., 89 (1967) 5182.
119 D.H. Williams, H. Budzikiewicz and C. Djerassi, J. Amer. Chem. Soc., 86 (1964) 284.
120 W. Carpenter, A.M. Duffield and C. Djerassi, J. Amer. Chem. Soc., 90 (1968) 160.
121 A.N.H. Yeo and D.H. Williams, J. Amer. Chem. Soc., 91 (1969) 3582.
122 A.N.H. Yeo, Chem. Commun., (1970) 1154.
123 A.N.H. Yeo, Chem. Commun., (1970) 987.
124 F.W. McLafferty and W.T. Pike, J. Amer. Chem. Soc., 89 (1967) 5953.
125 J. Diekman, J.K. MacLeod, C. Djerassi and J.D. Baldeschwieler, J. Amer. Chem. Soc., 91 (1969) 2069.
126 G. Eadon, J. Diekman and C. Djerassi, J. Amer. Chem. Soc., 91 (1969) 3986.
127 G. Eadon, J. Diekman and C. Djerassi, J. Amer. Chem. Soc., 92 (1970) 6205.
128 F.W. McLafferty, D.J. McAdoo, J.S. Smith and R. Kornfeld, J. Amer. Chem. Soc., 93 (1971) 3720.
129 D.J. McAdoo, F.W. McLafferty and T.E. Parks, J. Amer. Chem. Soc., 94 (1972) 1601.
130 G. Eadon, C. Djerassi, J.H. Beynon and R.M. Caprioli, Org. Mass Spectrom., 5 (1971) 917.
131 J.H. Beynon, R.M. Caprioli and T.W. Shannon, Org. Mass Spectrom., 5 (1971) 967.
132 K.B. Tomer and C. Djerassi, Org. Mass Spectrom., 6 (1972) 1285.
133 G. Eadon and C. Djerassi, J. Amer. Chem. Soc., 92 (1970) 3084.
134 Y.M. Sheikh, A.M. Duffield and C. Djerassi, Org. Mass Spectrom., 4 (1970) 273.
135 J.R. Dias, Y.M. Sheikh and C. Djerassi, J. Amer. Chem. Soc., 94 (1972) 473.
136 R.J. Liedtke, A.F. Gerrard, J. Diekman and C. Djerassi, J. Org. Chem., 37 (1972) 776.
137 A.F. Thomas and B. Willhalm, Helv. Chim. Acta, 50 (1967) 826.
138 D.R. Dimmel and J. Wolinsky, J. Org. Chem., 32 (1967) 411.
139 W.H. Pirkle, J. Amer. Chem. Soc., 87 (1965) 3022.
140 P. Brown and M.M. Green, J. Org. Chem., 32 (1967) 1681.
141 W.H. Pirkle and M. Dines, J. Amer. Chem. Soc., 90 (1968) 2318.
142 W.T. Pike and F.W. McLafferty, J. Amer. Chem. Soc., 89 (1967) 5954.
143 M.M. Bursey and L.R. Dusold, Chem. Commun., (1967) 712.
144 R.A.W. Johnstone, B.J. Millard, F.M. Dean and A.W. Hill, J. Chem. Soc., (1966) 1712.
145 J.H. Bowie and P.Y. White, J. Chem. Soc., B (1969) 89.
146 J.H. Bowie, R.G. Cooks, G.E. Gream and M.H. Laffer, Aust. J. Chem., 21 (1968) 1247.
147 O.H. Mattsson, Acta Chem. Scand., 22 (1968) 2479.
148 T.H. Kinstle, O.L. Chapman and M. Sung, J. Amer. Chem. Soc., 90 (1968) 1227.
149 D.H. Williams, H. Budzikiewicz, Z. Pelah and C. Djerassi, Monatsh. Chem., 95 (1964) 166.
150 M.K. Strong, P. Brown and C. Djerassi, Org. Mass Spectrom., 2 (1969) 1201.
151 R.T. Gray, R.J. Spangler and C. Djerassi, J. Org. Chem., 35 (1970) 1525.
152 G.D. Christiansen and D.A. Lightner, J. Org. Chem., 36 (1971) 948.
153 W.D. Weringa, Org. Mass Spectrom., 5 (1971) 1055.
154 G. Eadon and C. Djerassi, J. Amer. Chem. Soc., 91 (1969) 2725.
155 C. Fenselau, A.A. Baum and D.O. Cowan, Org. Mass Spectrom., 4 (1970) 229.
156 J.A. Ballantine and C.T. Pillinger, Org. Mass Spectrom., 1 (1968) 425.
157 R.J. Liedtke and C. Djerassi, J. Amer. Chem. Soc., 91 (1969) 6814.
158 J.L. Holmes, D. McGillivray and R.T.B. Rye, Org. Mass Spectrom., 7 (1973) 347.
159 C. Fenselau, J. Young, S. Meyerson, W. Landis, E. Selke and L.C. Leitch, J. Amer, Chem. Soc., 91 (1969) 6847.

160 A.G. Harrison, Org. Mass Spectrom., 3 (1970) 549.
161 C. Fenselau, J. Young, S. Meyerson, W. Landis, E. Selke and L.C. Leitch, Org. Mass Spectrom., 3 (1970) 689.
162 A. Venema, N.M.M. Nibbering and Th.J. de Boer, Org. Mass Spectrom., 3 (1970) 583.
163 H. Schwarz and F. Bohlmann, Org. Mass Spectrom., 6 (1972) 815.
164 D. Goldsmith and C. Djerassi, J. Org. Chem., 31 (1966) 3661.
165 J.L. Holmes and F. Benoit, Can. J. Chem., 47 (1969) 3611.
166 J. Seibl, Org. Mass Spectrom., 3 (1970) 417.
167 J.H. Beynon, B.E. Job and A.E. Williams, Z. Naturforsch., 20a (1965) 883.
168 S. Meyerson and J.L. Corbin, J. Amer. Chem. Soc., 87 (1965) 3045.
169 J.L. Holmes and F. Benoit, Org. Mass Spectrom., 4 (1970) 97, and references therein.
170 R.H. Shapiro, K.B. Tomer, R.M. Caprioli and J.H. Beynon, Org. Mass Spectrom., 3 (1970) 1333.
171 R.H. Shapiro, K.B. Tomer, J.H. Beynon and R.M. Caprioli, Org. Mass Spectrom., 3 (1970) 1593.
172 R.H. Shapiro and K.B. Tomer, Org. Mass Spectrom., 2 (1969) 1175.
173 R.H. Shapiro and K.B. Tomer. Org. Mass Spectrom., 3 (1970) 333.
174 F.W. McLafferty and R.S. Gohlke, Anal. Chem., 31 (1959) 2076.
175 F. Benoit, J.L. Holmes and N.S. Isaacs, Org. Mass Spectrom., 2 (1969) 591.
176 J.S. Shannon, M.J. Lacey and C.G. MacDonald, Org. Mass Spectrom., 5 (1971) 1391.
177 S.A. Benezra and M.M. Bursey, Org. Mass Spectrom., 6 (1972) 463.
178 R. Neeter and N.M.M. Nibbering, Org. Mass Spectrom., 5 (1971) 735.
179 R. Neeter and N.M.M. Nibbering, Tetrahedron, 28 (1972) 2575.
180 R.B. Fairweather and F.W. McLafferty, Org. Mass Spectrom., 2 (1969) 755.
181 J.S. Smith and F.W. McLafferty, Org. Mass Spectrom., 5 (1971) 483.
182 S. Meyerson and L.C. Leitch, J. Amer. Chem. Soc., 93 (1971) 2244.
183 J.L. Holmes and T.St. Jean, Org. Mass Spectrom., 3 (1970) 1505.
184 F. Benoit and J.L. Holmes, Org. Mass Spectrom., 6 (1972) 541.
185 F. Benoit and J.L. Holmes, Org. Mass Spectrom., 6 (1972) 549.
186 J.L. Holmes, Org. Mass Spectrom., 7 (1973) 341.
187 W.H. McFadden, L.E. Boggs and R.G. Buttery, J. Phys. Chem., 70 (1966) 3516.
188 P. Schulze and W.J. Richter, Int. J. Mass Spectrom. Ion Phys., 6 (1971) 131.
189 G.G. Smith and S.W. Cowley, Chem. Commun., (1971) 1066.
190 I. Howe and D.H. Williams, Chem. Commun., (1967) 733.
191 J. Deutsch and A. Mandelbaum, J. Amer. Chem. Soc., 92 (1970) 4288.
192 N. Dinh-Nguyen, R. Ryhage, S. Ställberg-Stenhagen and E. Stenhagen, Ark. Kemi, 18 (1961) 393.
193 N. Nakata and A. Tatematsu, Org. Mass Spectrom., 5 (1971) 1343.
194 J. Kossanyi, B. Furth and J.P. Morizur, Org. Mass Spectrom., 6 (1972) 593.
195 M.H. Wilson and J.A. McCloskey, J. Amer. Chem. Soc., 94 (1972) 3865.
196 C. Djerassi and C. Fenselau, J. Amer. Chem. Soc., 87 (1965) 5747.
197 F.W. McLafferty, Anal Chem., 29 (1957) 1782.
198 G.A. Smith and D.H. Williams, J. Amer. Chem. Soc., 91 (1969) 5254.
199 C.W. Tsang and A.G. Harrison, Org. Mass Spectrom., 3 (1970) 647.
200 W. Carpenter, A.M. Duffield and C. Djerassi, J. Amer. Chem. Soc., 89 (1967) 6164.
201 S.L. Bernasek and R.G. Cooks, Org. Mass Spectrom., 3 (1970) 127.
202 R. Smakman and Th.J. de Boer, Org. Mass Spectrom., 1 (1968) 403.
203 A.M. Duffield, H. Budzikiewicz and C. Djerassi, J. Amer. Chem. Soc., 87 (1965) 2920.
204 M. Katoh and C. Djerassi, Chem. Commun., (1969) 1385.
205 M. Katoh, D.A. Jaeger and C. Djerassi, J. Amer. Chem. Soc., 94 (1972) 3107.

206 G.W. Klein and V.F. Smith, J. Org. Chem., 35 (1970) 52.
207 J.K. MacLeod and C. Djerassi, J. Amer. Chem. Soc., 88 (1966) 1840.
208 F.W. McLafferty, M.M. Bursey and S.M. Kimball, J. Amer. Chem. Soc., 88 (1966) 5022.
209 P.D. Woodgate and C. Djerassi, Org. Mass Spectrom., 3 (1970) 1093.
210 A.N.H. Yeo and C. Djerassi, J. Amer. Chem. Soc., 94 (1972) 482.
211 M. Sheehan, R.J. Spangler and C. Djerassi, J. Org. Chem., 36 (1971) 3526.
212 P. Vouros and K. Biemann, Org. Mass Spectrom., 3 (1970) 1317.
213 H. Budzikiewicz, C. Djerassi and D.H. Williams, Mass Spectrometry of Organic Compounds, Holden-Day, San Francisco, 1967, pp. 96 et seq.
214 H. Budzikiewicz, C. Djerassi and D.H. Williams, Mass Spectrometry of Organic Compounds, Holden-Day, San Francisco, 1967, pp. 98 et seq.
215 L. Dolejs, P. Beran and J. Hradec, Org. Mass Spectrom., 1 (1968) 563.
216 M. Kraft and G. Spiteller, Monatsh. Chem., 99 (1968) 1839.
217 J.J. Kurland and R.P. Lutz, Chem. Commun., (1968) 1097.
218 B. Willhalm and A.F. Thomas, Org. Mass Spectrom., 1 (1968) 627.
219 N.M.M. Nibbering and Th.J. de Boer, Tetrahedron, 24 (1968) 1415.
220 N.M.M. Nibbering and Th.J. de Boer, Org. Mass Spectrom., 1 (1968) 365.
221 J.S. Shannon, Aust. J. Chem., 15 (1962) 165.
222 T.A. Molenaar-Langeveld and N.M.M. Nibbering, Tetrahedron, 28 (1972) 1043.
223 D.H. Williams, R.S. Ward and R.G. Cooks, J. Chem. Soc., B (1968) 522.
224 R.A.W. Johnstone and B.J. Millard, Z. Naturforsch., 21a (1966) 604.
225 L. Klasinc and H. Güsten, Z. Naturforsch., 27a (1972) 1681.
226 D. Van Raalte and A.G. Harrison, Can. J. Chem., 41 (1963) 3118.
227 A.G. Harrison and B.G. Keyes, J. Amer. Chem. Soc., 90 (1968) 5046.
228 C.W. Tsang and A.G. Harrison, Org. Mass Spectrom., 5 (1971) 877.
229 A.G. Harrison, A. Ivko and D. Van Raalte, Can. J. Chem., 44 (1966) 1625.
230 A.S. Siegel, Org. Mass Spectrom., 3 (1970) 1417.
231 T.J. Mead and D.H. Williams, J. Chem. Soc. Perkin II, (1972) 876.
232 Gy. Horvath and J. Kuszmann, Org. Mass Spectrom., 6 (1972) 447.
233 H. Budzikiewicz, C. Djerassi and D.H. Williams, Mass Spectrometry of Organic Compounds, Holden-Day, San Francisco, 1967, pp. 107–112.
234 R.H. Shapiro, S.P. Levine and A.M. Duffield, Org. Mass Spectrom., 5 (1971) 383.
235 J.H. Bowie, in D.H. Williams (Ed.), Mass Spectrometry, The Chemical Society, London, 1971, pp. 102–103.
236 M.M. Green and J.M. Schwab, Tetrahedron Lett., (1968) 2955.
237 H. Budzikiewicz, Z. Pelah and C. Djerassi, Monatsh. Chem., 95 (1964) 158.
238 R.S. Ward and D.H. Williams, J. Org. Chem., 34 (1969) 3373.
239 M.M. Green, R.J. Cook, J.M. Schwab and R.B. Roy, J. Amer. Chem. Soc., 92 (1970) 3076.
240 M.M. Green and R.B. Roy, J. Amer. Chem. Soc., 92 (1970) 6368.
241 C.E. Brion and L.D. Hall, J. Amer. Chem. Soc., 88 (1966) 3661.
242 L. Dolejs and V. Hanus, Collect. Czech. Chem. Commun., 33 (1968) 332.
243 R.T. Aplin, H.E. Browning and P. Chamberlain, Chem. Commun., (1967) 1071.
244 C.C. Fenselau and C.H. Robinson, J. Amer. Chem. Soc., 93 (1971) 3070.
245 H-Fr. Grützmacher, J. Winkler and K. Heyns, Tetrahedron Lett., (1966) 6051.
246 A. Buchs, Helv. Chim. Acta, 51 (1968) 688.
247 M.K. Strong and C. Djerassi, Org. Mass Spectrom., 2 (1969) 631.
248 J.L. Holmes and F. Benoit, Can. J. Chem., 49 (1971) 1161.
249 G.A. Singy and A. Buchs, Helv. Chim. Acta, 54 (1971) 537.
250 M.M. Bursey and T.A. Elwood, J. Amer. Chem. Soc., 91 (1969) 3812.
251 D.R. Dimmel and J. Wolinsky, J. Org. Chem., 32 (1967) 2735.
252 H. Kwart and T.A. Blazer, J. Org. Chem., 35 (1970) 2726.

253 K. Humski and L. Klasinc, J. Org. Chem., 36 (1971) 3057.
254 J.L. Holmes and D. McGillivray, Org. Mass Spectrom., 7 (1973) 559.
255 W. Benze and K. Biemann, J. Amer. Chem. Soc., 86 (1964) 2374.
256 H-Fr. Grützmacher and K-H. Fechner, Tetrahedron, 27 (1971) 5011.
257 P.M. Rylander, S. Meyerson, E.L. Eliel and J.D. McCollum, J. Amer. Chem. Soc., 85 (1963) 2723.
258 K.L. Rinehart, A.C. Buchholz and G.E. Van Lear, J. Amer. Chem. Soc., 90 (1968) 1073.
259 A.V. Robertson, M. Marx and C. Djerassi, Chem. Commun., (1968) 414.
260 A.V. Robertson and C. Djerassi, J. Amer. Chem. Soc., 90 (1968) 6992.
261 N.A. Uccella, I. Howe and D.H. Williams, Org. Mass Spectrom., 6 (1972) 229.
262 S. Hammerum and K.B. Tomer, Org. Mass Spectrom., 6 (1972) 1369.
263 R. Neeter, N.M.M. Nibbering and Th.J. de Boer, Org. Mass Spectrom., 3 (1970) 597.
264 D.G.I. Kingston and J.D. Henion, Org. Mass Spectrom., 3 (1970) 413.
265 P.D. Woodgate and C. Djerassi, Tetrahedron Lett., (1970) 1875.
266 D.A. Lightner, F.W. Sunderman, L. Hurtado and E. Thommen, Org. Mass Spectrom., 3 (1970) 1325.
267 N.A. Uccella, I. Howe and D.H. Williams, J. Chem. Soc., B (1971) 1933.
268 D.H. Williams and J. Ronayne, Chem. Commun., (1967) 1129.
269 R.J. Dickinson and D.H. Williams, J. Chem. Soc. Perkin II, (1972) 1363.
270 W.G. Cole, D.H. Williams and A.N.H. Yeo, J. Chem. Soc., B (1968) 1284.
271 P.M. Draper and D.B. MacLean, Can. J. Chem., 46 (1968) 1487.
272 P.M. Draper and D.B. MacLean, Can. J. Chem., 46 (1968) 1499.
273 C.K. Yu, D. Oldfield and D.B. MacLean, Org. Mass Spectrom., 4 (1970) 147.
274 W.J. Richter, J.M. Bursey and A.L. Burlingame, Org. Mass Spectrom., 5 (1971) 1295.
275 J.L. Holmes and F. Benoit, Org. Mass Spectrom., 3 (1970) 993, and references therein.
276 M.M. Bursey, Org. Mass Spectrom., 2 (1969) 907.
277 J.L. Holmes and F. Benoit, Chem. Commun., (1970) 1031.
278 G.E. Robinson, C.B. Thomas and J.M. Vernon, J. Chem. Soc., B (1971) 1273.
279 G.H. Lord and B.J. Millard, Org. Mass Spectrom., 2 (1969) 547.
280 N.M.M. Nibbering and Th.J. de Boer, Org. Mass Spectrom., 3 (1970) 487.
281 W. Carpenter, Y.M. Sheikh, A.M. Duffield and C. Djerassi, Org. Mass Spectrom., 1 (1968) 3.
282 R. Beugelmans, D.H. Williams, H. Budzikiewicz and C. Djerassi, J. Amer. Chem. Soc., 86 (1964) 1386.
283 N.C. Rol, Rec. Trav. Chim. Pays-Bas, 87 (1968) 321.
284 W. Heerma, J.J. de Ridder and G. Dijkstra, Org. Mass Spectrom., 2 (1969) 1103.
285 N.M.M. Nibbering and Th.J. de Boer, Tetrahedron, 24 (1968) 1435.
286 T.A. Molenaar-Langeveld, N.M.M. Nibbering and Th.J. de Boer, Org. Mass Spectrom., 5 (1971) 725.
287 A. Venema, N.M.M. Nibbering and Th.J. de Boer, Org. Mass Spectrom., 6 (1972) 675.
288 R.G. Cooks, R.S. Ward and D.H. Williams, Chem. Commun., (1967) 850.
289 B. Zeeh, Org. Mass Spectrom., 1 (1968) 315.
290 A.N.H. Yeo, R.G. Cooks and D.H. Williams, Org. Mass Spectrom., 1 (1968) 910.
291 H. Budzikiewicz, C. Djerassi and D.H. Williams, Mass Spectrometry of Organic Compounds, Holden-Day, San Francisco, 1967, pp. 336–353.
292 A. Prox and J. Schmid, Org. Mass Spectrom., 2 (1969) 121.
293 J.F. Biellmann and C.G. Hirth, Org. Mass Spectrom., 2 (1969) 723.
294 J.L. Holmes and F. Benoit, Org. Mass Spectrom., 5 (1971) 525.
295 C. Nolde, S.O. Lawesson, J.H. Bowie and R.G. Cooks, Tetrahedron, 24 (1968) 1051.
296 J.L. Holmes, Org. Mass Spectrom., 7 (1973) 335.
297 D.H. Williams, R.G. Cooks, J. Ronayne and S.W. Tam. Tetrahedron Lett., (1968) 1777.

298 H. Wynberg, R.M. Kellogg, H. Van Driel and G.E. Beekhuis, J. Amer. Chem. Soc., 89 (1967) 3501.
299 A. Padwa and R. Hartman, J. Amer. Chem. Soc., 88 (1966) 3759.
300 S.Meyerson and E.K. Fields, Org. Mass Spectrom., 2 (1969) 241.
301 F. de Jong, H.J.M. Sinnige and M.J. Janssen, Org. Mass Spectrom., 3 (1970) 1539.
302 A.S. Siegel, Tetrahedron Lett., (1970) 4113.
303 S. Meyerson and E.K. Fields, Org. Mass Spectrom., 1 (1968) 263.
304 W.D. Weringa, H.J.M. Sinnige and M.J. Janssen, Org. Mass Spectrom., 5 (1971) 1399.
305 M.E. Rennekamp, W.O. Perry and R.G. Cooks, J. Amer. Chem. Soc., 94 (1972) 4985.
306 P.J.M. Wakkers, M.J. Janssen and W.D. Weringa, Org. Mass Spectrom., 6 (1972) 963.
307 R.G. Cooks and S.L. Bernasek, J. Amer. Chem. Soc., 92 (1970) 2129.
308 A.M. Duffield, W. Carpenter and C. Djerassi, Chem. Commun., (1967) 109.
309 J.D. Henion and D.G.I. Kingston, Chem. Commun., (1970) 258.
310 K.B. Tomer and C. Djerassi, Org. Mass Spectrom., 7 (1973) 771.
311 A.S. Siegel, Org. Mass Spectrom., 3 (1970) 875.
312 W.H. McFadden and M. Lounsbury, Can. J. Chem., 40 (1962) 1965.
313 A.M. Duffield, S.D. Sample and C. Djerassi, Chem. Commun., (1966) 193.
314 W. Carpenter, A.M. Duffield and C. Djerassi, Chem. Commun., (1967) 1022.
315 M.M. Green, J. Amer. Chem. Soc., 90 (1968) 3872.
316 N.M.M. Nibbering and Th.J. de Boer, Tetrahedron, 24 (1968) 1427.
317 M.K. Hoffman and T.L. Amos, Tetrahedron Lett., (1972) 5235.
318 I. Howe and D.H. Williams, Chem. Commun., (1971) 1195.
319 J.L. Holmes and D. McGillivray, Org. Mass Spectrom., 5 (1971) 1339.
320 J.R. Hass, M.M. Bursey, D.G.I. Kingston and H.P. Tannenbaum, J. Amer. Chem. Soc., 94 (1972) 5095.

Chapter 4

ISOTOPES IN CARBANION REARRANGEMENTS

D.H. HUNTER

Department of Chemistry, University of Western Ontario, London, Ontario (Canada)

INTRODUCTION

Isotopes have played an important role in the study of most rearrangements of carbanions but naturally they have not been involved in all studies. To illustrate the value of isotopes it is necessary to view their use in perspective. Thus, it has been the author's intent not to emphasize those studies involving isotopes but to provide a fairly complete review of each type of carbanion rearrangement with the use of isotopes being presented in this context. Nonetheless, isotopes have been used as an excuse to limit some areas of discussion but hopefully not to the extent of unbalancing the presentation. The object of the review was to cover most of the selected topics in some detail rather than cover all possibilities in a broad or general fashion. While the selection of topics reflects the author's interests and biases, an attempt was made to emphasize areas of study that have developed recently.

I. STRUCTURES OF CARBANIONS

Before beginning a detailed look at the reactions of carbanions, it is desirable to develop an impression of the structures of some typical carbanions. The discussion of structure will begin by considering idealized carbanions in the gas phase and then looking at structures observed for organoalkalies in crystals. The perturbing influence of solvent on carbanion structure will then be discussed.

A. Gas Phase

In principle a minimum of complicating features will be encountered by observing discrete, isolated carbanions in the gas phase. Unfortunately, there are no structural observations on carbanions in the gas phase and our concepts of structure then come from analogy or from calculations. The methyl carbanion has received considerable attention. Cram [1a] proposed a trigonal pyramid rather than trigonal planar shape for the methyl carbanion by drawing an analogy with the isoelectronic ammonia molecule. Application of the Valence Shell Electron Pair Repulsion concepts developed by Gillespie [2] and Nyholm would result in a similar prediction although with the H–C–H bond angle being smaller than 109°. A bond angle close to 90° was suggested by Fort and Schleyer [3] in their detailed analysis of the methyl carbanion. The prediction of 90° was the result of the extrapolation of the observed bond angles for ammonia (107° for H–N–H) and the hydronium ion (117° for H–O–H) to the methyl carbanion. They also rationalized these observations and their prediction.

Ab initio molecular orbital calculations have also been attempted on the methyl carbanion using various basis sets. Kari and Csizmadia [4] came to the conclusion that the bond angles were sensitive to the basis set used and an H–C–H of about 105° was calculated. Owens and Streitwieser [5] found the H–C–H bond angle varied between 97.5° and 104.6° also depending upon the character of the minimal basis set. Thus, it would seem that, while neither analogy nor calculations lead to an exact shape for the methyl carbanion, there is certainly agreement that the methyl carbanion is non-planar with a barrier to inversion and that the H–C–H is probably less than 109°. Other structural features such as bond lengths and bond energies have not been discussed here although the ab initio calculations have provided data.

There is even less work on the structures of delocalized carbanions in the gas phase. It is generally felt that the shape of a carbanion should depend upon the substituents attached. If π-delocalization becomes important, presumably planarity will become more preferred. Some of the systems that are expected to be planar or near-planar are those stabilized by carbonyl, cyano, nitro and aryl groups, (e.g. fluorenyl and triphenylmethyl).

B. Solids

Almost all the observations on the properties and reactions of carbanions are not on the "free carbanion" but on the salts in various states. It is pertinent then to obtain an understanding of the structures of these salts both in solution and in the solid phase. It is tempting to use organo-alkalies as models for carbanion structure although the alkali metal cation would be expected to have a perturbing influence on the shape of the carbanion. Because of its size, lithium might be expected to have the greatest perturbing effect of the alkali metals; however lithium has also provided most of the examples of structures of organo-alkalies.

Recently, a number of diffraction studies were made of various organolithiums, and a sodium and a potassium compound. Weiss and co-workers have investigated the structure of methyllithium [6], methylpotassium [7] and ethylsodium [8] using X-ray powder data. Methylpotassium appeared to exist in the NiAs crystal form, ethylsodium appeared as sheets, while methyllithium existed in the interesting tetrameric form. With methyllithium, refinement of the data allowed the methyl hydrogens to be located. The methyl group does not appear to be planar trigonal but more pyramidal, although bond angles were not obtained with precision. Dietrich [9] has a crystal structure for tetrameric ethyllithium but again the hydrogens are only poorly defined.

Recently, detailed structures of some organolithiums complexed with tertiary amines have been published by Stucky and co-workers. Besides the structure of the 1-lithio derivative of bicyclo[1,1,0]butane [10] which must be quite rigid at the carbon, the first examples of delocalized carbanions have been obtained; benzyl [11] (*1*), triphenylmethyl [12] (*2*) and fluorenyllithium [13] (*3*) and the molecular geometries

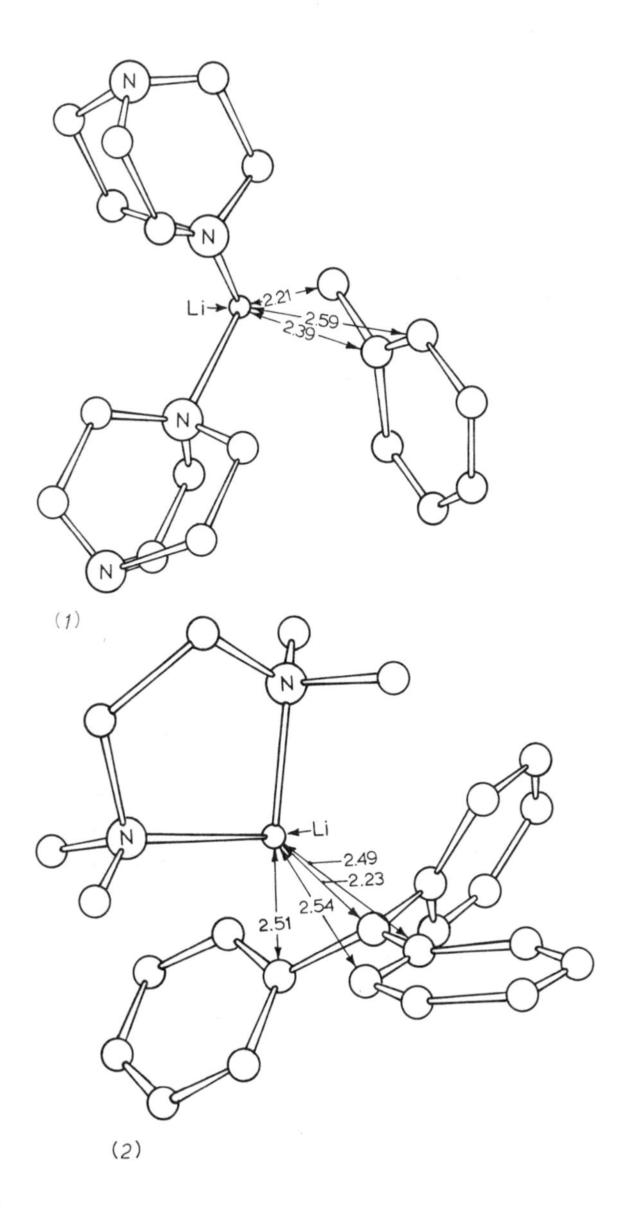

(1)

(2)

of the basic units are shown. A feature common to all three structures is that the lithium is not closely associated to just one carbon of the carbanion but to three, or four in the case of triphenylmethyllithium. In all three species studied the lithium is not symmetrically disposed above the carbanion as might be expected for strictly ionic bonding. The authors [13] interpret the bonding in terms of a three centre bond involving the lithium and the appropriate *p* orbitals of the carbanion.

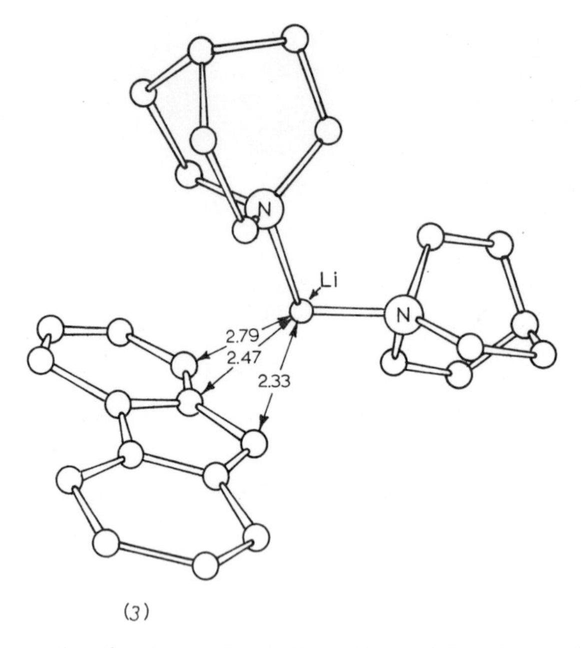

(3)

Another interesting feature of the triphenylmethyllithium structure (*2*) is illustrated in Fig. 1 which shows the bond angles around the substituted methyl carbanion. Each of the bond angles is approximately 120° the total being 358.1° indicating that within experimental error the carbanion is planar. The planar phenyl rings are twisted in propeller fashion about the central carbon. Unfortunately the usual difficulty of precisely locating protons in the presence of heavier atoms makes a similar analysis for (*1*) and (*3*) not sufficiently reliable.

Thus, while there are reservations about using crystal structures of organolithiums as models for carbanions, there is agreement between the observed shapes in the solid phase and the anticipated shapes in the gas phase. In summary, the crystal studies have provided examples of both planar and non-planar carbon moieties whose shapes are consistent with previous biases.

117.0±0.6
C
118.3±0.6
122.8±0.7

Fig. 1. Bond angles in triphenylmethyllithium [12].

C. Solution

Discussion of the behaviour of carbanions in solution can be arbitrarily divided into two classes: (1) short-lived carbanions in protic media, and (2) long-lived carbanions in aprotic media. This format will be used here.

1. Aprotic Media

There has been considerable activity in the study of the structure of organoalkalies in aprotic media in recent years. Recent books by Coetzee and Ritchie [14] and by Szwarc [15,16] attest to this activity and an attempt to detail this work here would be foolhardy. A point worth making, however, is that the importance of the role of the ion pairs is becoming more and more evident. Thus, the concept of structure of carbanions in solution must be expanded to include the various types of ion pairs, as well as dissociated ions, since different properties might well be expected for each.

Organo-lithiums provide dramatic examples of the range of structures which can be encountered in solution. As Table 1 shows, alkyllithiums exist not only with the ions paired but as aggregates [17,18]. The amount of polymerization is both substrate and solvent dependent and extends from hexamers to monomers. With a delocalized organolithium like fluorenyllithium in ether solvents, there is considerable evidence for the monomeric species existing in more than one form. Contact or intimate ion pairs, solvent separated ion pairs and dissociated ions have all been investigated [14–16]. Thus a range of structures from aggregates to dissociated ions is encountered with carbanions in aprotic media (Fig. 2) with each probably having a different chemical behaviour.

2. Protic Media

Base-catalyzed isotopic exchange is the most commonly used procedure for investigating carbanions in protic media (Fig. 3). Thus combinations of protium, deuterium, and tritium exchange between solvent and substrate can be used to monitor formation of short-lived carbanions and the relative rates of isotopic exchange at various sites is often related to kinetic and thermodynamic acidities.

TABLE 1

EXTENT OF POLYMERIZATION OF ORGANO-LITHIUMS[a]

	Solvent			
	Benzene	Cyclohexane	Diethyl Ether	Tetrahydrofuran
CH_3Li	6	6	4	4
$(CH_3)_3CLi$	4	4		
C_6H_5Li				2
$C_6H_5CH_2Li$	2			1

[a] From the data in refs. 17–18.

$$(\bar{R}\overset{+}{Li})_n \rightleftarrows \bar{R},\overset{+}{Li} \rightleftarrows \bar{R}\|Li^+ \rightleftarrows R^- + Li^+$$

various aggregates	contact or intimate ion-pair	solvent separated ion-pair	dissociated ions

Fig. 2. Possible equilibria for organo-lithiums in aprotic solvent.

$$R\text{-}H + \bar{B}: \rightleftarrows \left[R\ominus, HB\right] \xrightarrow[(TB)]{DB} \left[R\ominus, DB\right] \rightleftarrows R\text{-}D + B^-$$

Fig. 3. Simplified scheme for isotopic exchange.

The stereochemistry of isotopic exchange is used as a probe of carbanion structure. When short-lived carbanions are generated in solution, discussion of structure must again be extended to include not only the carbanion itself and the perturbing influence of the counterion but also perturbation by the surrounding solvent molecules. Cram [1b] has shown that the influence of the medium can be of major importance in determining the stereochemical behaviour of short-lived carbanions. With carbon acids that are suspected of giving nearly planar carbanions (e.g. fluorenyl), examples of reaction media are found which result in the carbon acid undergoing isotopic exchange with retention of configuration, with racemization, with net inversion of configuration as well as iso-inversion. Table 2 gives an illustrative example with a substituted fluorene (*4*) as substrate.

TABLE 2

EXAMPLES[a] OF ISOTOPIC EXCHANGE STEREOCHEMISTRY

(*4*) $\xrightarrow[ROH]{B:}$ product (D replaced by H)

Solvent	Base	T(°C)	k_e/k_α [b]
$(CH_3)_3COH$	NH_3	200°	>50
$(CH_3)_3COH$	$(CH_3)_3COK$	25°	1.0
CH_3COH	CH_3OK	25°	0.69

[a] See ref. lb for a more detailed discussion.
[b] Defined as the rate constant for exchange (k_e) over the rate constant for racemization (k_α). Thus $k_e/k_\alpha = 1$ implies racemization; $k_e/k_\alpha > 1$, net retention; $k_e/k_\alpha = 0.5-1$, net inversion.

These results add an extra air of uncertainty to attempts to deduce carbanion structure from carbon acid isotopic exchange stereochemistry. Observation of isotopic exchange with retention may well reflect the ability of the medium to retain its configuration relative to the carbanion, rather than the ability of the carbanion to retain its own configuration. Nonetheless there are some substrates which show a high degree of retention of configuration during isotopic exchange in a variety of media. Examples (Fig. 4) include the cyclopropyl anion [19a,b], the vinyl anion [20a–d] and interestingly, the α-sulfonyl carbanion [21]. Unlike the cyclopropyl and vinyl anions the α-sulfonyl carbanion's ability to maintain its configuration may be due not to a pyramidal shape but rather restricted rotations of a planar shape [21a–d]. It is this unique property which has contributed importantly to the analysis of the mechanism of the Ramberg–Bäcklund reaction [22].

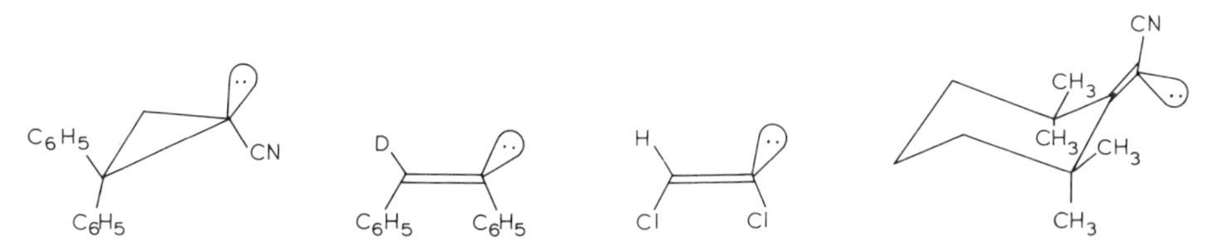

Fig. 4. Examples of carbanions that retain their configuration [19–20d].

II. PROTON TRANSFER REACTIONS

1,3 and 1,5 Proton transfers can be induced by base in systems which have allylic (or analogous) protons. Delocalized carbanions have been implicated as intermediates in these isomerizations and reprotonation of these ambident anions at the appropriate site leads to isomerization (Fig. 5). Besides providing an easy route for isomerization, the double bond plays the important role of greatly activating the proton thereby allowing use of a greater variety of bases and solvents.

Between the time of Zimmerman's review on carbanion rearrangement [23] in 1963 and the appearance of Cram's book in 1965, results on allylic isomerizations had increased dramatically. Thus, while Zimmerman did not include the topic, Cram devoted a full chapter [1c] to such proton transfers. Since then there have been two

$$\underset{\mathrm{H}}{\mathrm{C}}\!-\!\mathrm{C}\!=\!\mathrm{C} \;\underset{\mathrm{BH}}{\overset{\mathrm{B^-}}{\rightleftharpoons}}\; \left[\mathrm{C}\cdots\mathrm{C}\cdots\mathrm{C}\right]^- \;\underset{\mathrm{B^-}}{\overset{\mathrm{BH}}{\rightleftharpoons}}\; \mathrm{C}\!=\!\mathrm{C}\!-\!\underset{\mathrm{H}}{\mathrm{C}}$$

$$\underset{\mathrm{H}}{\mathrm{C}}\!-\!\mathrm{N}\!=\!\mathrm{C} \;\underset{\mathrm{BH}}{\overset{\mathrm{B^-}}{\rightleftharpoons}}\; \left[\mathrm{C}\cdots\mathrm{N}\cdots\mathrm{C}\right]^- \;\underset{\mathrm{B^-}}{\overset{\mathrm{BH}}{\rightleftharpoons}}\; \mathrm{C}\!=\!\mathrm{N}\!-\!\underset{\mathrm{H}}{\mathrm{C}}$$

$$\underset{\mathrm{H}}{\mathrm{C}}\!-\!\mathrm{C}\!\equiv\!\mathrm{C} \;\underset{\mathrm{BH}}{\overset{\mathrm{B^-}}{\rightleftharpoons}}\; \left[\mathrm{C}\cdots\mathrm{C}\!=\!\mathrm{C}\right]^- \;\underset{\mathrm{B^-}}{\overset{\mathrm{BH}}{\rightleftharpoons}}\; \mathrm{C}\!=\!\mathrm{C}\!=\!\mathrm{C}\!\diagdown\!\mathrm{H}$$

Fig. 5. Types of base-catalyzed 1,3-proton transfer reactions.

reviews, one [24] has emphasized allylic sulfides, sulfoxides and sulfones and the other [25] emphasized synthetic applications, and there has been continued interest in the topic. Isotopes have been used liberally to help elucidate the details of allylic and related isomerizations. This review will present examples in which isotopes have been used and will also include other aspects. For the purposes of discussion, systems analogous to allylic carbanions will also be included in this section: 2-aza-allylic carbanions, $\bar{C}$–N=C; propargyl carbanions, $\bar{C}$–C≡C; and dienyl and trienyl systems.

A. Intramolecularity

One of the first and truly unique features of allylic isomerizations was uncovered by observing the isomerization of 3-phenyl-1-butene (*5*) in *tert*-butyl alcohol-*O-d* con-

$$CH_3-\underset{H}{\overset{C_6H_5}{C}}-CH=CH_2 \xrightarrow[(CH_3)_3COD]{(CH_3)_3COK} \underset{54\%}{\overset{C_6H_5\quad H}{\underset{CH_3\quad CH_3}{C=C}}} + \underset{46\%}{\overset{C_6H_5\quad H}{\underset{CH_3\quad CH_2D}{C=C}}}$$

75°

(*5*) (*6*)

taining potassium *tert*-butoxide [26]. As illustrated, the isomerized product, 2-phenyl-*cis*-2-butene (*6*) contained only 0.46 atoms of deuterium per molecule even though the solvent was about 98% deuterated and the product did not exchange further under the reaction conditions. Thus 54% of the reacting molecules had not lost proton to solvent during rearrangement. Thus there was a 54% *intramolecular* 1,3 proton shift even in the presence of a large available deuteron pool.

A concurrent, independent study of proton transfer in 3-deuterioindene catalyzed by triethylamine in pyridine – 9 *M* deuterium oxide showed that an intramolecular 1,3 proton shift can occur with nearly 100% intramolecularity (case 27, Table 8). A large number of systems and reaction media have now been investigated and most of these results are summarized in Tables 3–8. The tables show the reactant and the product for which the percentage of the intramolecular component is reported. The solvent, base and temperature are also indicated along with the appropriate reference.

An intramolecular component in the proton shift is common to all the substrates and media studied even though magnitude of the contribution varies widely from extremes of 100% to 0.3%. Some generalizations seem possible on the basis of the data at hand. Tertiary amine bases (cases 13, 23, 25, 26, 27, 28 and 29) usually lead to a high percentage of intramolecularity in 1,3 or even 1,5 shifts (88–100%). Most alkoxide bases in alcohol-*O-d* solvent (cases 1, 3, 4, 8, 11, 12, 17, 18, 20, 21, 22, 23, 24 and 28) show similar and moderate percentages of intramolecularity (10–60%) even though the character of the medium and the reactive base change considerably. Cases 1 and 23 each present data on changes in medium, alcohol–alkoxide, without changes in sub-

TABLE 3

PERCENTAGE OF INTRAMOLECULARITY IN 1,3-PROTON TRANSFER REACTIONS OF PHENYLALKENES

Case	T(°C)	Solvent[d]	Base	Intramolecularity (%)[a]		Ref.
1		$CH_3-C(C_6H_5)(H)-CH=CH_2$	⟶	$(C_6H_5)(CH_3)C=C(H)(CH_3)$	+ $(C_6H_5)(CH_3)C=C(CH_3)(H)$	27
	75	$(CH_3)_3COD$	$(CH_3)_3COK$	51		
	50	$(C_2H_5)_3COD$	$(C_2H_5)_3COK$	56		
	50	DME–$(CH_3)_3COD$	$(CH_3)_3COK$	50	49	
	50	$(CH_3)_3COD$	$(CH_3)_4NOD$	45		
	144	$DOCH_2CH_2OD$	$DOCH_2CH_2OK$	32	33	
2		$CH_3-C(C_6H_5)(D)-CH=CH_2$	⟶	$(C_6H_5)(CH_3)C=C(H)(CH_2D)$		27
	75	$(CH_3)_3COH$	$(CH_3)_3COK$	23		
	50	$(C_2H_5)_3COH$	$(C_2H_5)_3COK$	17		
	50	DME–$(CH_3)_3COH$	$(CH_3)_3COK$	17		
	50	$(CH_3)_3COH$	$(CH_3)_4NOH$	6		
	145	$HOCH_2CH_2OH$	$HOCH_2CH_2OK$	12		
	25	DMSO–CH_3OH	CH_3COK	32		
3		$C_6H_5CH_2-C(C_6H_5)=CH_2$	⟶	$(C_6H_5)(H)C=C(C_6H_5)(CH_3)$	+ $(C_6H_5)(H)C=C(CH_3)(C_6H_5)$	28
	75	$(CH_3)_3COD$	$(CH_3)_3COK$	55	36	
4		$C_6H_5-CH_2-CH=CH_2$	⟶	$(C_6H_5)(H)C=C(H)(CH_3)$		29
	25	$(CH_3)_3COD$	$(CH_3)_3COK$	59		
5		$C_6H_5-CH_2-CH=CH_2$	⟶	$C_6H_5-CH=CH-CH_3$[b]		30
	[c]	CD_3SOCD_3	CD_3SOCD_2Na	80		
6		$(C_6H_5)(H)\,^{14}C=C(H)(CH_2C_6H_5)$	⟶	$(C_6H_5-{}^{14}CH_2)(H)C=C(H)(C_6H_5)$	+ $(C_6H_5-{}^{14}CH_2)(H)C=C(C_6H_5)(H)$	31
	[c]	$O(CH_2CH_2)O$ C_2H_5OH C_2H_5OT	$NaOC_2H_5$	99	97	

[a] Percentage does not refer to the percentage of either isomer formed but to the percentage of intramolecularity found for the isomer indicated. In some cases only one isomer was investigated and in others both isomers were isolated and analyzed for isotopic exchange.

[b] Not identified as *cis* or *trans*.

[c] Temperature not reported.

[d] DME = 1,2-dimethoxyethane; DMSO = dimethylsulfoxide.

TABLE 4

PERCENTAGE OF INTRAMOLECULARITY IN PROTON TRANSFER REACTIONS OF ALKENES

Case	T(°C)	Solvent	Base	Intramolecularity (%)[a]	Ref.
7		$CD_3CD_2CD_2{-}CD{=}CD_2$	→	$CD_3CD_2CD{=}CD{-}CD_3$[b]	32
	55	0.4 M $(CH_3)_3COH$ CH_3SOCH_3	$(CH_3)_3COK$	> 94	
8		$(CH_3)_3C{-}CH{=}C(CH_3)_2$	→	$(CH_3)_3C{-}CH_2{-}C({=}CH_2)CH_3$	33
	215	$(CH_3)_3COD$	$(CH_3)_3COK$	10	
9			→		34
	95	$CH_3CH_2C(CH_3)_2OD$	$CH_3CH_2C(CH_3)_2OK$	considerable	
10		D, D	→	D, H, D	35
	c	$(C_2H_5)_3COH$	$(C_2H_5)_3COK$	~ 90%	
	25	$CH_3SOCH_3{-}(C_2H_5)_3COH$	$(C_2H_5)_3COK$	< 7%	
11		$CH_3(CH_2)_{10}CH_2{-}O{-}CH_2{-}CH{=}CH_2$	→	$CH_3(CH_2)_{11}O{-}CH{=}CH{-}CH_3$ (H, H)	36
	80	$(CH_3)_3COD$	$(CH_3)_3COK$	30	

[a] Percentage does not refer to the percentage of the isomer formed but the percentage of intramolecularity found in its formation.
[b] Not identified as *cis* or *trans*.
[c] Temperature not reported.

strate, and clearly illustrate the minor effect of these media on the contribution of the intramolecular component. The insensitivity of intramolecularity to changes in both substrate and medium is illustrated by 16% intramolecularity in the reaction of 1-*tert*-butyl-3-methylindene (case 28) in methanol-*O-d* catalyzed by potassium methoxide combined with dicyclohexyl-18-crown-6-ether. Here the reactive base is probably dissociated methoxide ion. For comparison 25% intramolecularity is observed in the reaction of 1,3,3-triphenylpropyne with potassium *tert*-butoxide in *tert*-butyl alcohol

TABLE 5

PERCENTAGE OF INTRAMOLECULARITY IN ACETYLENE–ALLENE INTERCONVERSIONS

Case	T(°C)	Solvent	Base	Intramolecularity (%)[a]	Ref.
12		 $C_6H_5-CH(C_6H_5)-C\equiv C-C_6H_5$	⟶	$(C_6H_5)_2C=C=CH(C_6H_5)$	37a,37b
	27	$(CH_3)_3COD$	$(CH_3)_3COK$	25	
13		 $C_6H_5-CD(C_6H_5)-C\equiv C-C_6H_5$	⟶	$(C_6H_5)_2C=C=CD(C_6H_5)$	37a,37b
	28	CH_3OH	CH_3OK	18	
	28	$CH_3SOCH_3-(CH_3)_3COH$	$N(CH_2CH_2)_3N$	88	
	28	$CH_3SOCH_3-CH_3OH$	$N(CH_2CH_2)_3N$	88	
	28	$CH_3SOCH_3-CH_3OH$	$(CH_2)_4NH$	58	
14		$C_6H_5-CH_2-C\equiv CH$	⟶	$C_6H_5-CH=C=CH_2$	30
	b	CD_3SOCD_3	CD_3SOCD_2Na	>90	
15		$C_6H_5CD_2-C\equiv CD$	⟶	$C_6H_5-CD=C=CD_2$	30
	b	CH_3SOCH_3	CH_3SOCH_2Na	>90	
16		$C_6H_5CH=C=CH_2$	⟶	$C_6H_5-C\equiv C-CH_3$	30
	b	CD_3SOCD_3	CD_3SOCD_2Na	>95	

[a] Percentage does not refer to the percentage of the isomer formed but the percentage of intramolecularity found in its formation.
[b] Temperature not reported.

(case 12). Here the reactive base is ion paired and the dielectric of the medium is quite different.

One feature that does have a consistent effect on the percentage of the intramolecular component in 1,3 or 1,5 shifts is the position of the isotope. Three sets of data illustrate the isotope effects. The results obtained with 3-phenyl-1-butene (cases 1 and 2) show that consistently lower intramolecularity was observed when transferring a deuteron in a proton pool than when transferring a proton in a deuteron pool. The observation with 1,3-diphenylpropene (case 6) shows that almost 100% intramolecularity is observed when transferring a proton in a tritium pool although a direct com-

TABLE 6

PERCENTAGE OF INTRAMOLECULARITY IN 1,3 AND 1,5 PROTON TRANSFERS OF IMINES

Case	T(°C)	Solvent	Base	Intramolecularity (%)[a]	Ref.
17		$(H)(CH_3)(C_6H_5)C{-}N{=}C(D)C(CH_3)_3$	⟶	$(CH_3)(C_6H_5)C{=}N{-}CHD{-}C(CH_3)_3$	38
	75	$(CH_3)_3COD$	$(CH_3)_3COK$	13	
18		$(H)(CH_3)(C_6H_5)C{-}N{=}C(H)C(CH_3)_3$	⟶	$(CH_3)(C_6H_5)C{=}N{-}CH_2{-}C(CH_3)_3$	39
	75	$(CH_3)_3COD$	$(CH_3)_3COK$	46	
	75	$CD_3SOCD_3{-}CH_3OD$	CH_3OK	17	
19		$(D)(CH_3)(C_6H_5)C{-}N{=}C(H)C(CH_3)_3$	⟶	$(CH_3)(C_6H_5)C{=}N{-}CHD{-}C(CH_3)_3$	39
	75	$(CH_3)_3COH$	$(CH_3)_3COK$	8	
20		$C_6H_5{-}CH(CH_3){-}N{=}C(C_6H_4pOCH_3)(CH_3)$	⟶	$(C_6H_5)(CH_3)C{=}N{-}CH(CH_3){-}C_6H_4pOCH_3$	41
	100	$(CH_3)_3COD$	$(CH_3)_3COK$	49	
21		$C_6H_5CH{=}N{-}C(C_6H_5)(H){-}N{=}CHC_6H_5$	⟶	$C_6H_5CH{=}N{-}C(C_6H_5){=}N{-}CH_2C_6H_5$	42
	60	CH_3OD	CH_3OK	14	
22		$C_6H_5CH_2{-}N{=}C(C_6H_5){-}CH{=}CH{-}C_6H_5$	⟶	$C_6H_5CH{=}N{-}C(C_6H_5){=}CH{-}CH_2{-}C_6H_5$	43
	60	CH_3OD	CH_3OK	10	

[a] Percentages do not refer to the percentage of the isomer formed but to the percentage of intramolecularity found in its formation.

TABLE 7

PERCENTAGES OF INTRAMOLECULARITY IN 1,5-PROTON TRANSFER REACTIONS TO FORM TRIPHENYLMETHANES

Case	T(°C)	Solvent	Base	Intramolecularity (%)[a]	Ref.
23		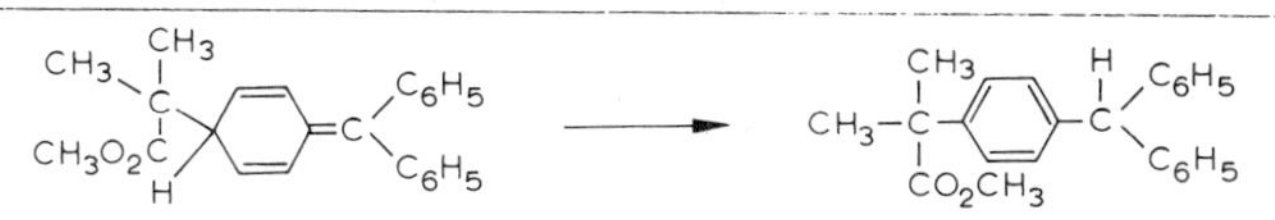			37a, 37b
	55	$DOCH_2CH_2OD$	$DOCH_2CH_2OK$	17	
	25	$THF-D_2O$	DONa	34	
	25	$DMSO-CH_3OD$	CH_3OK	40	
	25	CH_3OD	CH_3ONa	47	
	25	$(CH_3)_3COD$	$(CH_3)_3COK$	50	
	75	$(C_2H_5)_3COD$	$(C_3H_7)_3N$	98	
	75	$(C_2H_5)_3COD$ 0.1 *M* $(C_3H_7)_3NDI$	$(C_3H_7)_3N$	98	
	65	CH_3OD	$C_3H_7ND_2$	62	
24					44
	25	$(CH_3)_3COD$	$(CH_3)_3COK$	48	

[a] Percentages do not refer to the percentage of the isomer formed but to the percentage of intramolecularity found in its formation.

TABLE 8

PERCENTAGE OF INTRAMOLECULARITY IN 1,3-PROTON TRANSFER REACTIONS OF INDENES

Case	T(°C)	Solvent	Base	Intramolecularity (%)[a]	Ref.
25					45
	41	5 *M* $D_2O-C_5H_5N$	$CH_3(CH_2)_5ND_2$	60[b]	
26					46
	c	5.5 *M* $D_2O-C_5H_5N$	$(CH_3CH_2)_3N$	~100	

TABLE 8 (continued)

Case	T(°C)	Solvent	Base	Intramolecularity (%)[a]	Ref.
27			→		47
	[c]	9 *M* $D_2O-C_5H_5N$	$(CH_3CH_2)_3N$	~100	
28			→		48a, 48b
	50	$(CH_3)_3COD$	$N(CH_2CH_2)N$ $N(CH_2CH_2)_3NDI$	99.7	
	25	$CH_3SOCH_3-(CH_3)_3COD$	$N(CH_2CH_2)_3N$ $N(CH_2CH_2)_3NDI$	98.8	
	25	THF	$CH_3CH_2CH_2ND_2$	65	
	37	C_6H_6	$(CH_2)_5ND$	87	
	25	CH_3OD	CH_3OK	13	49
	25	CH_3OD	CH_3OK[d]	16	
	37	THF	$CH_3CH_2CH_2ND_2$	83	
	37	THF	$CH_3CH_2CH_2ND_2$[d]	90	
29			→		48a, 48b
	50	$(CH_3)_3COH$	$N(CH_2CH_2)_3N$	95	
	25	THF	$CH_3CH_2CH_2NH_2$	11	
	37	C_6H_5	$(CH_2)_5NH$	73	
	25	THF	$(CH_2)_5NH$	44	
	25	CH_3OH	CH_3OK	0.3	49
	100	$C_6H_6-C_6H_5OH$	C_6H_5OK	13	
	100	$C_6H_6-C_6H_5OH$	C_6H_5OK[d]	24	

[a] Percentages do not refer to the percentage of the isomer formed but to the percentage of intramolecularity found in its formation.
[b] Calculated from the available data in the paper.
[c] Temperature not reported.
[d] Added equivalent of dicyclohexyl-18-crown-6 ether.

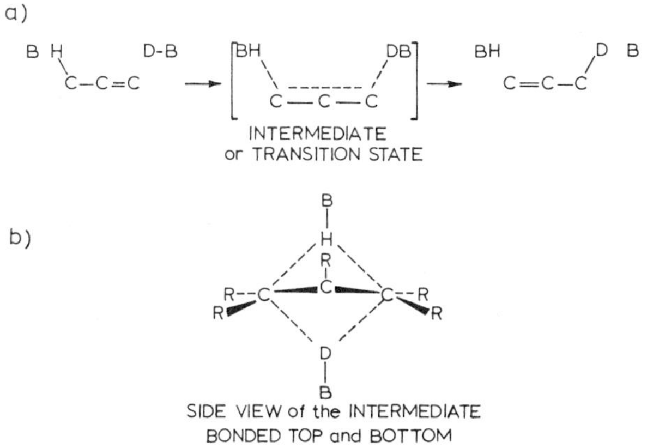

a)

INTERMEDIATE
or TRANSITION STATE

b)

SIDE VIEW of the INTERMEDIATE
BONDED TOP and BOTTOM

Fig. 6. Invalid models for 1,3-proton transfer reactions [27].

parison of media is not available. Cases 18 and 19 and cases 28 and 29 provide further examples of interchanging protons and deuterons in the medium and substrate.

The intervention of an intramolecular component in 1,3 and 1,5 proton transfers has important mechanistic implications and is itself sufficient evidence to rule out the concerted process (Fig. 6a) originally invoked [50a] for base-catalyzed aza-allylic rearrangements. A concerted process predicts 0% intramolecularity and this is rarely observed. This argument holds whether a transition state is involved in the transformation, or an intermediate with similar bonding. Another possible intermediate with a hydrogen bonded at the top face and a deuteron bonded at the bottom face (Fig. 6b), is ruled out since the intramolecularity should then be identical whether deuterium is in the substrate and proton is in the solvent or vice versa since the same intermediate would be formed in either case. As mentioned earlier, the position of the isotope does change the intramolecularity.

Figure 7 provides one of the simpler models for proton transfer not inconsistent with many of the observations on intramolecularity particularly since in an acceptable mechanism the proton being removed must be unique from those in bulk solvent. When the base is a tertiary amine the first formed intermediate would be an ion pair and the strong electrostatic attraction would result in high intramolecularity as is observed (vide supra). However, results have also been obtained with primary amines (cases 23, 25, 28 and 29) and secondary amines (cases 13, 28 and 29) and these show intramolecularities higher than would be expected if the protons (or deuterons) on nitrogen became scrambled. This suggests that the proton (or deuteron) removed from the substrate is held more specifically to the allylic carbanion than just by ion pairing but the isotope can still be exchanged. Presumably ion pairing interactions are important with alkoxide bases in low polarity solvents like *tert*-butyl alcohol but ion pairing need not be involved in the intermediate when bases like methoxide or deuteroxide

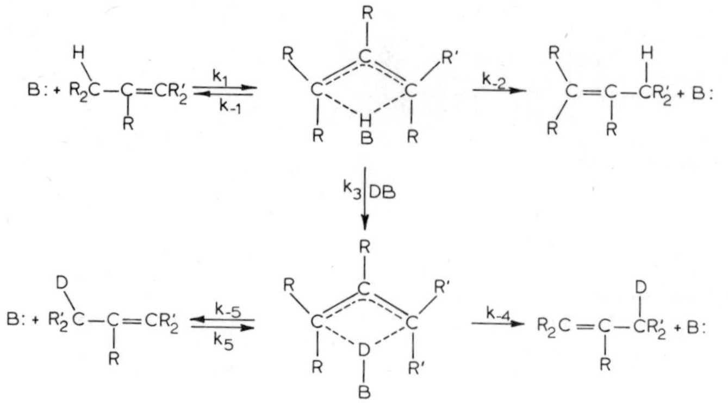

Fig. 7. Simple mechanistic model for 1,3 proton transfer reaction with intramolecular component [48a,b].

(case 23) are used. Again, however, the removed proton is still in a unique position compared to the bulk solvent. The origin of the isotope effects on intramolecularity is presumably complex and reflects a competition between the isotope effects for exchange and for collapse of the intermediate to product.

Some rather unique effects are encountered when dimsyl ion in dimethylsulfoxide was used with allylbenzene (case 5), propargylbenzene (cases 14 and 15) and phenylallene (case 16). The high intramolecularities, independent of the position of the deuterium (80–100%), suggest that again the proton being removed occupies a unique position and it has been proposed [30] that the dimsyl anion demonstrates ambident behaviour by removing the proton from the substrate with oxygen rather than with carbon (*7*). Unexpected intramolecularity was also observed in the dimsyl ion catalyzed *cis–trans* isomerization of *cis*-stilbene [20c] (*8*) where the recovered *trans*-stilbene (*9*) retained 32% of the original deuterium. While the intramolecularity in this

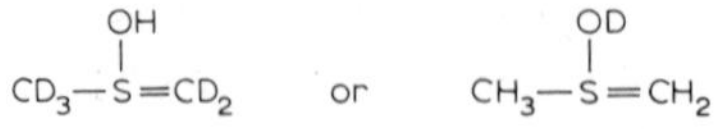

$$CD_3{-}\overset{\displaystyle OH}{\overset{|}{S}}{=}CD_2 \quad \text{or} \quad CH_3{-}\overset{\displaystyle OD}{\overset{|}{S}}{=}CH_2$$

(7)

cis–trans isomerization is lower, the vinyl carbanion must rotate with respect to the deuteron that has been removed increasing the opportunity for deuteron–proton scrambling. Proton abstraction by oxygen of the dimsyl anion provides an intriguing explanation for this result also.

The allylic isomerization of perdeuterio-1-pentene (case 7) in dimethylsulfoxide catalyzed by potassium *tert*-butoxide also proceeds with very high intramolecularity. While these results can be rationalized using *tert*-butoxide as the active base in a nearly aprotic medium (dimethylsulfoxide), the dimsyl anion again provides an interesting alternative as the active base.

$$\underset{(8)}{(D)(C_6H_5)C{=}C(D)(C_6H_5)} \xrightarrow[CH_3SOCH_3]{CH_3SOCH_2Na} \underset{30\%}{(D)(C_6H_5)C{=}C(C_6H_5)(D)} + \underset{70\%}{(D)(C_6H_5)C{=}C(C_6H_5)(H)}$$

(9)

The widespread involvement of intramolecularity in 1,3 and 1,5 proton shifts provides unambiguous evidence for the phenomenon of *internal return* in these delocalized carbanions. It is tempting to extend the experience obtained in the rearranging systems to those cases when isotopic exchange occurs without concurrent rearrangement. Thus intramolecularity could also provide a measure of the amount of carbanion formation without isotopic exchange (internal return) that is occurring in non-rearranging systems (e.g. fluorenyl, benzyl, triphenylmethyl and other more localized carbanions).

This approach leads to a number of generalizations applicable to all carbon acid isotopic exchange reactions: (1) in most reaction media, isotopic exchange will *not* measure all acts of carbanion formation; (2) since either tertiary amines in alcohol solvent or *tert*-butoxide and dimsyl anion in dimethyl sulfoxide show very high degrees of intramolecularity ($>$90%), isotopic exchange will probably measure only a small fraction of the acts of carbanion formation; (3) alkoxide–alcohol media may give a close but not complete measure of carbanion formation since moderate degrees of intramolecularity are observed (10–60%); (4) since intramolecularity in alkoxide alcohol is not highly sensitive to the specific medium, the fraction of internal return may also be insensitive to the particular alcohol employed.

Throughout this discussion the amount of internal return observed in 1,3 and 1,5 proton exchanges and shifts was used as a model for 1,1 proton exchange reactions. While this approach may prove to have some validity, it seems reasonable that the amount of internal return may often be lower in the rearranging systems than in the non-rearranging analogues since the movement of the medium with respect to the carbanion may provide a greater opportunity for drowning the removed isotope. This argument amplifies the conclusion that, while *relative* rates of isotopic exchange of carbon acids in a particular medium may be proportional to the actual relative rates of carbanion formation, the *absolute* values of the rate constants for the isotopic exchange process will almost always be lower than for the actual carbanion formation process.

B. Collapse Ratios

The factors that affect the preferred position of reaction is one of the interesting features of the chemistry of ambident species. Thus considerable attention has been devoted to studying the reactions at oxygen or carbon of enols and enolates [51].

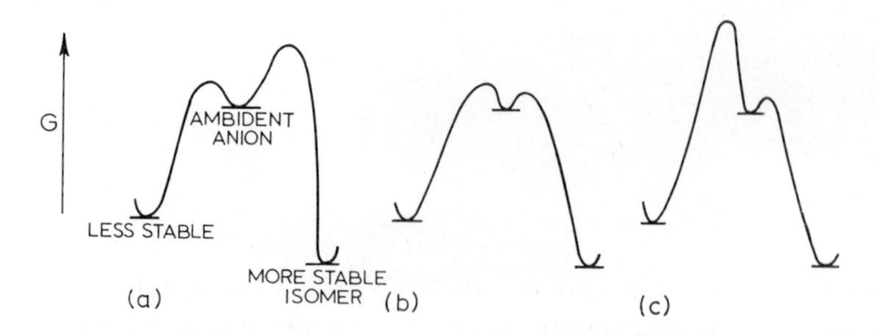

Fig. 8. Free energy profiles for reactions involving an intermediate [1d]. (Reproduced with permission of Academic Press.)

Studies of protonation of such species led to the generalization [50b] that "when a proton is supplied by acids to the mesomeric anion of weakly ionizing tautomers of markedly unequal stability, then the tautomer which is most quickly formed is the thermodynamically least stable; it is also the tautomer from which the proton is lost most quickly to bases". Cram [1d] pointed out exceptions to this generalization in his book and since then sufficient examples have appeared to justify an updating of the area.

There are three types of reactivity patterns for ambident anions that can be readily summarized in free energy profiles [1d] (Fig. 8) intended for reactions involving proton removal and reprotonation. Part (a) corresponds to the case where the less stable isomer forms the intermediate fastest but is also formed fastest from the intermediate as in the generalization mentioned above. In part (b) the less stable isomer remains the most reactive but is formed least often from the intermediate, while in (c) the less stable isomer is the least reactive and is formed least often.

There are three simple examples that serve to illustrate the possible free energy diagrams. Acetone exists primarily in the keto-form [55] and the enolate (*10*) is suspected to protonate and deprotonate more rapidly on oxygen than carbon thus fitting the energy curve of Fig. 8a. The prototropy of the substituted triphenylmethane (*11*) has been studied [37a,b] and the equilibrium greatly favours the triphenylmethane form (lies to the left) but, unlike enolates, protonation of the delocalized carbanion occurs to yield the thermodynamically preferred product as in Fig. 8b. Phenol (*12*) provides an example of the third type (c) since the enol form is greatly preferred at equilibrium but is almost certainly the kinetically most acidic form.

$$CH_3-\overset{O}{\overset{\|}{C}}-CH_3 \rightleftharpoons CH_3-C(\cdots O^-)(\cdots CH_2) \rightleftharpoons CH_3-C(OH)=CH_2$$

(*10*)

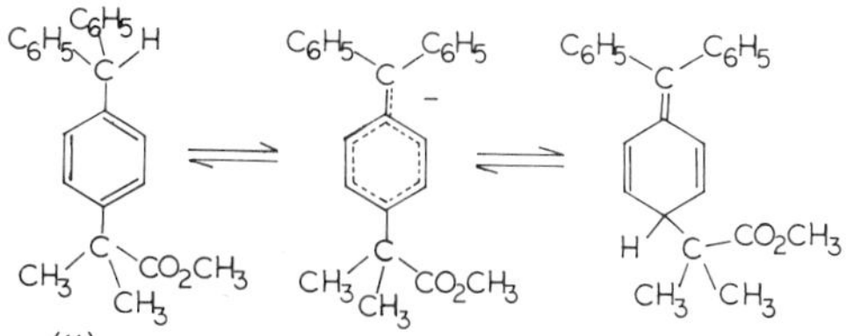

(11)

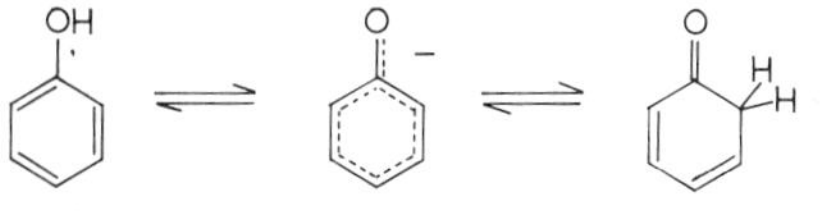

(12)

An analysis of the results of base-catalyzed hydrogen–deuterium exchange and isomerization also revealed more examples of the three types of energy profiles. Using the mechanism of Fig. 7 as a working model, the ratio k_{-4}/k_{-5} (assumed to equal k_{-2}/k_{-1}) provides a measure of the collapse preferences of the ambident carbanion on reprotonation. By comparing the rates of exchange and isomerization of a substrate, an estimate of the collapse ratio (k_{-2}/k_{-1} or k_{-4}/k_{-5}) can be obtained. A variety of allylic and aza-allyic anions have been analyzed in this way by following the relative rates of isomerization and of exchange in alkoxide–alcohol-*O-d*, and these are included in Tables 9, 10 and 11. The one feature that becomes apparent from these diverse results is that no single generalization or reactivity parameter will suffice to explain all the data.

Some interesting changes in collapse ratio with substitution at the *pseudo*-axial and *pseudo*-equatorial* positions (*13*) have been noted [56] and others are apparent for the

(13)

pseudo-equatorial *pseudo*-axial

reaction of some of the allylic and aza-allylic anions in *tert*-butyl alcohol-*O-d* catalyzed by potassium *tert*-butoxide (Fig. 9, p. 157) and other cases are also included in the following discussion. A comparison of allylic anions (*14*) with (*15*) and (*16*) with (*17*) shows that substitution of a phenyl group in the *pseudo*-equatorial position has only a small effect on the collapse ratio while *pseudo*-axial substitution ((*14*) with (*18*) or (*19*)) produces a larger effect. This may reflect a balancing of the opposing inductive and steric effects of the phenyl ring. In contrast a methyl group, with inductive and steric effects operating together, seems to unbalance the collapse ratios to a large degree ((*15*) with (*16*)

* Both the terms *pseudo*-axial, *pseudo*-equatorial and *cis, trans* have been used in labelling substitution in allylic anions.

TABLE 9

AMBIDENT ANIONS OF TYPE A IN FIG. 8

K_{eq}	Less stable isomer	Ambident anion	More stable isomer	Ref.
10^{+6}	HO–C(CH$_3$)=CH$_2$	O⋯C(CH$_3$)⋯CH–H	CH$_3$–C(=O)–CH$_3$	55
10^{+7}	CH$_2$=N(OH)–O	O=N(⋯O)⋯CH–H	CH$_3$–NO$_2$	52
large	cyclopentenyl–C(H)(CO$_2$Et)CO$_2$Et	cyclopentyl⋯C(CO$_2$Et)CO$_2$Et	cyclopentylidene=C(CO$_2$Et)CO$_2$Et	53
4	C$_6$H$_5$–CH(CH$_3$)–N=C(CH$_3$)–C(CH$_3$)$_3$	C$_6$H$_5$–C(CH$_3$)⋯N⋯C(CH$_3$)–C(CH$_3$)$_3$ (200 ↓ C, 1 ↓ C)	(C$_6$H$_5$)(CH$_3$)C=N–CH(CH$_3$)–C(CH$_3$)$_3$	56
15	C$_6$H$_5$–CH(CH$_3$)–N=CH–C(CH$_3$)$_3$	C$_6$H$_5$–C(CH$_3$)⋯N⋯CH–C(CH$_3$)$_3$ (5 ↓ C, 1 ↓ C)	(C$_6$H$_5$)(CH$_3$)C=N–CH$_2$C(CH$_3$)$_3$	39
2.4	(CH$_3$)$_2$C=CH–C(CH$_3$)$_3$	H$_2$C⋯C(CH$_3$)⋯CH–C(CH$_3$)$_3$ (3 ↓ C, 1 ↓ C)	CH$_2$=C(CH$_3$)–CH$_2$–C(CH$_3$)$_3$	33
$>10^2$	fluorenyl(H)–CH=CH–C$_6$H$_5$	fluorenyl C⋯CH⋯CH–C$_6$H$_5$ (2 ↓ C, 1 ↓ C)	fluorenylidene=CH–CH$_2$–C$_6$H$_5$	61

and (*20*) with (*21*)). Substitution at the central atom does not appear critical if the groups are *trans* as indicated by the small effect of the phenyl ring when comparing (*18*) with (*19*). The effect of the methylsulfinyl, methylsulfonyl and trimethylammonium groups (Table 10) suggests an important role for inductive effects. The consequence of the results obtained to date is to reveal that both steric and inductive effects play a role in determining the collapse ratios to the extent that thermodynamic control of product ratios is not common. The principle of least motion has also been in-

TABLE 10

AMBIDENT ANIONS OF TYPE B IN FIG. 8

K_{eq}	Less stable isomer	Ambident anion	More stable isomer	Ref.
large	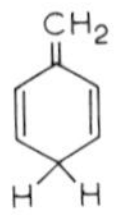 CH_2; H H	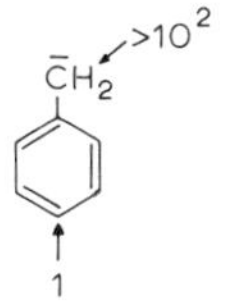 $\bar{C}H_2$ (>10^2); 1	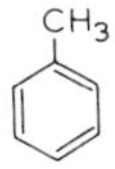 CH_3	58
large	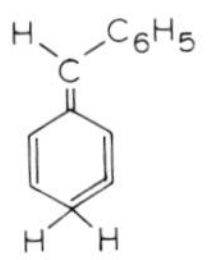 H, C_6H_5, C; H H	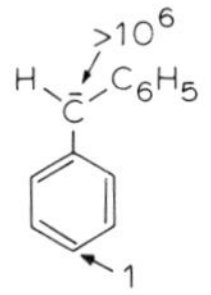 H, C_6H_5, $\bar{C}$ (>10^6); 1	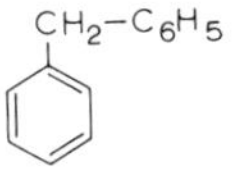 $CH_2-C_6H_5$	57
200	$C_6H_5-CH_2-CH=CH_2$	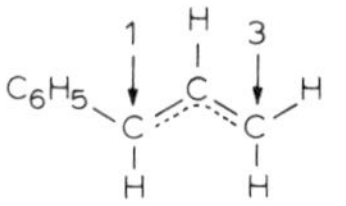 1, H, 3; C_6H_5, C, C, C, H; H, H	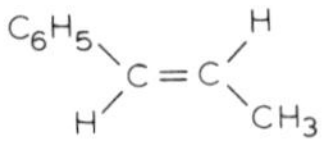 C_6H_5, H; C=C; H, CH_3	29
45	$C_6H_5-CH_2-CH=CH_2$	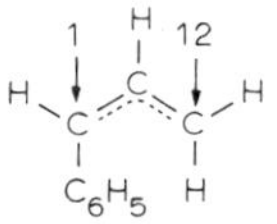 1, H, 12; H, C, C, C, H; C_6H_5, H	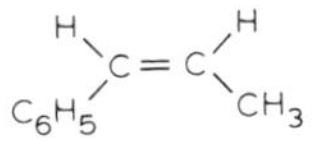 H, H; C=C; C_6H_5, CH_3	29
>10^3	$C_6H_5-CH(CH_3)-CH=CH_2$	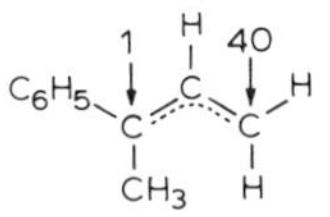 1, H, 40; C_6H_5, C, C, C, H; CH_3, H	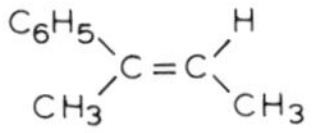 C_6H_5, H; C=C; CH_3, CH_3	27
40	$C_6H_5-CH_2-C(C_6H_5)=CH_2$	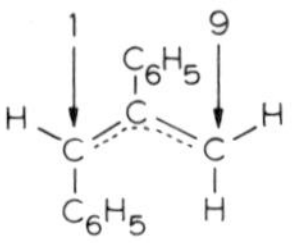 1, 9; C_6H_5; H, C, C, C, H; C_6H_5, H	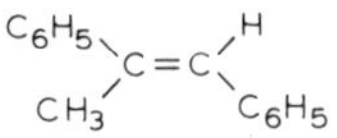 C_6H_5, H; C=C; CH_3, C_6H_5	28
10	$C_6H_5CH_2-C(C_6H_5)=CH_2$	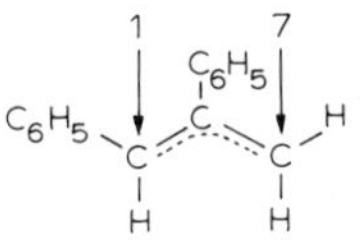 1, 7; C_6H_5; C_6H_5, C, C, C, H; H, H	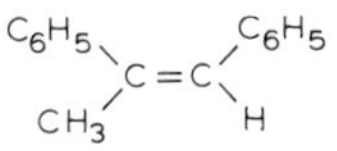 C_6H_5, C_6H_5; C=C; CH_3, H	28
large	$C_8H_{17}OCH_2-CH=CH_2$	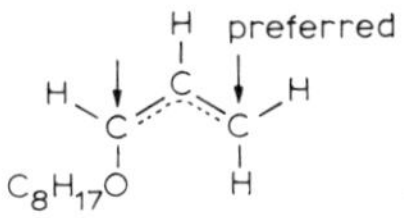 H, preferred; H, C, C, C, H; $C_8H_{17}O$, H	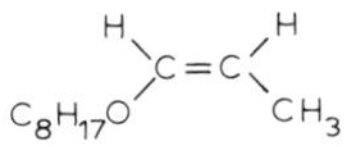 H, H; C=C; $C_8H_{17}O$, CH_3	36

TABLE 10 (continued)

K_{eq}	Less stable isomer	Ambident anion	More stable isomer	Ref.
24	$C_9H_{19}CH_2CH=CH-SOCH_3$	$C_9H_{19}-CH \cdots CH \cdots CH-SOCH_3$ (1, 1000)	$C_9H_{19}CH=CH-CH_2SOCH_3$	29, 36
>100	$C_3H_7CH_2-CH=CH-SO_2CH_3$	$C_3H_7-CH \cdots CH \cdots CH-SO_2CH_3$ (1, 10^4)	$C_3H_7CH=CH-CH_2SO_2CH_3$	59
60	$(CH_3)_2C=CH-\overset{+}{N}(CH_3)_3$ I^-	$H_2C \cdots C(CH_3) \cdots CH-\overset{+}{N}(CH_3)_3$ (1, 3)	$CH_2=C(CH_3)-CH_2-\overset{+}{N}(CH_3)_3$ I^-	60

TABLE 11

AMBIDENT ANIONS OF TYPE C IN FIG. 8

K_{eq}	Less stable isomer	Ambident anion	More stable isomer	Ref.
large	2,4-cyclohexadienone (6,6-H,H)	phenoxide ($C_6H_5O^-$)	phenol (C_6H_5OH)	
1.6	$C_6H_5-CH(CH_3)-CH=CH-C_6H_5$	$C_6H_5-C(CH_3) \cdots CH \cdots CH-C_6H_5$ (1, 35)	$(C_6H_5)(CH_3)C=CH-CH_2C_6H_5$	54
2.1	$(CH_3)(C_6H_5)C=CH-CH_2C_6H_5$	$CH_3-C(C_6H_5) \cdots CH \cdots CH-C_6H_5$ (1, 40)	$C_6H_5-CH(CH_3)-CH=CH-C_6H_5$	54
2.0	$C_6H_5-CH(CH_3)-N=C(CH_3)-C_6H_4OCH_3$	$C_6H_5-C(CH_3) \cdots N \cdots C(CH_3)-C_6H_4OCH_3$ (1, 3)	$(C_6H_5)(CH_3)C=N-CH(CH_3)-C_6H_4OCH_3$	41

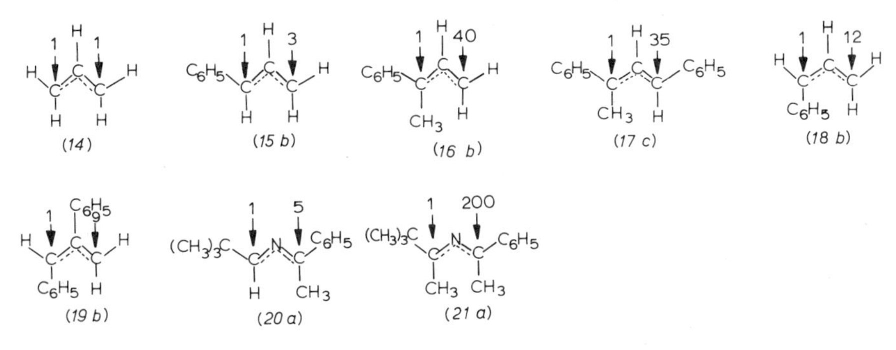

Fig. 9. Collapse ratios for selected carbanions classified by free energy profile (see Fig. 8).

voked to explain and predict positions of protonation [68] but it is not universally successful. For example, the observed collapse preference of the phenylallyl anion proves to be the opposite of that predicted.

Calculated collapse ratios have been used to further refine the mechanistic model for the allylic isomerization [48a] of the substituted indenes. The amine catalyzed exchange and isomerization of 3-*tert*-butyl-1-methylindene ((*22-h*) and (*22-d*)) to 1-*tert*-butyl-3-methylindene (*23*) (K_{eq} = 7.6 at 25°) has been studied using primary (*n*-propylamine) and secondary (piperidine) amines. The results pertaining to collapse ratios are shown in Table 12 and for both bases the carbanion collapsed preferentially to yield starting material, reflecting steric effects. Different collapse ratios were observed with the primary and secondary amine but the primary amine also showed a different calculated collapse ratio depending upon the position of the isotope.

TABLE 12

VARIATION OF THE POSITION OF REACTION OF THE 1-METHYL-3-*TERT*-BUTYLINDENYL ANION WITH BASE [48a]

(D)H, CH3 — k_{-1} — k_{-2} — H
CH3 CH3 CH3 (22) ⇌ ⇌ (23)

Isotope	Base	Solvent	k_{-1}/k_{-2}
D	n-PrNH$_2$	THF	1.1
H	n-PrND$_2$	THF	2.3
D	N-H (piperidine)	THF	6.9
H	N-D (piperidine)	THF	6.7

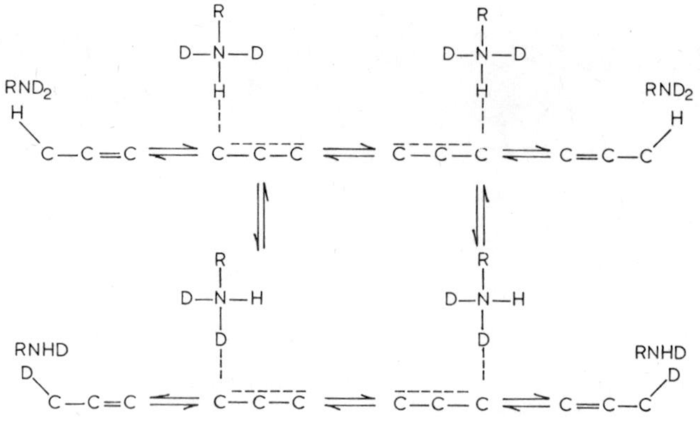

Fig. 10. Mechanistic model for 1,3 proton transfer reactions with unsymmetrical intermediates [48a,b].

A change in collapse ratio with position of isotope is inconsistent with the simplified kinetic scheme of Fig. 7 where for the purposes of calculation of the collapse ratio, the ratios k_{-1}/k_{-2} and k_{-5}/k_{-4} are equated. Another level of sophistication was introduced into the isomerization mechanism to rationalize the results with *n*-propylamine (Fig. 10). In the extreme version of this mechanism, hydrogen bonding occurs to just one site at a time at either end. If conduction of the proton from one end of the carbanion to the other is faster than both collapse and isotopic exchange then the kinetics of the scheme of Fig. 10 simplifies to that of Fig. 7. However, if collapse rates, isotopic exchange rates and proton conduction rates are of similar magnitude then isotope effects on these rates could appear as a change in the calculated collapse ratio. The secondary amine seems to be adequately described by the former scheme while the primary amine seems to require the latter. These results suggest a change in conduction rates with amine which may be a consequence of the bases existing in different states of aggregation. The general applicability of either of these mechanistic models to 1,3 and 1,5 proton transfer reactions is still an unsettled question.

The collapse preferences of pentadienyl anions have also received attention and here the evidence has been gained primarily from the ratios of products of quenching of organolithiums by water. Analysis of isotopic exchange and isomerization experiments has also been used. While the two procedures are not directly comparable, similar intermediates may be involved in the collapse of the carbanion with water in an ether solvent or with alcohol in an alcohol solvent. Cyclohexadienyl anions have been postulated as intermediates in the Birch reduction and as a general rule under these conditions protonation occurs preferentially at the central carbon to yield cyclohexa-1,4-dienes [62,63].

Bates et al. have generated an interesting variety of pentadienyl anions and investigated the product distribution when these anions were quenched with water, methyl iodide and ethylene oxide [64]. Figure 11 contains these and two other examples

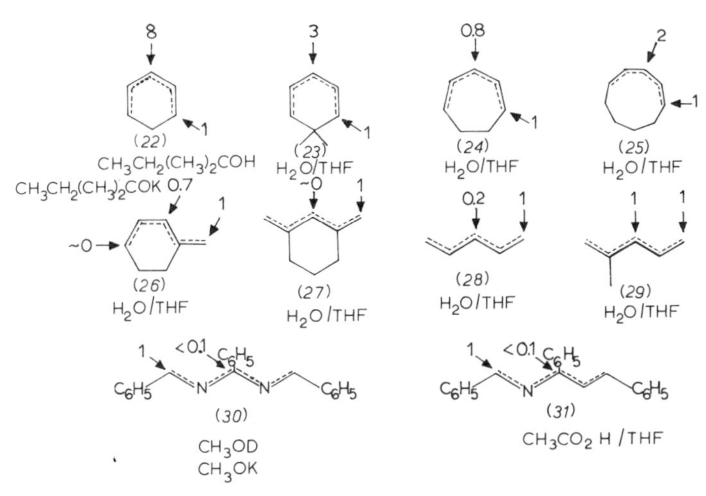

Fig. 11. Collapse ratios for pentadienyl anions [42,43,64].

((*30*) and (*31*)) with the collapse ratios and the medium indicated. The cyclohexadienyl anions [34,65], (*22*) and (*23*), both show a preference for protonation at the central atom consistent with the proposed scheme for the Birch reductions. However, in the larger and probably distorted seven- [66] and nine-membered [67] rings ((*24*) and (*25*)) this preference disappears.

Anions (*22*), (*26*) and (*27*) provide models for the U, sickle and W shapes of the pentadienyl anions and a changing pattern of collapse ratios is observed [34]. But it is not clear whether the changes are a consequence of the shapes or the degree of substitution (primary versus secondary). The open-chain anions (*28*) and (*29*) do not show high-selectivity [34] whereas the anions (*30*) [42] and (*31*) [43] show exclusive reaction at the terminal position. The geometries of the products of quench of (*28*) and (*29*) show that the anions exist in two or more geometries of similar stability (presumably W and sickle) and the collapse ratios then reflect the behaviour of more than one shape. The anions (*23*), (*24*), (*25*) and (*28*) show different collapse ratios upon reaction with methyl iodide and ethylene oxide showing not surprising sensitivity to the nature of the quenching species.

Preferential collapse at the central atom is a feature of just the cyclohexadienyl anion and with allylic and aza-allylic anions a wide range of collapse preferences is observable. While inductive and steric effects, and the principle of least motion [68] may be playing a role here, Bates et al. have noticed an interesting relationship between the position of protonation and charge distribution [64]. Carbon-13 shifts were used as a measure of electron density [69]. Similar conclusions have been reached in a study of some 1,3-diphenyl allyl anions [70] but details are yet to be reported.

C. Asymmetric Induction

The stereochemistry of 1,3 proton shifts catalyzed by base, besides shedding further light on the details of the reaction mechanism, has also provided examples of facile *intramolecular suprafacial 1,3 proton transfers.* To date the stereochemical studies have been restricted to two types of substrates; alkyl-substituted indenes and aryl-substituted imines (methylene azomethines). Because of their differing acidities the indenes have been studied primarily with amines as bases while alkoxides have been used to generate the aza-allylic anions as intermediates.

Bergson and Wiedler [71] were the first to observe stereoselectivity, noting that the rearrangement of (–)-1-methyl-3-isopropylindene (*32a*) in pyridine solvent catalyzed by butylamine produced optically active (+)-1-isopropyl-3-methyl-indene (*33a*). At

(32) ⇄ (33)

	a	*b*†
R	$(CH_3)_2CH$-	$(CH_3)_3C$-

† In ref. 72, the isomer of opposite configuration was used.

that time, neither the absolute or relative configurations of (*32a*) or (*33a*) were known. The racemization of the equilibrium mixture consisting of 20% (*32a*) and 80% (*33a*) was considerably slower than the isomerization rates and thus the isomerizations of (*32a*) → (*33a*) must be occurring with high stereoselectivity. The studies were extended [72] to the methyl-*tert*-butylindenes ((*32b*) and (*33b*)) and the sensitivity to base and solvent were investigated. Both (*32a*) and (*32b*) behaved similarly with the tertiary amine 1,4-diazabicyclo[2.2.2]octane, DABCO, as base in pyridine and DABCO gave much the same results as butylamine. However, with dilute potassium *tert*-butoxide in *tert*-butyl alcohol racemization of (*33b*) occurred significantly faster than isomerization. The stereoselective proton transfers were postulated to be suprafacial since the starting materials (*32*) and the products (*33*) had opposite optical rotations and this assumption was later confirmed [73,74].

Cram and co-workers have looked in detail at the reaction of (*32b*) and (*33b*) having both protium and deuterium at the 1-position [48a,b, 49, 74]. Optically pure isomers of known maximum rotation and absolute configuration were used. The stereospecificity of the isomerizations were deduced by interrupting the isomerization reaction of (*32b*) and then separating (*32b*) and (*33b*) which were analyzed separately for racemization. Provided (*32b*) and (*33b*) are not being racemized rapidly, the *stereospecificity* is revealed by the magnitude and sign of rotation of (*33b*). Use of deuterated substrate or solvent provides information on *intramolecularity* and *collapse ratios*, which have been discussed previously.

Some selected results obtained by starting with (*32b*) and isolating (*33b*) are re-

TABLE 13

EXCHANGE AND RACEMIZATION DURING ISOMERIZATION OF 1-*TERT*-BUTYL-3-METHYLINDENE (*32b*) TO 3-*TERT*-BUTYL-1-METHYLINDENE (*33b*) [48,49,74]

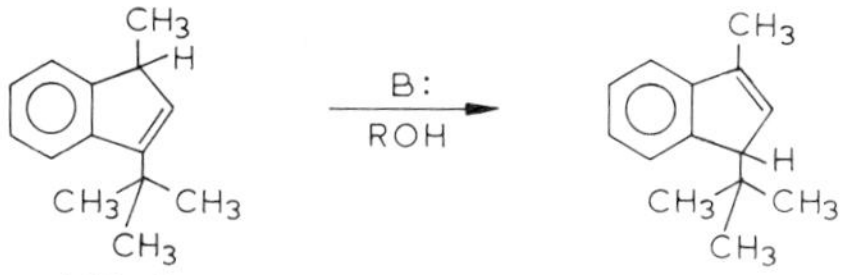

Base[a]	Solvent[b]	Recovered (*33b*) Racemization (%)	Exchange (%)
DABCO	$(CH_3)_3COD$	0	0
DABCO	1 *M* $(CH_3)_3COD$ DMSO	0	1
PMG	$(CH_3)_3COD$	2	0
$(CH_2)_5ND$	C_6H_6	0	11
$CH_3CH_2CH_2ND_2$	THF	0	35
CH_3OK	CH_3OD	100	87

[a] DABCO = 1,4-diazabicyclo[2.2.2]octane; PMG = pentamethylguanidine.
[b] DMSO = dimethylsulfoxide; C_6H_6 = benzene; THF = tetrahydrofuran.

corded in Table 13 and show the effect of varying the base and the solvent on both the extent of isotopic exchange during the 1,3 proton shift and the amount of racemization. With primary, secondary and tertiary amines the isomerization is completely stereospecific showing an exclusive suprafacial 1,3 shift in all the solvents used even though up to 35% exchange was observed with the *n*-propylamine. The reaction in methanol–methoxide stands in marked contrast showing complete racemization even with a 13% intramolecular component. These data are consistent with the results of Bergson and Wiedler (vide supra) and have the advantage of more closely defining the magnitude of the stereospecificity.

With these results the mechanistic models of Figs.7 and 10 can be also elaborated to include stereochemistry. With amine bases and indenes as substrates, the proton that has been abstracted remains on the same face of the anion and exchange of isotope also seems to occur on the same face of the anion. With methoxide in methanol-*O-d* the CH_3OH which is generated by proton removal appears to remain in the vicinity of the carbanion (13% intramolecularity) but has the opportunity to become symmetrical with respect to the carbanion, suggesting a solvent cage is formed.

As mentioned earlier, imines have also provided opportunities for observing asymmetric induction in a 1,3 proton transfer. Three systems have been studied to date and are shown in Fig. 12 along with the calculated stereospecifity and percentage of ex-

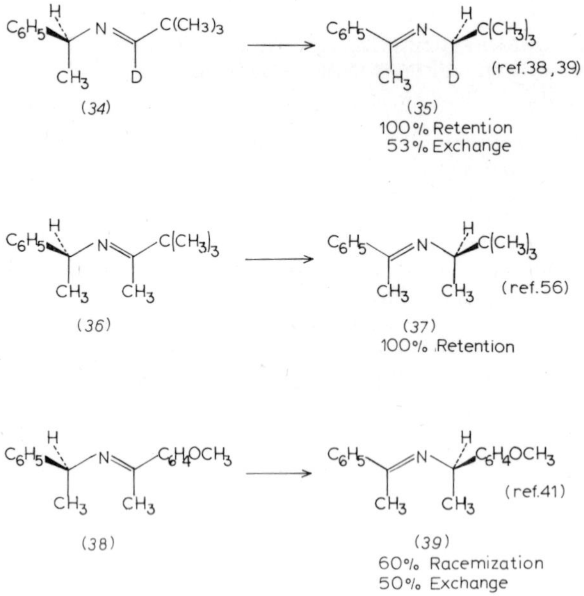

(34) → (35) (ref.38,39)
100% Retention
53% Exchange

(36) → (37) (ref.56)
100% Retention

(38) → (39) (ref.41)
60% Racemization
50% Exchange

Fig. 12. Stereochemistry and isotopic exchange for imine isomerizations in $(CH_3)_3COD$–$(CH_3)_3COK$.

change when the reaction was run in *tert*-butyl alcohol-*O-d* with potassium *tert*-butoxide. In the case of the (*34*)→(*35*) interconversion [38,39] and the (*36*)→(*37*) interconversion [56] the starting imine ((*34*) and (*36*)) was racemizing at a rate similar to the isomerization rate, a result of the collapse ratio favouring starting material (see Table 9). Within experimental error after correction for the prior racemization, both systems underwent a 100% suprafacial 1,3 proton shift in both the intramolecular and intermolecular component. Thus, all reaction was occurring on one face of the delocalized carbanion. Interestingly, the stereospecificity disappeared when dimethylsulfoxide-*tert*-butyl alcohol was used as solvent for the potassium *tert*-butoxide catalyzed reaction of (*36*)→(*37*) and recovered (*37*) was completely racemized.

To rationalize the results with the imines an important role has been given to the potassium cation which is associated with the base in *tert*-butyl alcohol but dissociated

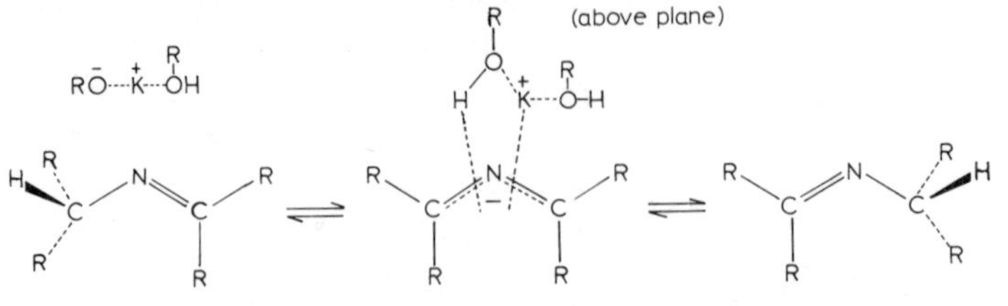

(above plane)

(*40*)

in dimethylsulfoxide. Proton abstraction from one conformation of imine by the potassium *tert*-butoxide ion pair in *tert*-butyl alcohol produces a carbanion ion-paired to potassium which in turn is solvated by the just-formed alcohol molecule (*40*). The presence of the potassium cation preserves the asymmetry of the intermediate and collapse to product occurs only on the side ion-paired to potassium to avoid formation of a product separated ion-pair in this low dielectric medium. It is also to be expected that the protons on alcohols solvating the potassium should be more acidic than those on alcohols in bulk solvent, thus also favouring reaction on the same side as the potassium. The addition of dimethylsulfoxide to the medium results in dissociation of the base and the organizing role of the potassium cation disappears resulting in racemization. Similar reasoning can be applied to the results with the indenes. With amines as bases, ion-pairs composed of alkyl–ammonium cations and indenyl anions show the same stereochemical behaviour as with ion-pairs of potassium cations and aza-allylic anions; all reaction occurs on just one face of the carbanion. With dissociated methoxide ions as base all asymmetry is lost.

The results obtained [41] with the imines (*38*)→(*39*) stand in contrast to the previous cases since under conditions where (*34*) and (*36*) react with complete retention, (*38*) shows about 60% racemization. The authors argue in favour of a twisted aza-allylic anionic intermediate, presenting arguments against a variety of other possible explanations. Thus, in spite of the close structural similarity between the anionic intermediates in the (*36*)→(*37*) and (*38*)→(*39*) interconversions, the presence of the second aryl ring may stabilize a twisted form with resultant loss of asymmetric induction.

The ramifications of a *suprafacial intramolecular 1,3 proton shift* in both the indenes and imines is worth considering in light of orbital symmetry predictions on sigmatropic processes [75]. The highest occupied molecular orbital in a simple allylic anion is antisymmetric about the plane perpendicular to the plane of the carbanion (Fig. 13). On this basis a concerted suprafacial 1,3 hydrogen shift is considered a forbidden thermal process. While this generalization need not be directly applicable to base-catalyzed 1,3 shifts, orbital symmetry would seem to make some demands upon the reaction pathways. Two applications are illustrated in Fig. 14 which show possible geometries of intermediates or transition states in base-catalyzed 1,3 proton shifts.

Figure 14a is meant to represent the orbitals in a state where a proton is symmetrically disposed below the face of the allylic anion. Such a diagram obviates the conclusion that overlap of the highest occupied molecular orbital of the allylic anion and the 1*s* orbital of hydrogen will not provide significant stabilization. To the extent that this

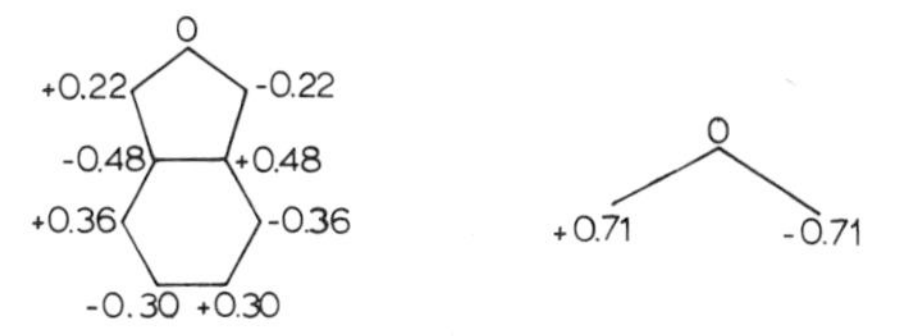

Fig. 13. Hückel coefficients for highest occupied molecular orbital of the indenyl and allyl anion.

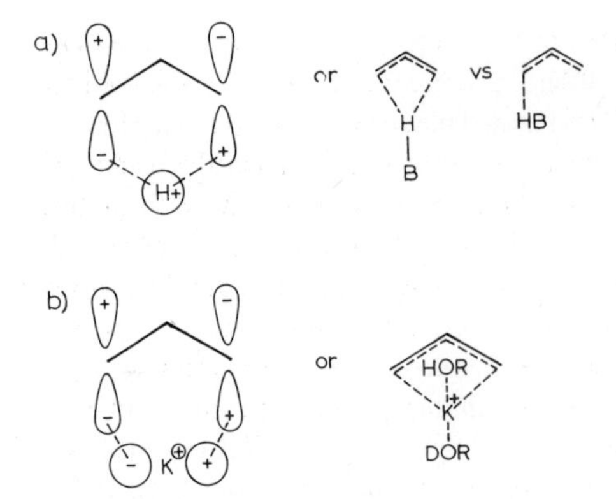

Fig. 14. Orbital representation for possible intermediates in base-catalyzed 1,3 proton transfer reactions.

may be used as a model for the case when a base molecule is also attached to the hydrogen, then an unsymmetrical geometry would seem preferable to a symmetrical geometry for the intermediate. (See mechanistic schemes in Figs. 7 and 10 and accompanying discussion.) Interestingly, a symmetrical species involving a potassium cation (Fig. 14b) would seem better adapted for overlap since vacant $4p$ orbitals of appropriate symmetry are available for bonding. As discussed earlier this potassium cation bonded to one face of the allylic carbanion could then serve to guide reaction to that same side of the carbanion.

III. GEOMETRICAL ISOMERISM IN DELOCALIZED CARBANIONS

Geometrical isomerism in carbanions seems to arise in two different circumstances; the *cis–trans* isomerism of vinyl anions and *cis–trans* isomerism associated with restricted rotation in allylic and related anions. The behaviour of vinyl anions has been discussed briefly in Section I C. The discussion in this section will be developed around the behaviour of allylic and vinylogous delocalized carbanions and will begin with consideration of the evidence available on the geometries adopted by delocalized anions. Results on the barriers to rotation in delocalized anions will also be discussed.

A. Geometries of Delocalized Carbanions

Evidence on the preferred geometries of delocalized carbanions has come both from studies of short-lived carbanionic intermediates in protic media and from studies of long-lived carbanions in aprotic media; e.g. organo-lithiums in tetrahydrofuran. As discussed in Section II a variety of substituted allylic anions have been generated as intermediates during base-catalyzed isomerization and isotopic exchange reactions. Commonly, when more than one geometrical isomer can be produced as the product

TABLE 14

COMPARISON OF THE KINETIC AND THERMODYNAMIC RATIO OF *CIS* AND *TRANS*-2-ALKENES PRODUCED FROM 1-ALKENES WITH BASE CATALYSIS [76]
(CH_3SOCH_3 AT 55°)

1-Alkene	*cis/trans* Initial ratio	Final ratio
$CH_3CH_2-CH=CH_2$	47	0.25
$CH_3CH_2CH_2-CH=CH_2$	11	0.23
$(CH_3)_2CHCH_2-CH=CH_2$	3	0.23
$(CH_3)_3CCH_2-CH=CH_2$	0.25	<0.001

in these reactions, the kinetic product ratio of *cis* and *trans* isomers is different from the thermodynamic ratio. A striking example [76] of this phenomenon comes from a study of the isomerization of 1-alkenes to 2-alkenes in dimethyl sulfoxide–potassium *tert*-butoxide at 55° (Table 14). Thus, the less stable *cis*-2-alkenes are produced initially in far greater abundance than would be expected upon the basis of thermodynamic control. This observation has, of course, led to speculation on the source of this unexpected behaviour and both ground state conformational preferences and factors affecting carbanion stability have been considered.

The preferential formation of *cis*-2-butene from 1-butene has been rationalized in terms of the anticipated conformational preference of 1-butene [77] (*41*). Isomeriza-

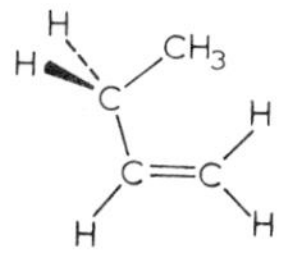

(*41*)

tion and isotopic exchange results for 2,4,4-trimethyl-2-pentene (*42*) in potassium *tert*-butoxide–*tert*-butyl alcohol at 215° bear on this suggestion [33]. Treatment of (*42*) with base can lead either to a *cis*-allylic anion (*43*) or a *trans*-allylic anion (*44*)

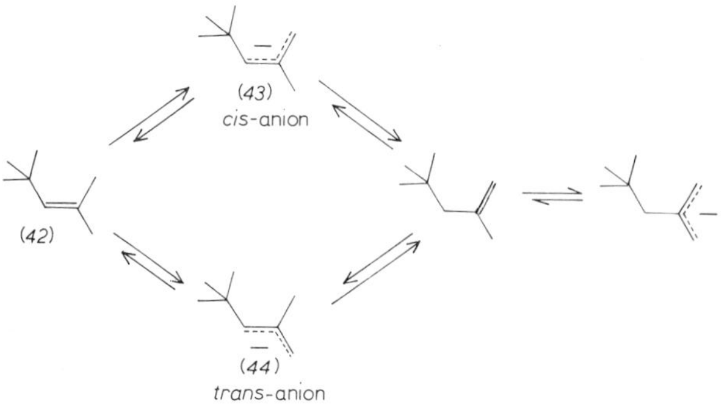

(*43*) *cis*-anion

(*42*)

(*44*) *trans*-anion

and the relative rates of formation of these two anions can be deduced from the initial rates of isotopic exchange in the methyl groups of (*42*). The methyl group *cis* to the *tert*-butyl group exchanged faster than the *trans*-methyl group by at least a factor of 10. In this rigid system where both conformations exist equally, the *cis*-anion is still kinetically preferred and this suggests an effect beyond ground state conformations.

It has also been proposed [78] that the product distribution of the 2-alkenes reflects the carbanion stability and that *cis*-substituted allylic anions (*45*) are more stable than *trans*-substituted allylic anions (*46*). Among the explanations offered to rationalize

cis-anion (45) trans-anion (46)

the stability of *cis*-anions are dipole moments [79], non-bonded attractive interactions [80a,b], steric effects on solvation of the anions [1c] and electronic stability as obtained by extended Hückel calculations [81]. It is probably safe to say that for these reactions no clear cut decision has been made as to which is the correct explanation.

The assumption that *cis*-alkyl-substituted allylic anions are more stable than *trans*-anions has only recently been corroborated for allyl-alkalies with n.m.r. studies. A low temperature n.m.r. spectrum of methylallyl lithium (*47*), prepared from 2-butene and *n*-butyl lithium in *N,N,N',N'*-tetramethylethylenediamine, showed a spectrum consistent with more than 80% of the *cis*-form [82]. Two conflicting n.m.r. studies have been reported of (*47*) in diethylether. Seyferth and Jula [83] report obtaining 60% *trans*-product and 40% *cis*-product when the lithium compound (*47*) was reacted with trimethylsilyl chloride but the n.m.r. signals for (*47*) are broad, poorly resolved and give no evidence of two isomers. On the other hand, the Russian group [84] found two isomers in similar amounts in the n.m.r. but resolution seems sufficiently poor to be able to deduce much structural detail. The latter samples yielded temperature dependent spectra.

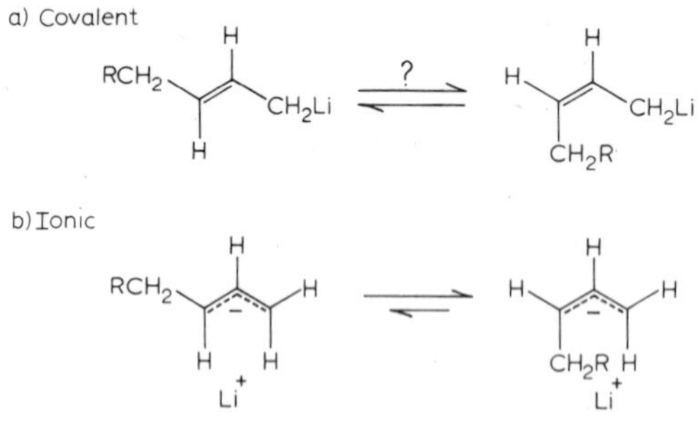

a) Covalent

b) Ionic

Fig. 15. Proposed structures for methylallyllithium [82–84] (*47*, R=H) and neopentylallyllithium (*48*, R=$(CH_3)_3C$–) [185a–d].

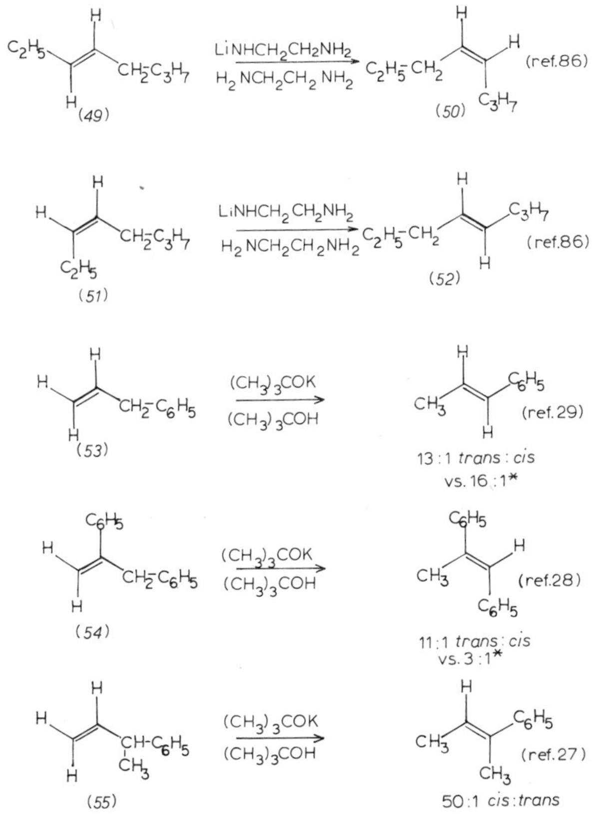

C_2H_5–CH=CH–CH_2–C_3H_7 (49) $\xrightarrow[H_2NCH_2CH_2NH_2]{LiNHCH_2CH_2NH_2}$ C_2H_5–CH_2–CH=CH–C_3H_7 (50) (ref.86)

H(H)C=C(H)–CH_2–C_3H_7 with C_2H_5 (51) $\xrightarrow[H_2NCH_2CH_2NH_2]{LiNHCH_2CH_2NH_2}$ C_2H_5–CH_2–CH=CH–C_3H_7 (52) (ref.86)

(53) H₂C=CH–CH_2–C_6H_5 $\xrightarrow[(CH_3)_3COH]{(CH_3)_3COK}$ CH_3–CH=CH–C_6H_5 (ref. 29)

13 : 1 *trans* : *cis*
vs. 16 : 1*

(54) $H_2C=C(C_6H_5)$–CH_2–C_6H_5 $\xrightarrow[(CH_3)_3COH]{(CH_3)_3COK}$ CH_3–C(C_6H_5)=CH–C_6H_5 (ref.28)

11 : 1 *trans* : *cis*
vs. 3 : 1*

(55) $H_2C=CH$–CH(CH_3)–C_6H_5 $\xrightarrow[(CH_3)_3COH]{(CH_3)_3COK}$ CH_3–CH=C(CH_3)–C_6H_5 (ref.27)

50 : 1 *cis* : *trans*

Fig. 16. Kinetic products of allylic isomerization. (*Thermodynamic ratio.)

Neopentylallyl lithium (*48*) has been studied [85a–d] in cyclohexane, toluene and tetrahydrofuran. While two geometries for (*48*) were observed in the hydrocarbon solvents and quenching yielded a 3 to 1 ratio of the *trans*/*cis*-4,4-dimethyl-2-pentenes, the proton n.m.r. spectra were interpreted in terms of the covalent structures of Fig. 15 [R = $-C(CH_3)_3$] or their equivalent since the terminal vinylic protons appeared as singlets. In tetrahydrofuran at $-4°$, two ionic geometries (Fig. 15b) were observed with a *cis* to *trans* ratio of 1.4.

Some other examples of kinetic control in hydrocarbon isomerizations are shown in Fig. 16. An interesting change in selectivity is exhibited by *trans* and *cis*-3-octene [86] where *trans*-3-octene (*49*) yields essentially only *cis*-2-octene (*50*) but *cis*-3-octene (*51*) favours *trans*-4-octene (*52*). With phenyl-substituted alkenes the preferred kinetic product seems also to be the thermodynamic product as illustrated for allylbenzene [29] (*53*), α-benzyl-styrene [28] (*54*) and 3-phenyl-1-butene [26,27] (*55*). A variety of heterosubstituted alkenes have also been studied and are recently reviewed [25].

TABLE 15

PROPOSED GEOMETRIES AND PROTON COUPLING CONSTANTS OF PENTADIENYL, HEPTATRIENYL, AND NONATETRAENYL CARBANIONS

	Cation/Solvent	J_{gem}[a]	J_{vic}[a]					
	T(°C)	1,2	1,3	2,3	3,4	4,5	5,6	6,
(56)	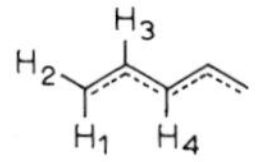 Li/THF–hexane[87]							
	15		16	9				
	40[b]		13	13	11			
	K/NH_3[88]	3.4	15.1	9.0	11.5			
(57)	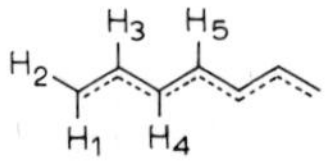 Li/THF–hexane [66]							
	–50	2	16	8	12	12		
	K/NH_3 –60° [89]	3.1	15.8	9.5	11.2	12.4		
(58)	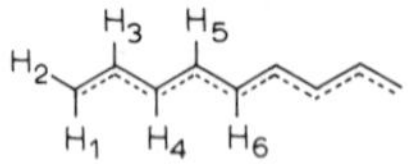 Li/THF-hexane [90]	2	17	8	12	12	12	
(59)	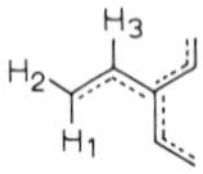 Li/THF–hexane [91]	2	16	10				
(60)	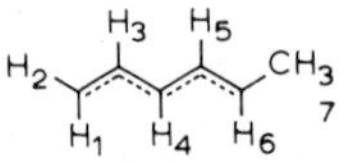 K/NH_3 –60 [92]	3.5	15.2	9.0	11.9	11.2	13.7	6.
↓ c								
(61)	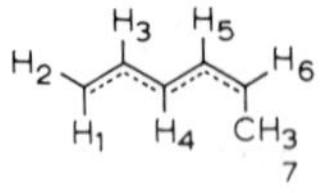 K/NH_3 –20 [80b]	3.5	15.1	8.8	11.5	11.5	8.7	6.
	Li/THF–hexane [93] –30				12.0	11.5	9.5	
(63)	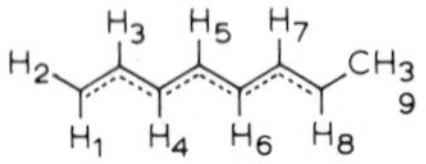 K/NH_3 –60 [94]	3.2	15.7	9.3	11.3	12.2	12.6	1
				7,8 = 14.0 8,9 = 6.5				
⇅ d								
(64)	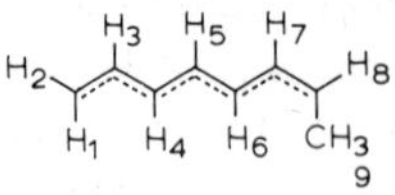 K/NH_3 0 [94]	3.2	15.7	9.4	11.2	12.4	12.5	1
				7,8 = 9.8 8,9 = 6.5				
(62)	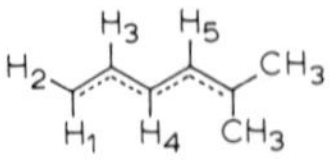 Li/THF–hexane [93] –37				12.5	11.0		

[a] Coupling constants in Hertz.
[b] Above coalescence temperature and therefore average coupling constants are observed.
[c] No (60) observed at equilibrium in NH_3 but 38% (60) in THF–hexane at +80°.
[d] K_{eq} = 4 at 0°.

TABLE 16

PROPOSED GEOMETRIES AND PROTON COUPLING CONSTANTS FOR PHENYL ALLYLIC CARBANIONS

	Cation/Solvent T(°C)	J_{gem}[a] 1,2	J_{vic}[a] 1,3	2,3	3,4	
(65)	Li/THF 5 [95]	3.2	15.4	9.4	12.2	
	K/NH_3 –40 [96]	3.5	15.4	9.3	12.2	
(66)	K/NH_3 –20 [96]	3.6	16.4	10.4		
(67)	Li/THF 38 [97a,97b]		13		13	
	K/NH_3 –20 [96]		12.9		12.9	
(68)	K/NH_3 –20 [96]	2.7				1,4 = 1.6
(69)	K/NH_3 [98]	3.5	15.2	9.4		
(70)	K/NH_3 –20 [96]	3.3				2,4 = 1.2
(71)	K/NH_3 –20 [80b,99]	6.5	13.8	1.2	11.4	
⇅ b						
(72)	K/NH_3 –20 [80b,99]	6.6	1.2	9.8	12.5	
(73)	K/NH_3 [98]		13.5			
(74)	Li/THF [97a,97b] [b]		c			

TABLE 16 (continued)

		Cation/Solvent T(°C)	J_{gem}[a] 1,2	J_{vic}[a] 1,3	2,3	3,4
(75)	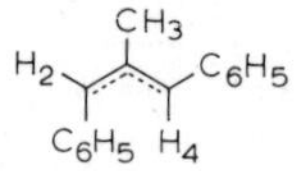	Li/THF [97a,97b][b]		*d*		
(76)	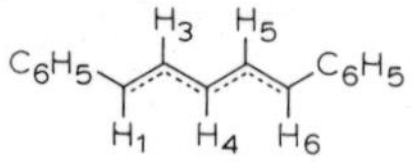	Li/Et_2O 38 [100]		14.4		11.2

[a] Coupling constants in Hertz.
[b] K_{eq} = 5 at −20°.
[c] Geometry based upon chemical shifts.
[d] Geometry based upon nonequivalence of H_2 and H_4.

TABLE 17

PROPOSED GEOMETRIES AND PROTON COUPLING CONSTANTS FOR OXYGEN-SUBSTITUTED CARBANIONS

		Cation/Solvent T(°C)	J_{gem}[a] 1,2	J_{vic}[a] 1,3	2,3	3,4	4,5	5,6	6,7
(77)	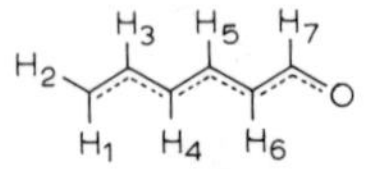	K/NH_3 −60 [101]	2.4	16.7	9.9	10.6	14.3	11.3	10.3
	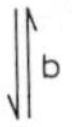								
(78)	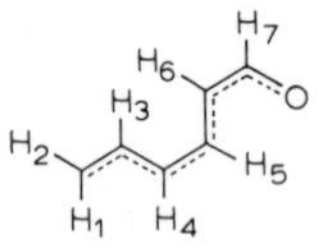	K/NH_3 −60 [102]	2.5	16.9	9.8	11.2	10.2	12.1	4.4
(79)	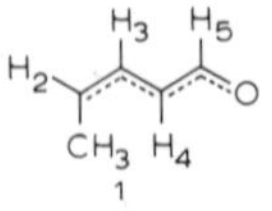	K/NH_3 −20 [96]	6.7	1.5	10.4	11.2	10.5		
(80)	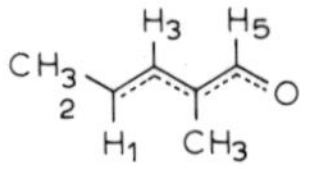	K/NH_3 [98]	6.4	14.8					
(81)	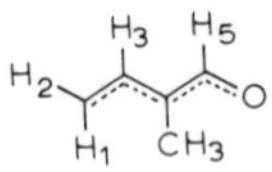	K/NH_3 [98]	3.0	16.5	10.3				
(82)	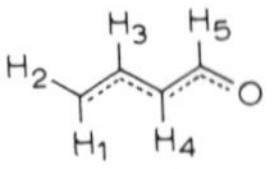	K/NH_3 [88]	3.2	16.5	10.7	10.6	10.5		

TABLE 17 (continued)

	Cation/Solvent T(°C)	J_{gem}[a] 1,2	J_{vic}[a] 1,3	 2,3	 3,4	 4,5	 5,6	 6,7
(*83*) H_3, H_2, O, C_6H_5	K/NH_3 –20 [96] K/DMSO [96]			5.0				
⇅ c								
(*84*) H_3, C_6H_5, O, H_1	K/NH_3 –20 [96] K/DMSO [96]		11.2					
(*85*) CH_3, H_2, O, C_6H_5	K/NH_3 [96]							
(*86*) H_3, C_6H_5, OCH_3, H_1, H_4	K/NH_3 –60 [103]		11.6		11.2			
↓ d								
(*87*) H_3, C_6H_5, H_4, H_1, OCH_3	K/NH_3 –60[103]		12.1		5.5			

[a] Coupling constants in Hertz.
[b] No interconversion observed between (*77*) and (*78*).
[c] N.m.r. spectrum shows a 3:2 mixture of (*84*) to (*83*) but this need not be the equilibrium mixture.
[d] No (*86*) observed at equilibrium.

Phenyl-substituted allyl lithiums and pentadienyl lithiums have proven to be more amenable to n.m.r. study than allyl and alkyl lithiums since they destroy ether solvents with less enthusiasm. A number of studies have been directed at determining the geometrical preferences of open chain carbanions using the magnitude of vicinal proton–proton coupling constants as a criterion for geometry. In Tables 15, 16 and 17 the data on coupling constants and the proposed geometry for a variety of delocalized open chain carbanions are collected. The cation, solvent and temperature at which the n.m.r. spectra were obtained is indicated along with the calculated geminal and vicinal proton coupling constants. An attempt has been made to be systematic in the labelling of protons so that comparisons can be made down a column. The vicinal coupling constant between protons assigned as *cis* fall in the range 8.0–10.4 Hz and those assigned as *trans* fall in the range 11–17 Hz. Chemical shift arguments about geometry have also been applied in a number of cases. Carbon-13 n.m.r. has recently been applied to the question of the geometry of some of the anions [64,93] and in particular supports the assigned geometry for (*56*).

The geometrical assignments by proton n.m.r. have been checked by quenching experiments in only three cases ((*56*), (*67*) and (*76*)) but in the case of (*76*) the two methods were in disagreement, with n.m.r. indicating all-*trans* but quenching providing a 1,5-diphenyl-1,3-pentadiene with a *cis* double bond. In the other two cases ((*56*) and (*67*)), the products of quench were consistent with the n.m.r. assignment but both revealed minor amounts of a second geometry not seen by proton n.m.r. Carbon-13 n.m.r. has also shown [93] the presence of a second geometry for (*56*) in minor amounts and this also is the case for (*60*), (*61*) and (*62*).

Some pairs of substituted anions are related as geometrical isomers ((*60*) and (*61*), (*63*) and (*64*), (*71*) and (*72*), (*77*) and (*78*), (*83*) and (*84*), (*86*) and (*87*)) and the equilibrium composition was determined for some of these pairs. The position of the equilibrium is indicated by arrows on the table and is detailed in the footnotes to the tables. For most of the other anions, it is possible that the carbanion has adopted the most stable geometry available since the barriers to interconversion are often sufficiently low although the enolates may be an exception to this generalization (vide infra).

It is interesting to note that the pentadienyl (*56*), heptatrienyl (*57*) and nonatetraenyl (*58*) anions seem to adopt the extended planar shapes in spite of theoretical predictions [81] although the quenching and carbon-13 n.m.r. results suggest the preference is not large. The trivinylmethyl anion (*59*) seems to suggest a further preference for the "sickle" shape over the "U". The pentadienyl and heptatrienyl anions with a terminal methyl group ((*60*) and (*61*), (*63*) and (*64*)) seem to prefer a shape in which the methyl is in the *pseudo*-axial position. If this phenomenon can be extrapolated to the allyl anion, it would support the assumption based upon kinetic results that *cis*-substituted allyl anions (*45*) are more stable than *trans* (*46*). Further correlations between preferred anion geometries and kinetic results are apparent from a comparison of the data in Fig. 16 and anions (*65*), (*68*) and (*69*).

The phenyl- and methyl-substituted allylic anions of Table 16 reveal stereochemical preferences. While a phenyl ring apparently prefers the *pseudo*-equatorial position ((*65*) and (*67*)) a methyl group seems to show a preference for the *pseudo*-axial position ((*71*) and (*72*), K_{eq} = 5) which can be offset by a bottom 1,3 interaction [54] with a second methyl group (*73*). Also the preference of a phenyl group for the *pseudo*-equatorial position can be offset by a top 1,2 interaction with a phenyl (*68*) or a methyl ((*70*) and (*75*)) group. The preference of a phenyl for the *pseudo*-equatorial position and a methyl for the *pseudo*-axial position is further illustrated in anions (*69*), (*73*) and (*74*) although the evidence for the positioning of the methyl and phenyl groups is not strong. There certainly seems to be justification for quenching experiments on a number of these anions.

The oxygen-substituted carbanions of Table 17 show characteristics similar to the hydrocarbon anions. In anions (*77*), (*78*), (*83*) and (*84*) the preferred geometry has not been demonstrated experimentally. In the methoxy-substituted-phenylallyl anions (*86*) and (*87*), the methoxy group apparently prefers the *pseudo*-axial position, behaving in this manner much like a methyl group ((*71*) and (*72*)). Unfortunately, it is not certain whether bonding to a cation may be involved here (*88*) as has been pro-

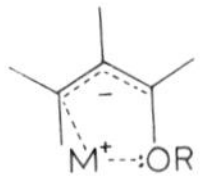

M^{+}···:OR

(*88*)

posed to rationalize kinetic results with allyl ethers [24,25,103,104a–d] although it is likely that the carbanions exist as ion pairs in both studies. If the anions (*86*) and (*87*) are not in such a special environment then the preference for a *cis*-geometry appears inconsistent with an interpretation in terms of dipole moments [79] and has a bearing upon the other models put forward to rationalize the kinetic observations with 1-alkenes (vide supra).

B. Barriers to Rotation in Delocalized Carbanions

One of the most important facts to come from the study of the geometry of phenylallylic and polyenylic carbanions is the preference for planarity consistent with the existence of barriers to rotation around partial double bonds. Attempts to estimate the

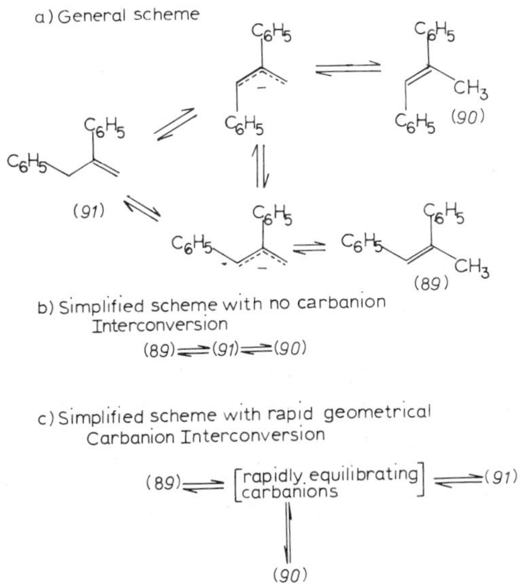

a) General scheme

C_6H_5 C_6H_5 CH_3 (*90*) C_6H_5 (*91*) (*89*)

b) Simplified scheme with no carbanion Interconversion

(*89*) ⇌ (*91*) ⇌ (*90*)

c) Simplified scheme with rapid geometrical Carbanion Interconversion

(*89*) ⇌ [rapidly equilibrating carbanions] ⇌ (*91*)

(*90*)

Fig. 17. Kinetic models for the reaction [28] of α-benzylstyrene (*91*) and *cis*- and *trans*-α-methylstilbenes ((*89*) and (*90*)). (a) General scheme; (b) simplified scheme with no carbanion interconversion; (c) simplified scheme with rapid geometrical carbanion interconversion.

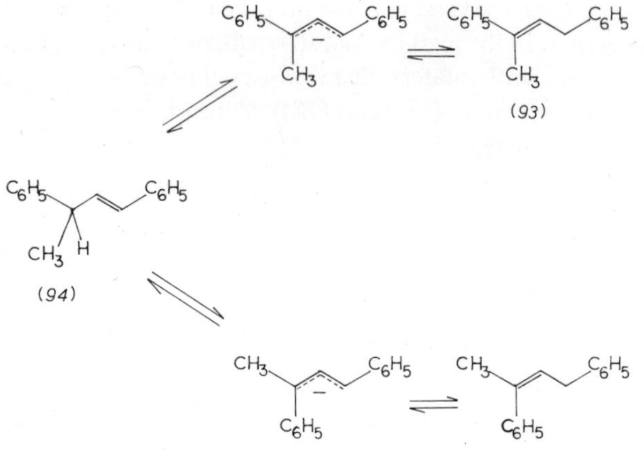

Fig. 18. Mechanism for the interconversion [54] of *cis*- and *trans*-1,3-diphenyl-2-butene ((*92*) and (*93*)) and *trans*-1,3-diphenyl-1-butene (*94*).

magnitude of the barriers to rotation have been made on carbanions both as short-lived species formed during base-catalyzed 1,3 proton transfer reactions in protic media, and as long-lived organo-alkalies in aprotic media. There are two studies of 1,3 proton transfers catalyzed by potassium *tert*-butoxide in *tert*-butyl alcohol which provide evidence about the geometrical stability of phenyl-substituted allylic anions.

One of these was a study of the kinetics of interconversion of *cis* and *trans*-α-methylstilbene ((*89*) and (*90*)) and α-benzylstyrene (*91*) (Fig. 17) [28]. It was shown that the results could be accommodated by the kinetic scheme (b) of Fig. 17 but not scheme (c). Thus geometric interconversion of these allylic anions could not compete with the rate at which the anions were reprotonated by solvent. Similar results were obtained using the 1,3-diphenylbutenes [54] (Fig. 18) where the kinetic results were best accommodated by non-interconverting carbanions. These results served to show that geometric interconversions were not rapid in *tert*-butyl alcohol. In fact, the rates of geometric interconversion in aprotic media are sufficiently slow to be appropriate for study by variable temperature n.m.r. techniques.

The temperature dependence of the proton n.m.r. spectrum of 1.5 F allyllithium (*95*) [105] in tetrahydrofuran serves to illustrate the approach (Fig. 19). At room temperature and above the spectrum appeared as an AX_4 pattern; a doublet centred at 2.28 for the end protons of the allylic system and a quintet centred at 6.48 for the central proton. When the sample was cooled, broadening, followed by coalescence near $-50°$, followed by sharpening again at $-87°$ produced a spectrum best described as an $AA'BB'C$ pattern. At the low temperatures the terminal protons are no longer equivalent on the n.m.r. time scale but appears as two types. These observations are interpretable in terms of a process in which the terminal protons are changing environment at a temperature-dependent rate. These spectra may then be used to obtain kinetic

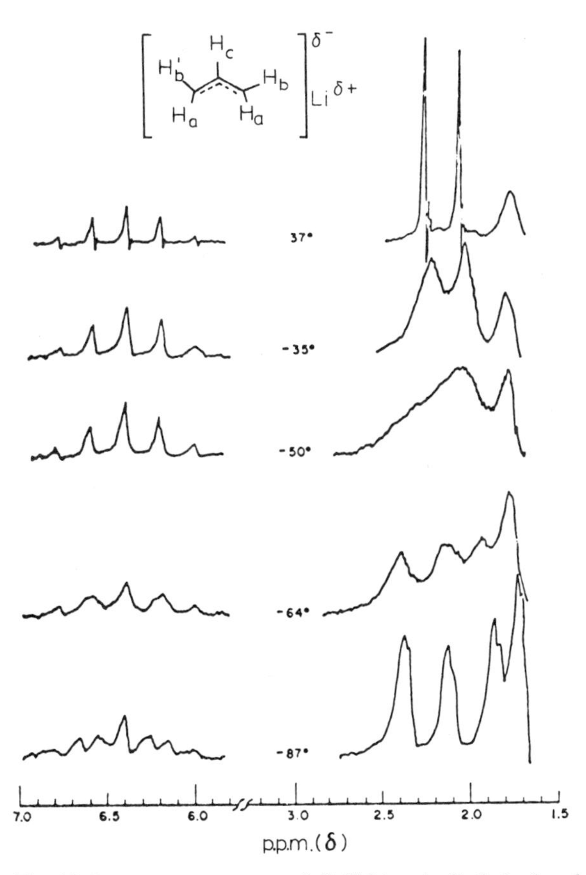

Fig. 19. Proton n.m.r. spectra of allyllithium in diethyl ether-d_{10} at various temperatures [105]. (Reprinted by permission of the American Chemical Society.)

parameters for the system (ΔG^*, ΔH^*, ΔS^*, E_a) and for this particular case the E_a = 10.5±2 kcal mol^{-1}.

A similar approach has been used for a variety of delocalized carbanions and these are gathered in Figs. 20 and 21. Figure 20 contains data for geometrical isomerizations that do not produce a change in structure (isothermic) and Fig. 21 contains examples of exothermic isomerizations. Included in the figures are the solvent and cation used and where available the Gibb's free energy of activation (ΔG^*) for the interconversion process and the temperature for which it was calculated. In the cases of anions (*74*) and (*75*) the authors did not obtain activation parameters but reported only coalescence temperatures. These n.m.r. results are interpreted as a measure of the rate of rotation around the bond indicated by the arrow. The activation-free energies range from 10–21 kcal mol^{-1} or a change in rate of about 8 powers of 10 showing a large

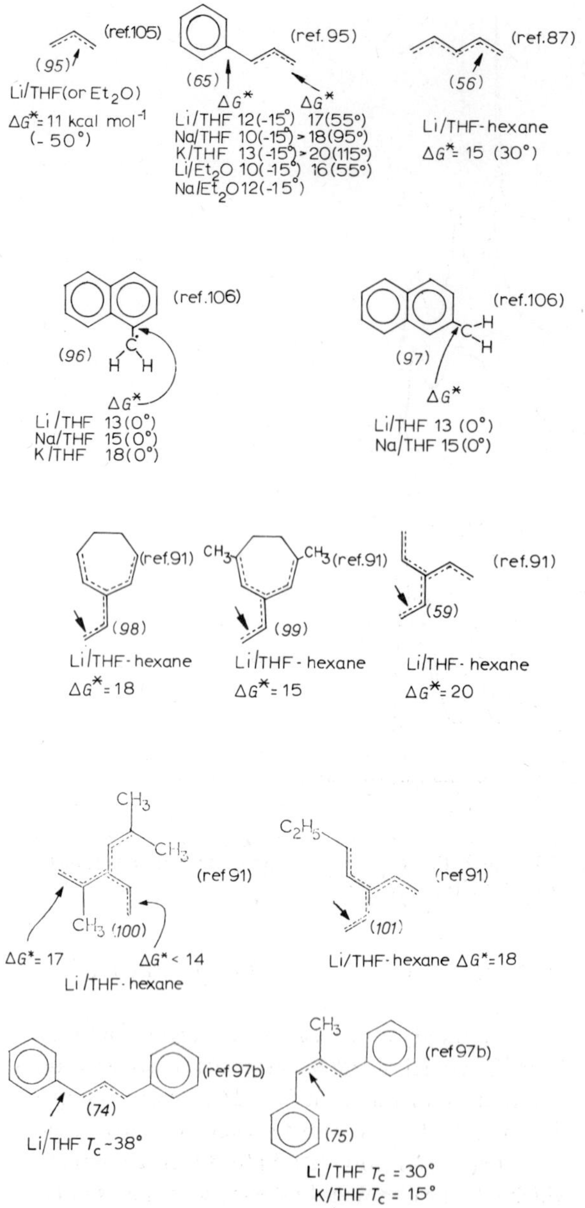

(ref.105)
(95)
Li/THF(or Et_2O)
ΔG^*= 11 kcal mol^{-1}
(- 50°)

(ref. 95)
(65)
ΔG^* ΔG^*
Li/THF 12(-15°) 17(55°)
Na/THF 10(-15°) > 18(95°)
K/THF 13(-15°) > 20(115°)
Li/Et_2O 10(-15°) 16(55°)
Na/Et_2O 12(-15°)

(ref.87)
(56)
Li/THF-hexane
ΔG^*= 15 (30°)

(ref.106)
(96)
C
H H
ΔG^*
Li/THF 13(0°)
Na/THF 15(0°)
K/THF 18(0°)

(ref.106)
C-H
H
(97)
ΔG^*
Li/THF 13 (0°)
Na/THF 15(0°)

(ref.91)
(98)
Li/THF-hexane
ΔG^*= 18

CH_3 CH_3 (ref.91)
(99)
Li/THF - hexane
ΔG^*= 15

(ref.91)
(59)
Li/THF- hexane
ΔG^*= 20

CH_3
CH_3
(ref 91)
CH_3 (100)
ΔG^*= 17 ΔG^* < 14
Li /THF-hexane

C_2H_5
(ref 91)
(101)
Li/THF-hexane ΔG^*=18

(ref 97b)
(74)
Li/THF T_c ~38°

CH_3
(ref 97b)
(75)
Li /THF T_c = 30°
K/THF T_c = 15°

Fig. 20. Free energies of activation (kcal mol^{-1}) or coalescence temperatures for isothermic geometrical isomerization of delocalized organo-alkalies.

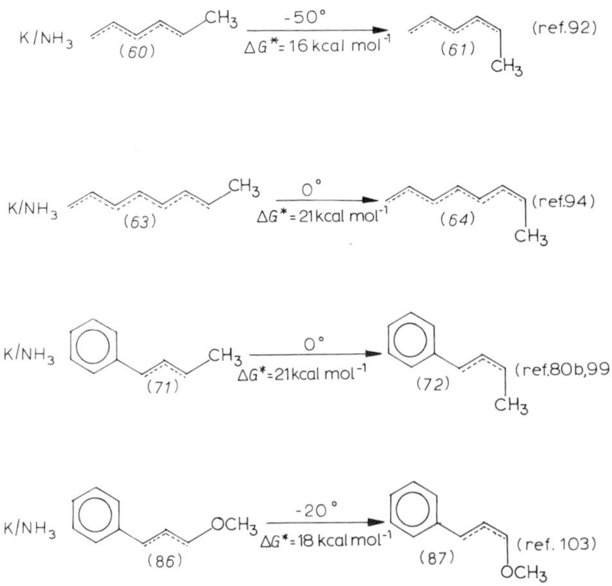

K/NH$_3$ (60) CH_3 $\xrightarrow[\Delta G^{*}=16\ \text{kcal mol}^{-1}]{-50^{\circ}}$ (61) CH_3 (ref.92)

K/NH$_3$ (63) CH_3 $\xrightarrow[\Delta G^{*}=21\ \text{kcal mol}^{-1}]{0^{\circ}}$ (64) CH_3 (ref.94)

K/NH$_3$ (71) CH_3 $\xrightarrow[\Delta G^{*}=21\ \text{kcal mol}^{-1}]{0^{\circ}}$ (72) CH_3 (ref.80b,99)

K/NH$_3$ (86) OCH_3 $\xrightarrow[\Delta G^{*}=18\ \text{kcal mol}^{-1}]{-20^{\circ}}$ (87) OCH_3 (ref. 103)

Fig. 21. Free energies of activation for exothermic geometrical isomerization of delocalized organopotassiums.

sensitivity to structure and medium. Even the same rotation in the same molecule (*96*) can change rate by 3 to 4 powers of 10 on changing just the cation.

There are a number of examples showing barriers to rotation around the terminal bond in both allylic ((*95*), (*65*), (*75*), (*71*) and (*86*)) and polyenylic ((*56*), (*98*), (*99*), (*59*), (*100*), (*101*), (*60*) and (*63*)) anions and the absolute rate constants at coalescence vary from about 10^2 sec^{-1} (at -50° for (*95*)) to 10^{-4} sec^{-1} (at 0° for (*63*)). Interestingly, barriers to rotation are also observed for bonds connecting conjugated aryl rings to allylic anions ((*65*), (*96*), (*97*) and (*74*)) and these barriers are of the same magnitude as those in the conjugated system itself. There is only one example of an estimate of a barrier to rotation around an inside carbon—carbon bond but the spectra could not be completely analyzed [87]. Carbon-13 n.m.r. spectra [93] of the lithium salts of anions (*56*), (*60*), (*61*) and (*62*) in THF—hexane also show temperature dependent spectra consistent with barriers to rotation around inner bonds.

Interpretation of the barriers to rotation requires a knowledge of the mechanism of the process and there are basically two types of mechanisms that have been proposed. In one of these the rotation occurs in the delocalized carbanion itself and thus the barriers would be a measure of partial double bond character in the carbanion. This picture could be sensitive to the cation if the reactive species were an ion pair or π-bonded species. In the other mechanism the carbanion exists primarily as a delocalized or π-bonded species but reacts as a σ-bonded species. The σ-bonding serves to localize the negative charge and reduce the barriers to rotation. Presumably if this mechanism pertains then the barriers need not reflect partial double bond character in the delocalized anion but rather will be related to the preferred position of σ-bonding.

At the present time, it is not certain whether either or both of these mechanistic models is valid but some features of the reactions can be identified. The observation of barriers to rotation and n.m.r. spectra of these carbanions show that a σ-bonded organo-alkali is not the major species in solution*. Several authors have thus concluded that the carbanions exist as ionic species in tetrahydrofuran [87,95,97b,105,106] and in liquid ammonia [80b]. Burley and Young [97b] have obtained spectroscopic evidence that the lithium salts of (*74*) and (*75*) exist as solvent-separated ion pairs and the potassium salt is a contact ion pair. Kronzer and Sandel [106] decided that the lithium, sodium and potassium salts of (*96*) and (*97*) are monomeric and exist as contact ion pairs in tetrahydrofuran. Further the polarization of the carbanion as revealed by the proton n.m.r. suggest that the lithium contact ion pair may be best described in terms of interaction between the methylene carbon and the metal and π-allyl bonding may be involved. In contrast, allyllithium (*95*) in ether exists as an aggregate of n_{app} > 10 but still shows the properties of an ionic or π-bonded molecule. On the basis of this sort of evidence, it seems reasonable to expect that most of the organo-lithiums in Fig. 20 will exist as ion pairs.

There is no direct evidence available on the character of the ions in liquid ammonia (Fig. 21) but studies of more common electrolytes in liquid ammonia have shown that ion pairing is common at low dilution and aggregation occurs in more concentrated solution [107]. Heiszwolf et al. [80b] found that addition of excess potassium ion (130%) did not change the rate of isomerization of (*60*) to (*61*) and deduced that the isomerization must therefore be occurring in the ionic species rather than through σ-bonded species. But as Kronzer and Sandel [106] have pointed out, an effect of added cation would be expected only if the major species in solution is dissociated ions and not ion pairs.

The exact mechanism of rotation remains in doubt and hence so does the interpretation of the barriers to rotation.

IV. ELECTROCYCLIC REACTIONS

Cyclization reactions of carbanions have been of synthetic and mechanistic importance for a long time (e.g. the Dieckmann and Thorpe reactions). It is only more recently that examples of the cyclization of delocalized carbanions have appeared and many of these fall into the class of reactions now known as electrocyclic reactions. An electrocyclic reaction can be defined as the formation of a single bond between the

---

* In contrast, the spectra reported for neopentylallyllithium in cyclohexane and toluene have been interpreted in terms of σ-bonded species [85a–d], (see Fig. 15 and the accompanying discussion) as have the spectra for 2-butenyl-1-lithium [83,84] in ether solvents.

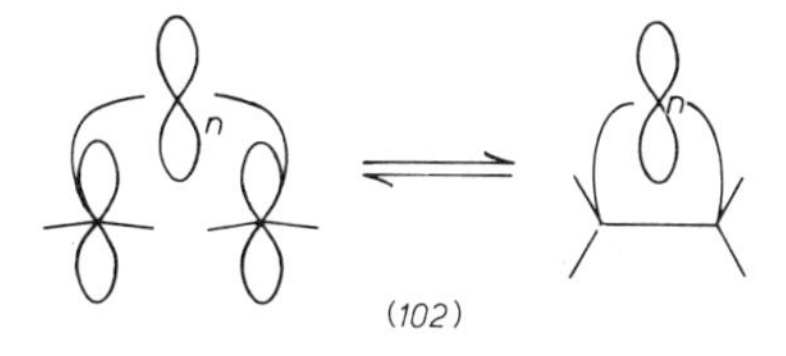

termini of a linear conjugated π system and also the converse process (*102*). Examples are available of electrocyclic reactions involving three-, five-, seven- and nine-membered rings and they provide information on both reactivity and stereoselectivity.

Orbital symmetry has been invoked [75] to make predictions about the preferred stereochemistry of concerted electrocyclic processes and the stereochemical motions have been classified as conrotatory and disrotatory. As a reminder of the stereochemical pathways for thermal cyclization of polyenylic anions, the symmetry-allowed motions for reactions involving three-, five-, seven- and nine-membered rings are summarized in Fig. 22. Stereochemical results are not available for all these systems and further discussion of the stereochemistry of electrocyclic reactions will await the specific examples.

A. Allyl–Cyclopropyl Interconversion

All of the examples of allyl–cyclopropyl rearrangements involve ring-opening of a cyclopropyl anion rather than ring-closure. In all cases studied, the reaction should be highly exothermic since besides relieving ring strain the product anions are more stable. In fact, the products of some of the reactions are better described as aromatic

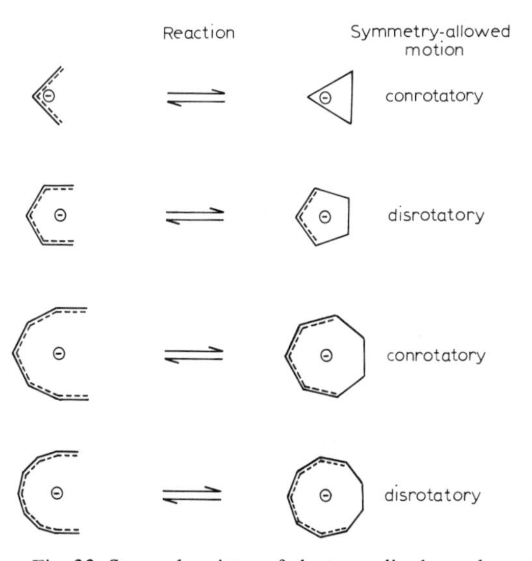

Fig. 22. Stereochemistry of electrocyclic thermal reactions of carbanions [75].

anions than allylic anions but are included here since the same mechanistic generalities apply.

There are a few examples of an electrocyclic reaction which fit both the cyclopropyl and allyl criterion directly. Reaction of dimethyl 1-methyl-3,3-diphenylcyclopropane-1,2-*trans*-dicarboxylate [108] (*103*) with sodium hydride in DMF at 20° produced a red-coloured solution which on acidification (hydrochloric acid/methanol)

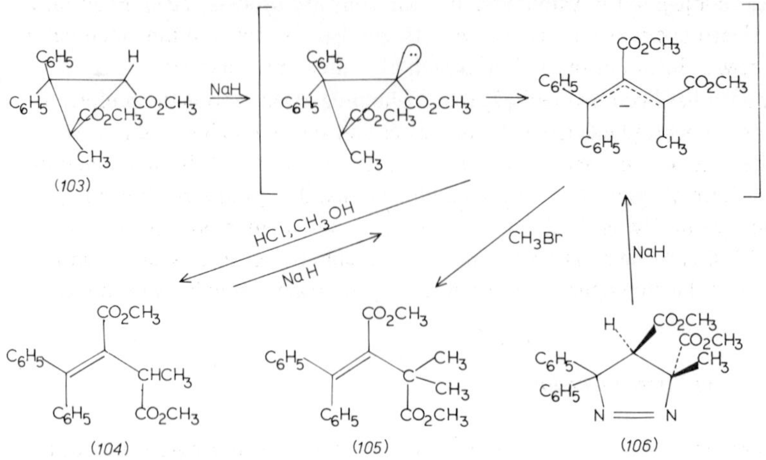

(*103*) (*104*) (*105*) (*106*)

yielded the diester (*104*) in 70% yield and (*105*) could be obtained upon reaction of the red solution with methyl bromide. The red colour can be regenerated from (*104*) upon reaction with sodium hydride and interestingly also from the reaction of (*106*) with sodium hydride [109]. The results suggest that the initially formed cyclopropyl anion rapidly opens at 20° to the allylic carbanion of unknown geometry.

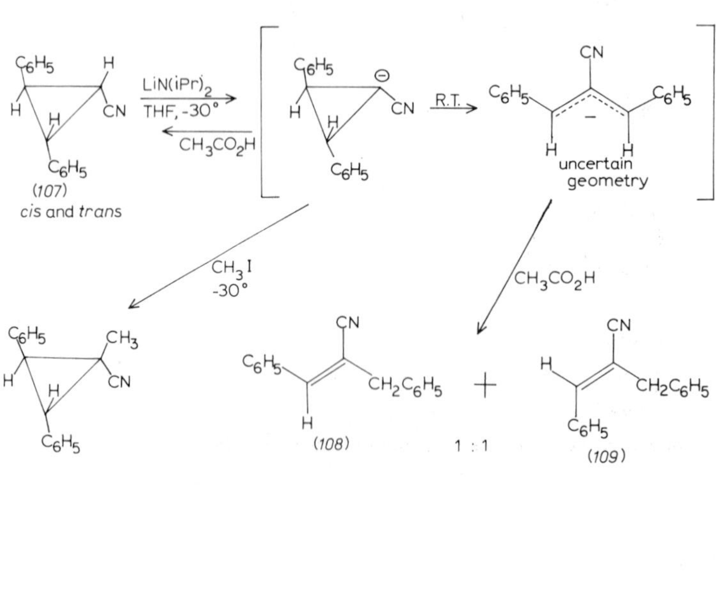

(*107*) cis and trans

uncertain geometry

(*108*) 1 : 1 (*109*)

Analogously treatment of *cis* or *trans*-2,3-diphenyl-1-cyanocyclopropane (*107*) with lithium diisopropylamide in THF at –30° or –78° produced the cyclopropyl anion as shown by quenching with acetic acid and methyl iodide [110,111] and some *cis–trans* interconversion was also observed. If the solution was allowed to warm to room temperature a red colour developed. Treatment of the red solution with acetic acid yielded a 1:1 mixture of the two olefins (*108*) and (*109*). Apparently, this cyclopropyl anion ring opens more slowly than the anion derived from (*103*). This is perhaps a reflection of the product stability upon the activation energy for the ring-opening.

C_6H_5 C_6H_5 H H H C_6H_5 (*110*) — *n*-Bu Li, TMEDA, Hexane, R.T. → [C_6H_5 C_6H_5 ⊖ H H C_6H_5 ↓ C_6H_5 C_6H_5 H H C_6H_5 uncertain geometry] — D_2O → C_6H_5 C_6H_5 CHD H C_6H_5 ; C_6H_5 H CHD C_6H_5 C_6H_5 (*111*)

A close relative to (*107*), 1,2,3-triphenylcyclopropane (*110*), was also observed to undergo facile ring-opening upon treatment with *n*-butyl lithium-*N,N,N′,N′*-tetra-methyl-ethylenediamine in hexane at room temperature [112]. Upon treatment of the purple solution so obtained with deuterium oxide the triphenylpropenes, 90% of (*111*) and 10% of its isomer, were obtained. Under these conditions the cyclopropyl anion was not observed and it appeared that ring-opening was complete to yield allylic anions of yet unknown geometry.

Ring-opening reactions have also been observed [113] for the dianions derived from the *cis* and *trans*-2,3-diphenylcyclopropane-1-carboxylic acids upon treatment with excess lithium di-isopropylamide in THF at 0°, although different bonds were cleaved in the *cis* and *trans* isomer. Thus reaction of (*112*) apparently produced the di-anion (*113*) as the transient intermediate on the way to the allyl di-anion which could also be produced from the olefinic acid (*115*). Quenching of the allyl anion of unknown geometry yielded (*114*) and (*115*) in temperature dependent relative amounts. At 0° the quenched mixture was about 40% (*114*) and 60% (*115*) while quenching at –78° yielded about 10% of (*114*) and 90% of (*115*).

C_6H_5 CO_2H (112) C_6H_5 $\xrightarrow[THF, 0°]{LiN(iPr)_2}$ [C_6H_5 CO_2^- (113) C_6H_5 → C_6H_5 CO_2^- H H C_6H_5 uncertain geometry] $\xrightarrow{CH_3OH, H^+}$ C_6H_5 CO_2H $CH_2C_6H_5$ H (114) + H CO_2H $CH_2C_6H_5$ C_6H_5 (115) ($\xrightarrow[THF]{LiN(iPr)_2}$ back to intermediate)

While the *trans*-compound (*112*) reacted to open the carbon–carbon bond between the phenyl-bearing carbons, the geometrical isomer (*116*) reacted to open the other bond as indicated by the structure of the product olefins (*117*) and (*118*). The change in the preferred direction of ring-opening may reflect a change with geometry in both

C_6H_5 C_6H_5 CO_2H (116) $\xrightarrow[THF]{LiN(iPr)_2}$ [C_6H_5 C_6H_5 CO_2^- → C_6H_5 C_6H_5 H H CO_2^- uncertain geometry] $\xrightarrow{CH_3OH, H^+}$ C_6H_5 C_6H_5 CH_2CO_2H H (117) + C_6H_5 H CH_2CO_2H C_6H_5 (118) ($\xrightarrow[THF]{LiN(iPr)_2}$ back to intermediate)

the rate of cyclopropyl anion formation and the relative rates of ring-opening, since complex kinetics were observed for the reaction of both substrates. Again the ratio of (*117*) and (*118*) obtained on quenching was temperature dependent, but both at 25° and –78° (*117*) was by far the major product (~ 90% yield).

These examples should be contrasted with the cyclopropyl anions [114–116b] (*119*), (*120*) and (*121*) for which no ring-opening has been observed under the mild

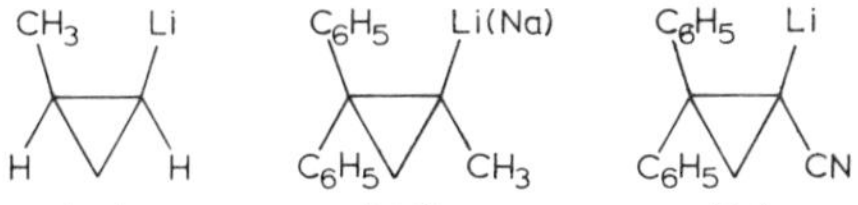

(*119*) (*120*) (*121*)

conditions used. On the basis of exothermicity of the ring-opening [117a] and the estimated pK_a's of the carbanions [1a–e], the ring-opening of (*119*), (*120*) and probably (*121*) would be exothermic, although theoretical calculations [117b] predict a high activation energy. Apparently a 1,3-distabilized allyl anion must be formed for facile ring-opening. The effect of the extra carbomethoxy or phenyl groups must be on the activation energy for ring-opening process rather than just on the thermodynamics.

An example of an aziridine anion–aza-allylic anion rearrangement, an analogue for the all-carbon system, has provided direct evidence on the stereochemistry of the ring-opening process. Kauffmann et al. [118] observed that solutions of *N*-lithio-2,3-*cis*-diphenylaziridine (*122*) in THF turned red when kept at 40–60° and the proposed

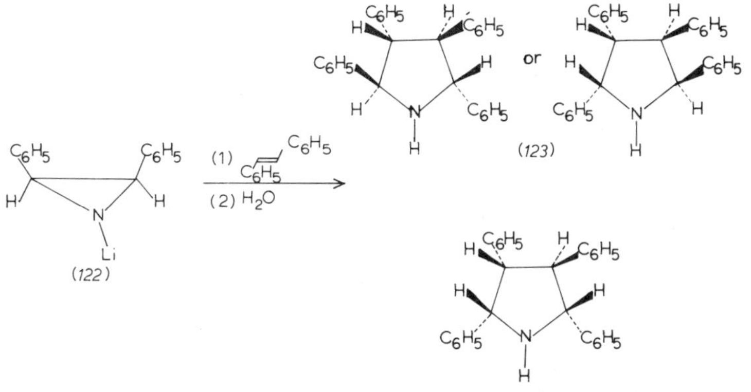

(*122*) (*123*) (*124*)

aza-allylic anion could be trapped with *trans*-stilbene to produce a 2,3,4,5-tetraphenylpyrrolidine. When the ring-opening reaction was run in the presence of *trans*-stilbene, a 73% yield of one isomer of (*123*) was obtained and 11% of (*124*) was produced. In contrast, when the *trans*-stilbene was added after ring-opening, only (*124*) was isolated. These results were interpreted as indicating that the initial ring-opened aza-allylic anion has the *cis–trans* geometry (*125*) and this anion can be trapped by *trans*-stilbene to produce (*123*). On standing (*125*) converts to the *trans–trans* geometry (or *cis–cis*) (*126*) which is trapped to yield (*124*). The initial production of the *cis–trans* aza-allylic anion (*125*) is consistent with orbital symmetry predictions of a conrotatory electrocyclic reaction.

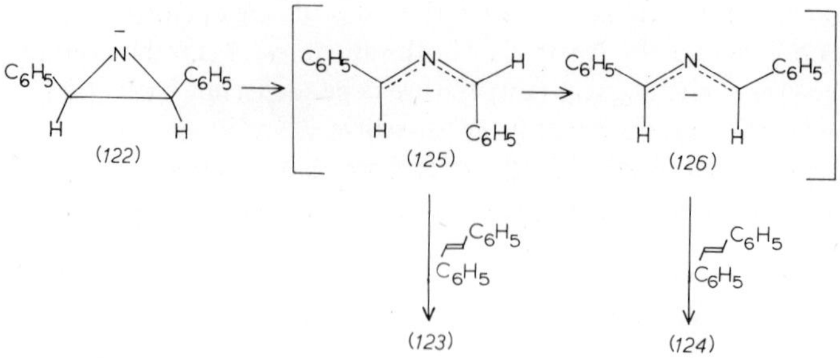

The ring-opening of the potassium salt from *anti*-9-methoxy-*cis*-bicyclo[6.1.0]nona-2,4,6-triene (*127*) is also consistent with a conrotatory ring-opening. Boche et al.

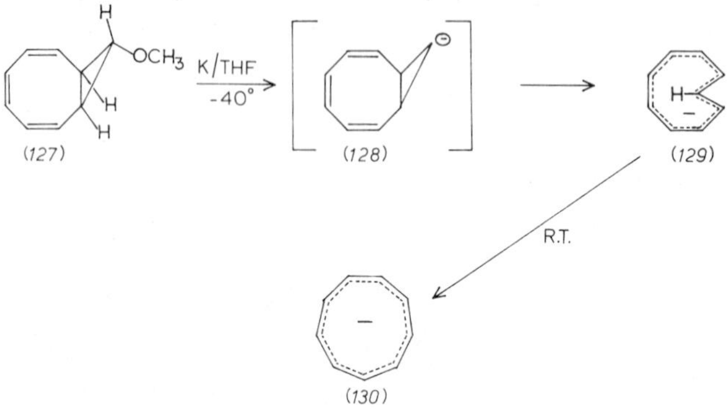

[119] followed the reaction of (*127*) in THF-d_8 at –40° by n.m.r. and found that the spectrum of the first observed product was consistent with potassium *trans, cis, cis, cis*-cyclononatetraenide (*129*). The proton pointing inside the aromatic ring was particularly unique appearing at δ = –3.5 p.p.m. Upon warming to room temperature (*129*) was replaced by a new species whose spectrum (singlet, δ = 7.13 p.p.m.) was consistent with the all *cis*-anion (*130*). The results seem consistent with initial formation of the cyclopropyl anion (*128*) which opens in a conrotatory manner to make the seemingly badly strained *trans, cis, cis, cis*-anion (*129*) as the initial product.

In contrast to these two cases there are also examples of electrocyclic ring-openings which apparently do not take the symmetry allowed pathway. Wittig et al. [120] reacted the *exo*-7-bromocycloprop(a)acenaphthylene (*131*) with *n*-butyl lithium in diethylether at 0° or THF at –70°. The *exo*-cyclopropyl lithium (*132*) was characterized by reaction with deuterium oxide, methyl iodide, iodine, carbon dioxide, benzophenone and mercuric chloride. Upon heating to room temperature in ether or THF the anion (*132*) seemed to decompose the solvent yielding cycloprop(a)acenaphthylene (*134*) (H for D). If hexane was used as solvent and the solution warmed to 100°, the first reaction was formation of the *endo*-cyclopropyl anion (*133*) characterized by obtaining a 59% yield of (*134*) on quenching with deuterium oxide. After heating at 100° for 24 h a 43% yield of (*135*) was obtained as well as 48% of the all protio counterpart of (*134*). These results indicate that a cyclopropyl anion can open in a distrotatory manner if the conrotatory pathway is not available.

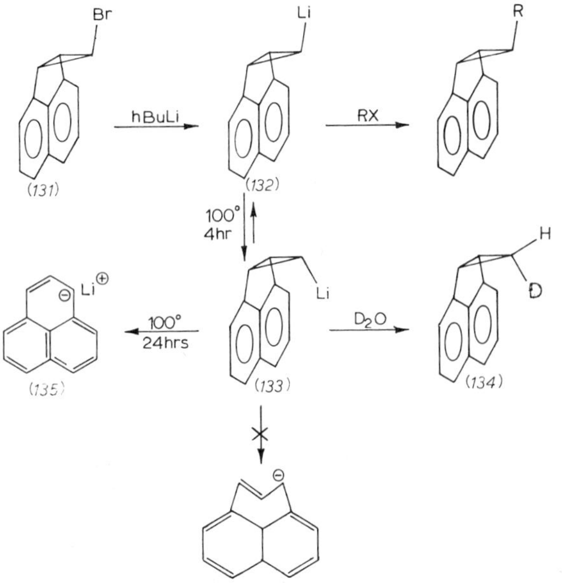

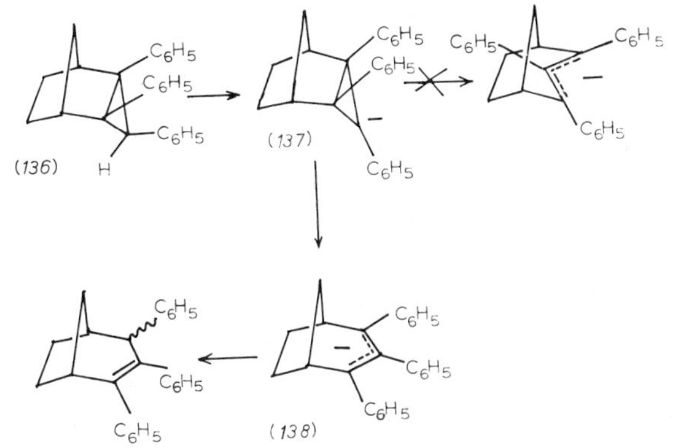

Londrigan and Mulvaney [121] looked at the ring-opening of a bicyclic analogue (*136*) and (*139*) of the triphenylcyclopropyl system (*110*) mentioned above. They found that both (*136*) and (*139*) produced ring-opened products upon treatment with potassium *tert*-butoxide in dimethylsulfoxide at 70° for 20 h. Again it seems reasonable to propose the cyclopropyl anions, (*137*) and (*140*), as initial intermediates which ring-open to yield allylic (*138*) and bishomocyclo-pentadienyl (*141*) anions, respectively. In both cases, the symmetry-allowed conrotatory motion would lead to a very high energy product and it is not clear that this would actually be an exothermic process. Instead the product of net disrotatory motion is formed producing a more stable anion.

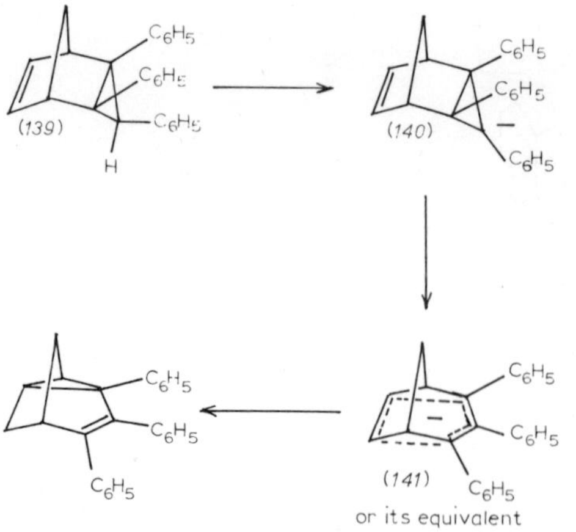

A ring-constrained analogue of the *cis* and *trans*-2,3-diphenyl-cyclopropane-1-carboxylic acids (*111*) and (*116*) has also been investigated [113]. Reaction of (*142*) under the conditions used for (*112*) and (*116*) lead to a mixture of both types of ring-opened products ((*143*) and (*144*)) which are analogues to the products from (*112*) and (*116*). There are two senses in which the reaction of (*142*) is unique: first, whereas (*112*) and (*116*) reacted with exclusive cleavage of different cyclopropyl bonds, (*142*) reacted with competitive cleavage of both types of bonds and secondly, the competitive cleavage of the two types of bonds is between a disrotatory and presumably forbidden process (to form (*144*)) and a process to which either stereochemistry is available (to form (*143*)). The complex kinetic orders observed for the reaction show that the rate of reaction does not reflect just one step. Thus, although the product distribution provides no support for the orbital symmetry prediction of preferred conrotatory ring-opening, neither does it provide evidence against the prediction.

A kinetically simpler case has been investigated [111] as a source of reaction rate evidence on orbital symmetry control. The rates of reaction of *cis*-2,3-diphenyl-1-cyanocyclopropane (*cis*-(*107*)), *trans*-2,3-diphenyl-1-cyanocyclopropane (*trans*-(*107*)) and 7a-,7b-dihydrocycloprop(a)acenaphthylene-*anti*-7-carbonitrile (*145*) with lithium diisopropylamide or *tert*-butylamide in tetrahydrofuran were measured. All three substrates were converted to the corresponding cyclopropyl anions at −78° as revealed by quenching with deuterium oxide. Upon warming from −30° to −10°, the anions from

(*107*) opened and the *pseudo*-first order rates were measured as a function of temperature. In contrast to (*107*), (*145*) required warming to 36° and the ring-opened product was obtained in about 25% yield. It was found that the anion from *trans*- (*107*) reacted 42 times faster than *cis*-(*107*) at –25° and *cis*-(*107*) extrapolated to 10^4 faster

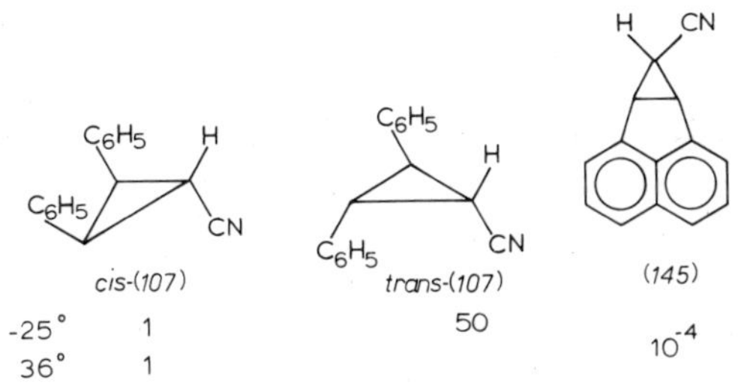

Fig. 23. Relative rates of ring-opening catalyzed by lithium diisopropylamide and lithium *tert*-butylamide in tetrahydrofuran [111].

than (*145*) at 36° (Fig. 23). Since (*145*) should produce a more stable product anion than (*107*), it is argued that the reduced reactivity of (*145*) reflects its inability to follow the allowed conrotatory pathway. This would then indicate that orbital symmetry control is energetically important. While these cases of disrotatory opening might be viewed as violating orbital symmetry rules [75], they should better be considered as illustrating that the orbital symmetry allowed pathway is not the only pathway available for many reactions. When the symmetry allowed reaction becomes energetically unfeasible (e.g. no longer exothermic), then other reactions can and still do occur.

B. Pentadienyl–Cyclopentenyl Interconversion

In a sharp contrast to the allyl–cyclopropyl anion rearrangement, the examples with the pentadienyl–cyclopentenyl system are of ring-closure reactions rather than ring-opening reactions. There are not many reported examples with the five-carbon system. The pentadienyl anion has been prepared [87,88,93] (see Table 15, Section 3) but its cyclization has not been observed, although thermochemical estimations suggest that the reaction should be spontaneous and bond rotation appears facile [87,93].

However, when the pentadienyl anion is constrained in an eight-membered ring, cyclization seems to occur with ease. Thus, when 1,3-octadiene (*146*) (or 1,4 and 1,5)

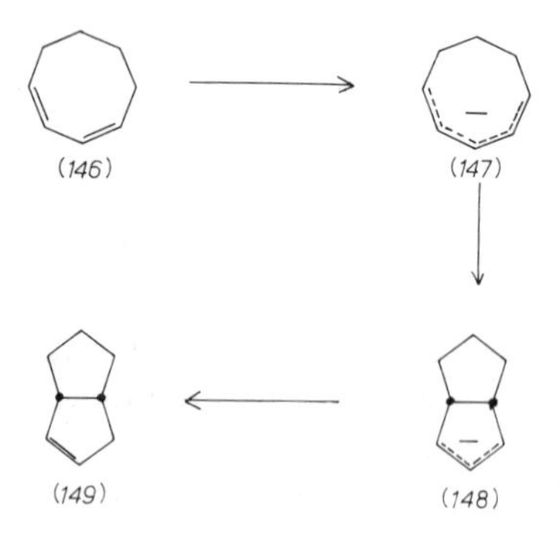

was treated with phenyl potassium [122], potassium hydride [123], *n*-butyl lithium [67], or potassium amide [124], it is eventually converted to *cis*-bicyclo[3.3.0]oct-2-ene (*149*), in good yield. The cyclooctadienyl anion (*147*) has been characterized by proton n.m.r. [67,124] and its cyclization followed (25° and 35°). The bicyclic anion (*148*) was also characterized by n.m.r. [67] but proved to be unstable and produced (*149*). The conversion of (*147*) to (*148*) has occurred in a net disrotatory manner consistent with orbital symmetry preferences but it should be noted that the alternative conrotatory motion would be very prohibitive in energy.

An analogous but more complicated rearrangement has been observed for the nine-membered ring [125]. Treatment of (*150*) with *n*-butyl lithium in hexane leads to two

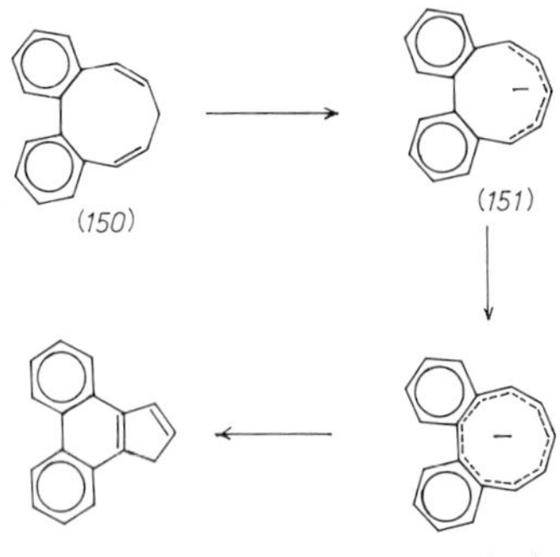

(150) (151) (152) (153)

nonatetraenyl anions (*151*) and (*152*) of which (*151*) was produced initially and then rearranged to (*152*). These were assigned non-planar (*151*) and planar (*152*) conformations. When solutions of (*152*) were quenched with water, the oxidized cyclized compound (*153*) was obtained in about 25% yield. It is not clear at what point the cyclization occurs or what is the oxidizing agent. The six- and seven-membered ring anions have been prepared and n.m.r. studies of the cycloheptadienyl (*154*) [66,124] and

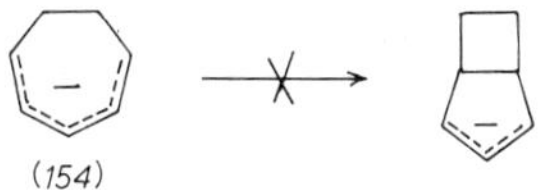

(154)

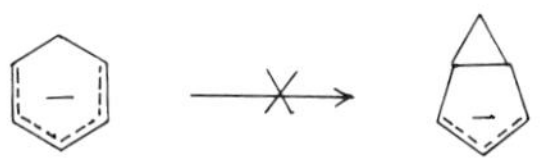

(155) (156)

cyclohexadienyl anion (*155*) [87,124] show no tendency for these smaller ring anions to cyclize.

Rearrangements of hexadienyl anions have provided permissive evidence for the formation of anions like (*156*) as unstable intermediates. Treatment [126] of 6,6-diphenylbicyclo[3.1.0]hex-2-ene (*157*) with refluxing *tert*-pentyl alcohol–potas-

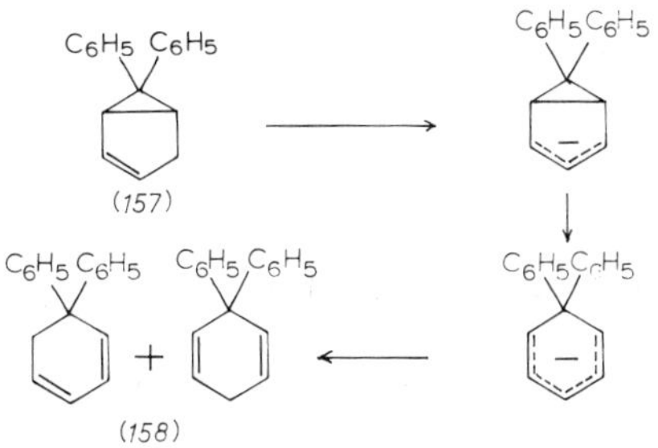

C_6H_5 C_6H_5 (157) C_6H_5 C_6H_5

C_6H_5 C_6H_5 + C_6H_5 C_6H_5 (158) C_6H_5 C_6H_5

sium *tert*-pentoxide leads to equilibrium amounts of the diphenylcyclohexadienes, (*158*) suggesting a disrotatory ring-opening. Reaction [127] of 5-methyl-5-phenyl-1,3-cyclohexadiene (*159*) with alkali amides in liquid ammonia leads to a complex mixture

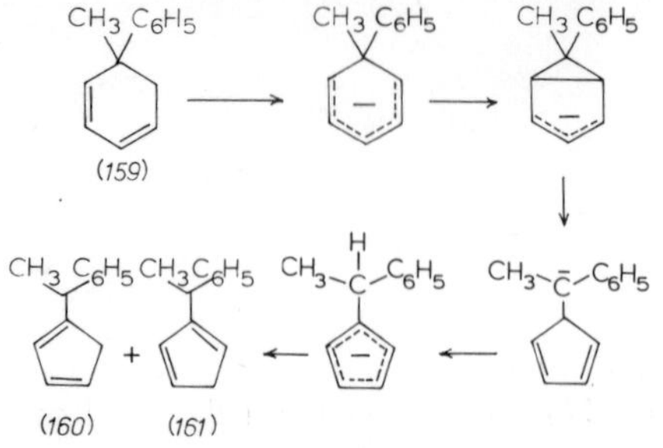

of products among which are the ring contracted compounds (*160*) and (*161*). These have been rationalized as arising via an electrocyclic ring closure followed by a ring-opening, both in a disrotatory manner.

There are two examples of ring-closures involving nitrogen substituted pentadienyl anions; one is a 2,4-diazapentadienyl system and the other is 2-azapentadienyl anion and both provide stereochemical information. When 1,3,5-triphenyl-4-aza-1,3-pentadiene (*162*) was treated [43] with lithium tetramethylpiperidide at $-78°$ in THF a

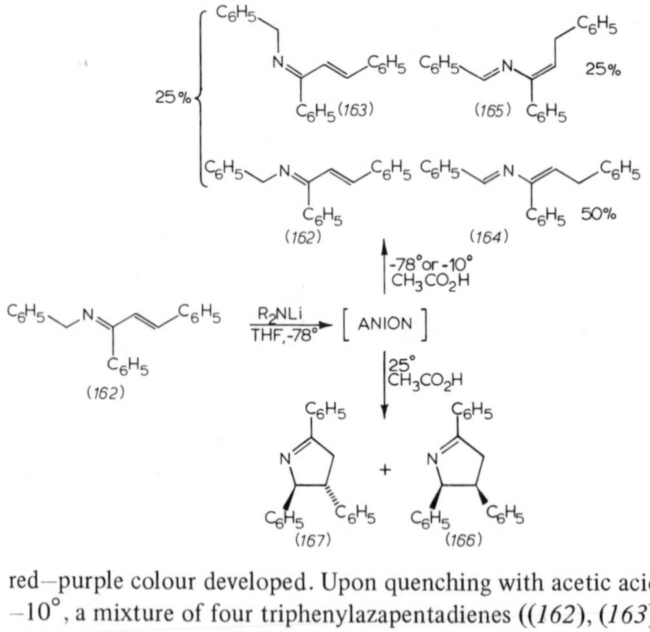

red–purple colour developed. Upon quenching with acetic acid at either $-78°$ or $-10°$, a mixture of four triphenylazapentadienes ((*162*), (*163*), (*164*) and (*165*)) is obtained but upon allowing the solution to warm to room temperature for four hours before quenching, only the *cis* and *trans* cyclized products ((*166*) and (*167*)) are obtained and in nearly equal amounts. The quenching results at low temperature indicate two or more geometries of anion are present in similar concentrations. The room temperature results suggest that the geometries of open chain anions proceed at similar rates to cyclized products. It was shown that the *trans*-pyrroline (*167*) is greatly preferred at equilibrium over the *cis*-pyrroline (*166*) and that no *cis–trans* isomerization is occurring under the reaction conditions. The actual geometries of open chain anions that are present are not known but the "W" and "sickle" shapes seem the best candidates.

The 2,4-diazapentadiene system was studied both thermally and photochemically [128a,b]. When hydrobenzamide (*168*) was treated with phenyl lithium in THF at

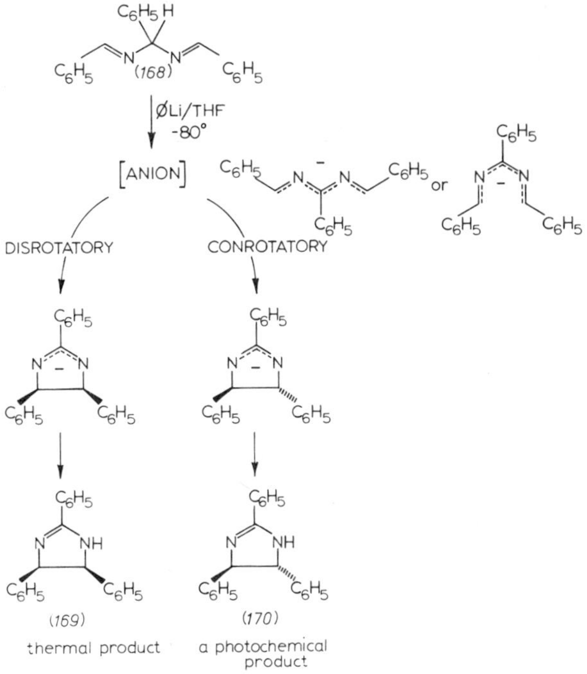

C_6H_5 H
C_6H_5 (168) C_6H_5
ØLi/THF -80°
[ANION] C_6H_5 N N C_6H_5 or N N
C_6H_5 C_6H_5 C_6H_5
DISROTATORY CONROTATORY
(169) thermal product
(170) a photochemical product

–78°, a blue colour developed (λ_{max} 555, 592 nm) which faded to light yellow on warming to room temperature. Upon quenching with acetic acid, only the *cis* cyclic compound (amarine, (*169*)) was obtained and in high yield. The *cis* compound proved to be thermodynamically less stable than the *trans* compound (isoamarine, (*170*)) with 4% (*169*) and 96% (*170*) at equilibrium. Isoamarine (*170*) could be produced however, when the blue solution was irradiated in the long wavelength region. The percentage of (*170*) in the cyclized products increased with decreasing temperature. These results suggest a carbanionic intermediate which reacts thermally to produce exclusively the thermodynamically less stable product (*169*) and photochemically to generate amounts of the *trans* product (*170*). The selective formation of the thermodynamically less stable product has been interpreted in terms of orbital symmetry control in cyclization from a "W" (or "U") shaped open chain.

There are experimental results which bear on the relative rates of cyclization of these two systems. The rates of 1,3 and 1,5 proton transfer and of cyclization have been measured for both systems in methanol–potassium methoxide at 60°. On the basis of the proton transfer reaction rates, it appears that both (*162*) and (*168*) produce carbanions at about the same rate. In contrast potassium *tert*-butoxide in *tert*-butyl alcohol at 120° is required to cyclize (*162*) while (*168*) cyclizes readily at 60° in methanol–potassium methoxide. This implies a large difference in the rate at which the carbanions cyclize and this is estimated to be about 6 powers of 10. This difference can be rationalized in terms of the stability of the cyclized anions (*171*) and (*172*) since most of the charge should reside on the terminal atoms of the delocalized system.

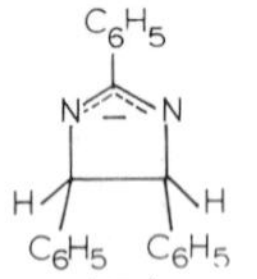

(*171*) 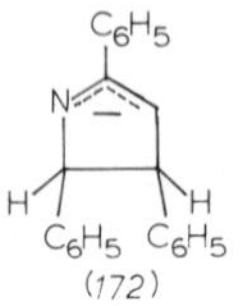 (*172*)

A comparison of the electrocyclization stereochemistry of the 2-azapentadienyl anion and the 2,4-diazapentadienyl anion is intriguing since (*162*) produced both geometries of cyclized product, (*166*) and (*167*) in similar amounts while (*168*) produced only the *cis* product (*169*). This change with aza-substitution may reflect a competition between conrotatory and disrotatory modes from a U-shaped anion due to lack of molecular symmetry in the monoaza anion. Alternatively, the change in stereochemistry may be a consequence of disrotatory electrocyclization but from different geometries of open chain anion in the two systems.

C. Heptatrienyl–Cycloheptadienyl Interconversion

Unlike the three and five carbon systems, the heptatrienyl anion to cycloheptadienyl rearrangement has been observed in the parent system but unfortunately none of the systems which have been studied provide any stereochemical results. Treatment [89] of 1,3,5-heptatriene (*173*) with potassium amide in liquid ammonia at a low temperature leads to an anion whose spectrum at –60° was interpreted in terms of the all-*trans*-heptatrienyl anion (see Table 15). Upon warming to 0° conversion to a new species occurred ($t_{1/2} \approx 60$ min) whose n.m.r. spectrum was interpreted in terms of the cycloheptadienyl anion (*175*). Products of quenching with protons, methyl iodide and carbon dioxide are reported to confirm this assignment although details have not

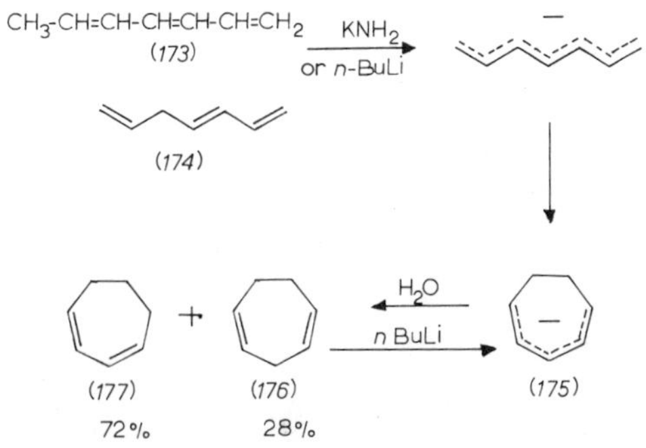

CH_3-CH=CH-CH=CH-CH=CH_2 (173) → KNH_2 or *n*-BuLi

(174)

(177) 72% + (176) 28% ← H_2O / *n* BuLi → (175)

been given. Likewise reaction [66] of 1,3,6-heptatriene (*174*) with *n*-butyl lithium in THF–hexane at –50° yielded the all-*trans*-heptatrienyl anion which cyclized at –30° with a half-life of 13 min to the cycloheptadienyl anion (*175*). The identity of (*175*) was confirmed by generation from 1,4-cycloheptadiene (*176*) and by quenching with water to yield 28% of (*176*) and 72% of 1,3-cycloheptadiene (*177*).

Three groups have reported on the cyclization of octatrienes to methyl substituted cycloheptadienes. Treatment [129] of 1,3,6 and 1,3,7-octatrienes ((*178*) and (*179*))

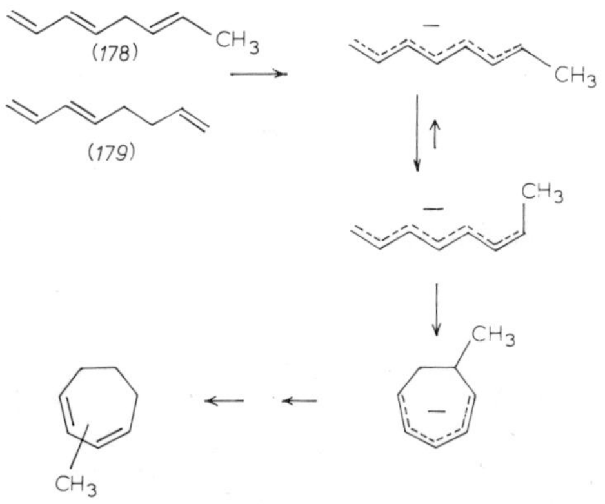

(178) CH_3

(179)

CH_3

CH_3

CH_3

CH_3

with piperidinopotassium first produced proton transfer reactions but then produced methylcycloheptadienes (primarily the 1-methyl and 2-methyl-1,3-cycloheptadienes, (*180*)). Evidence for carbanionic intermediates in this type of rearrangement was obtained by n.m.r. studies of the open chain anion and its ring-closure product. A half-life of 300 min at –30° was observed [66] for the lithium salt in THF–hexane while the potassium salt in liquid ammonia cyclized completely upon standing for several hours at room temperature [92]. With the amide bases further proton transfers apparently occur in the cyclized products to yield the observed products.

Further methyl-substitution on the cycloheptatriene reveals [66] a sensitivity to substitution of both the cyclization and proton-abstraction step. While (*174*) and (*178*) yield open chain anions at –50° upon reaction with *n*-butyl lithium in THF–

	R_1	R_2	R_3	R_4
(174)	H	H	H	H
(178)	CH_3	H	H	H
(181)	H	CH_3	H	CH_3
(182)	H	CH_3	CH_3	CH_3

hexane, (*181*) requires warming to –30° and (*182*) does not react until room temperature. When (*182*) reacts, the open chain anion is not observed and instead the cyclized product forms. The rates of cyclization were observed for (*174a*), (*178a*) and (*181a*) and at –30° half-lives were (*174a*)→(*174b*), 13 min; (*178a*)→(*178b*), 30 min; and (*181a*)→(*181b*), 200 min.

Substitution [90] of a vinyl group at the 6-position of cycloheptatrienyl anions results in a change of the position of equilibrium with the open chain form now being favoured (nonatetraenyl anion). Thus when (*183*) was reacted with *n*-butyl lithium in

THF–hexane at –30° the product appeared to be the nonatetraenyl anion (*184*). Protonation followed by hydrogenation yielded undecane. In an analogous manner, reaction of (*185*) yielded tridecane. Both results suggest a rapid opening of the cycloheptatrienyl ring to form the more stable nonatetraenyl anion (*184*) (vide infra).

The possible photochemical conversion of a bicyclic pentadienyl anion to a monocyclic heptatrienyl anion in a disrotatory manner was investigated [130]. Irradiation of the potassium salt of the bicyclo[5.1.0]octadienyl anion (*186*) at –70° with a 500 watt high pressure mercury lamp resulted in the appearance of the cyclooctatetraene

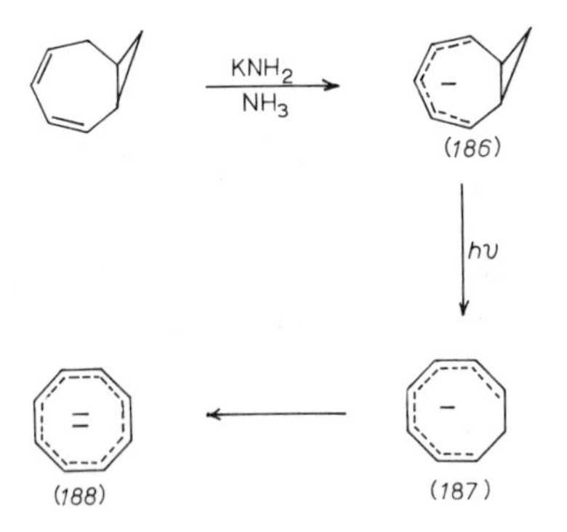

di-anion (*188*). This is permissive evidence for photochemical ring-opening to the cyclooctatrienyl anion (*187*) since it has previously been shown [131] that (*187*) is unstable relative to the di-anion (*188*) and cyclooctatetraene.

D. Nonatetraenyl–Cyclononatrienyl Interconversion

There is only one report [90] of an attempt to observe a nonatetraenyl–cyclononatrienyl interconversion and interestingly the open chain anion (*189*) could not be persuaded to cyclize even on heating to 70° nor could the presumed cyclononatrienyl

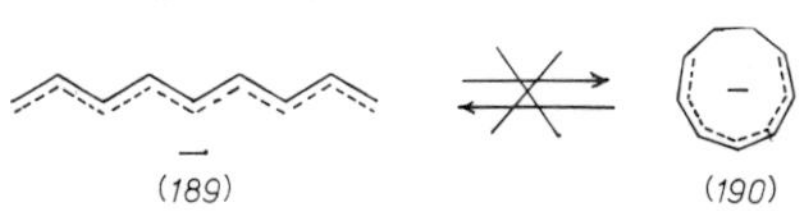

(189) (190)

anion (*190*) be persuaded to open upon heating to 90°. Both preferred to decompose to products of unknown structure. Reactions were run in THF–hexane.

E. Enolate–Cyclooxaenyl Interconversion

There are two examples of oxygen analogues to the carbon system. When 2,5-dihydrofuran (*191*) was treated with base (potassium amide in liquid ammonia at –60°

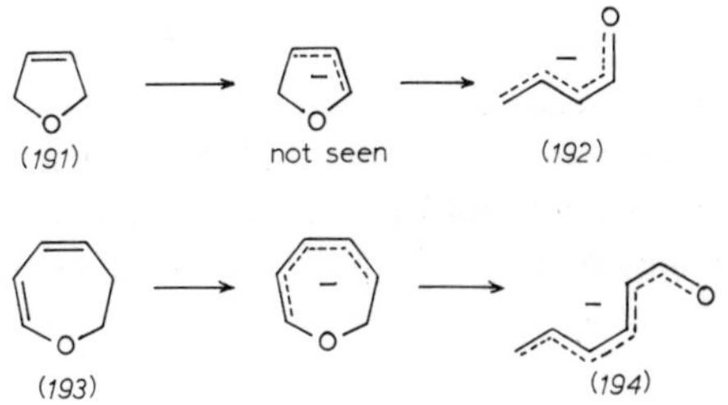

(191) not seen (192)

(193) (194)

[132] or *n*-butyl lithium in THF–hexane at 35° [133] the "sickle" shaped open chain anion (*192*) is produced. The "sickle" shape has not been observed to convert to the "W" shape even upon heating. The vinylogous system, 2,3-dihydrooxepin (*193*), when treated with potassium amide in liquid ammonia at –60°, also produced the bent open chain anion, (*194*) [102].

Both these systems contrast with the carbon analogues which prefer the cyclized form. This presumably is because in the open chain enolate form, significant charge can be placed on the oxygen.

V. HOMOCONJUGATION AND RESULTANT REARRANGEMENTS

Associated with many of the usual π-conjugated systems are, in principle, a further series of homoconjugated species. A π-conjugated system of *p*-orbitals is usually defined as a series of nearly coplanar or overlapping *p*-orbitals on adjacent directly bonded atoms (*195*) while in a homoconjugated system the adjacent *p*-orbitals are not on

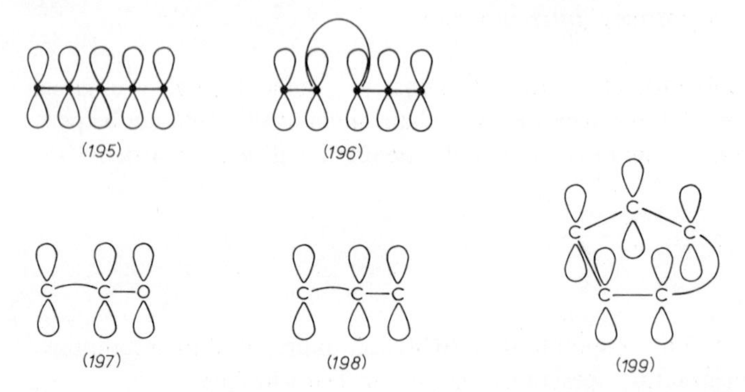

atoms directly bonded to one another but have one or more atoms intervening (*196*). Since in principle the size, position, type, and number of atom bridges can vary, a large variety of homoconjugated species is possible. In practice however, there are restrictions on each of the variables and the examples are more limited. If the further restriction of considering only carbanions is imposed, a discussion of the variety of species that have been studied becomes feasible.

Homoenolic (*197*), homoallylic (*198*) and homocyclopentadienyl (*199*) anions are the systems which have received the most attention. Rearrangements in a variety of substrates have been observed and serve as one of the best criteria for the intervention of homoconjugated systems. Isotopes have been used both for labelling positions and as a monitor for carbanion formation by observing isotopic exchange. The discussion will be organized according to the type of ion involved; homoenolates, homoallylic anions, homocyclopentadienyl anions and homocyclopropenyl anions.

A. Homoenolates

Carbonyl groups exert a profound stabilizing influence on a carbanionic centre when placed in direct conjugation (enolates). For example, the estimated pK_a's for the methyl hydrogens of propane and acetone are 42 and 21, respectively [1]. Since both inductive effects and the π-conjugated system play a major role in stabilizing enolate anions and since the effect is large, the opportunity for homoconjugation to be energetically important seems high. Thus the presence of an appropriately oriented carbonyl group could lead to a homoconjugated anion ((*200*)→(*201*)) whose character

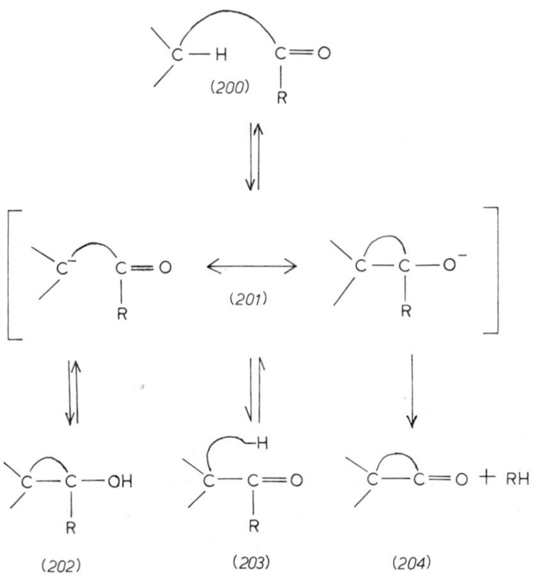

(200)

(201)

(202) (203) (204)

can be described in terms of two contributing structures. The relative importance of each contributor should depend upon the detailed structure of the molecule involved, even resulting in two distinct ions rather than just one. The intervention of the closed structure provides an opportunity for cyclization (*202*) and rearrangement (*203*) as well as isotopic exchange ((*200*), where H is D). It is worth noting that a second rearrangement (or cleavage, (*204*)) is also possible.

One of the first pieces of evidence suggesting the formation of a homoenolate intermediate was the result of a study [134a,b] of the base-catalyzed racemization and isotopic exchange of camphenilone. When optically active (+)-camphenilone (*205*) was treated at 185° with 0.7 *M* potassium *tert*-butoxide in *tert*-butyl alcohol for various periods of time and then recovered, it was found to be partially racemized. When the same experiments were carried out using *tert*-butyl alcohol-*O-d*, the recovered (*205*) had racemized but had also incorporated deuterium; up to 3 atoms of deuterium per molecule. Furthermore, the rate of racemization and the rate of appearance of

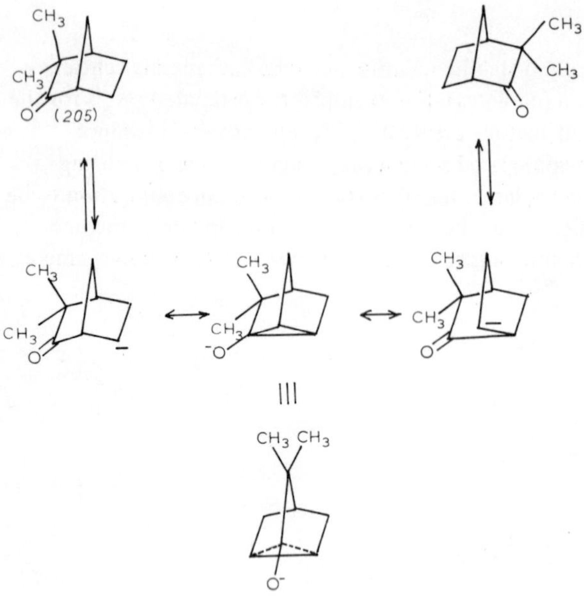

(205)

(206)

molecules with more than one deuterium per molecule were equal. These results could be nicely accommodated by postulating the formation of a symmetrical homoenolate (*206*). Provided formation of this intermediate (or its equivalent) was accompanied by isotopic exchange, the rates of racemization and appearance of molecules with one deuterium or more would be equal. Also this process would lead to the incorporation of up to 3 atoms of deuterium per molecule since the 1 and 6-positions become scrambled.

Bicyclic and tricyclic systems have now provided further examples of rearrangements consistent with formation of homoenolates (Fig. 24). In all of these systems the reaction medium was potassium *tert*-butoxide in *tert*-butyl alcohol at 185° or higher, i.e. roughly the conditions used for the racemization and exchange of camphenilone (*205*). Attempts to use the more basic solution of potassium *tert*-butoxide in dimethyl sulfoxide at lower temperatures have failed due to the onset of side reactions [134a,b]. All of these structurally similar systems point to the common involvement of homoenolate anions in the rearrangement processes.

Further permissive evidence for the involvement of homoenolates has been obtained for 2-norbornanone (*206*), camphor (*207*) and the half-cage ketones ((*208*) and (*209*)) (Fig. 25). The appropriate homoenolates were generated from the homoenols or enol acetates and under the reaction conditions were found to generate the ketones at a much faster rate than exchange or rearrangement was occurring in the parent ketones. Thus the homoenolates generated in this way behave in a manner consistent

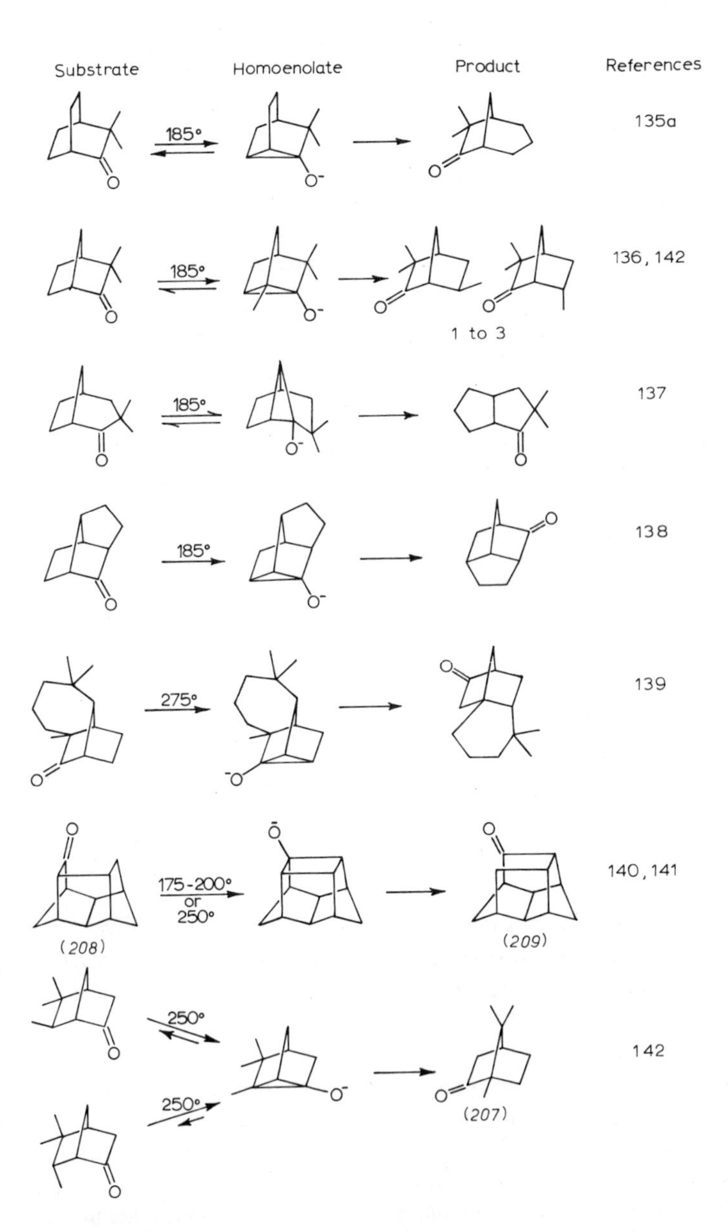

Fig. 24. Examples of homoenolic rearrangements in bicyclic and tricyclic systems in $(CH_3)_3COH$–$(CH_3)_3COK$.

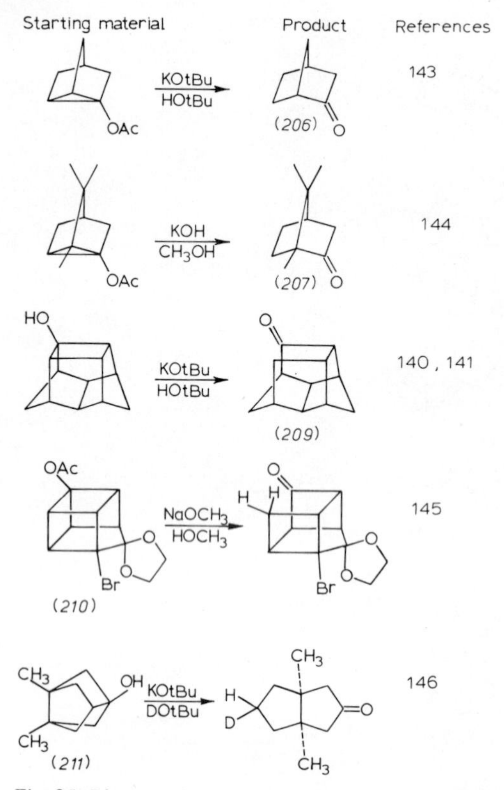

Fig. 25. Ring-opening reactions of some homoenols and homoenolacetates.

with the role proposed for them in homoenolization. The homocubane derivative (*210*) and the tricyclic carbinol (*211*) provide further examples of ring-opening reactions. Such ring-opening reactions are also proving to be of interest particularly from a stereochemical point of view.

The establishment of a connection between the racemization and the exchange in (*205*) and the formation of a homoenolate anion makes it seem reasonable to look at isotopic exchange itself as a measure of homoenolate anion formation in related systems. This approach is especially useful where no rearrangement is occurring either due to return of the homoenolate to starting material or due to formation of a symmetrical homoenolate (e.g. (*206*)). When (*205*) was treated with potassium *tert*-butoxide in *tert*-butyl alcohol-*O-d* for extended periods of time, incorporation of more than 3 atoms of deuterium per molecule was noted [147]. The possible positions of exchange were tested by treating (*212*) and (*213*) under the isotopic exchange conditions for (*205*). No exchange was noted and on this basis it was concluded that the methyl groups had been incorporating the deuterium.

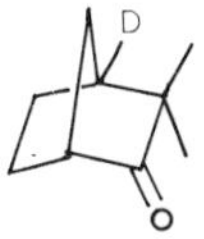

(212) 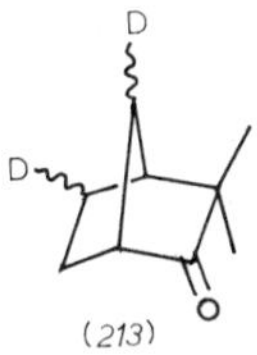 (213)

Recently carbon-13 n.m.r. has provided positive evidence for methyl hydrogen–deuterium exchange [142] in several related systems. The carbon-13 n.m.r. technique should prove most valuable for if the carbon signals can be assigned, the extent of deuteration at each position can be monitored. This technique has been applied to the bicyclic ketones ((*214*), (*215*), (*216*) and (*217*)) included in Fig. 26. The authors were able to assign approximate relative rates of isotopic exchange per exchangeable hydrogen at a variety of positions and these are shown along with the homoenolate anions that would be responsible for the exchange. Some of these systems can also rearrange and were previously included in Fig. 24. The exchange data for the methyl hydrogens is intriguing from two points of view. Firstly, there appears to be a high selectivity in the exchange reactions remembering that the methyls in (*214*) are equivalent and that the homoenolate (*218*) from (*217*) results in interconversion of the methyls. Secondly, no ring enlargement products have been found from the postulated homoenolates for methyl exchange. However, such products may not be stable under the reaction conditions. Another feature of note is the large variation in reactivity of the bridgehead hydrogens.

The exchange of (*219*), (*220*) and (*221*) have been followed by mass spectrometry and while methylene exchange was observed for (*219*), no exchange was seen in (*220*) and (*221*). These results suggest that the inductive effect of the carbonyl group is not sufficient in itself to induce isotopic exchange. These negative results then point to the necessity for homoenolization in the isotopic exchange processes*.

Rigid bicyclic systems have not provided the only examples of rearrangements consistent with formation of homoenolates. The alkaline hydrolysis of some 3-acetoxy-Δ^1-pyrazolines has produced open chain ketones [150] whose structure can be rationalized in terms of homoenolate intermediates. For example, 3,5,5-trimethyl-3-acetoxy-Δ^1-pyrazoline (*222*) yielded pinacolone (*223*) as the major product and methyl isobutyl ketone (*224*) as the minor product. This same product distribution was observed upon hydrolysis of 1,2,2-trimethyl-1-acetoxycyclopropane (*225*). These results can be interpreted in terms of a homoenolate intermediate (*226*) which is formed faster than the open chain anion is captured by solvent. Similar rearrangements were observed for (*222*) with the 3-methyl group replaced by an ethyl or an 1-isobutenyl group.

A γ-diketone has also provided an example of a rearrangement consistent with homoenolate formation; actually two homoenolate anions appeared involved. Reaction [151a,b] of 4-hydroxy-2,2-diphenyl-3-pentenoic acid lactone (*227*) with phenyl lithium leads to the formation of three products which are open chain ketones; 1,2,2-

* A reinvestigation of (*221*) has shown that exchange does occur under homoenolization conditions [135b].

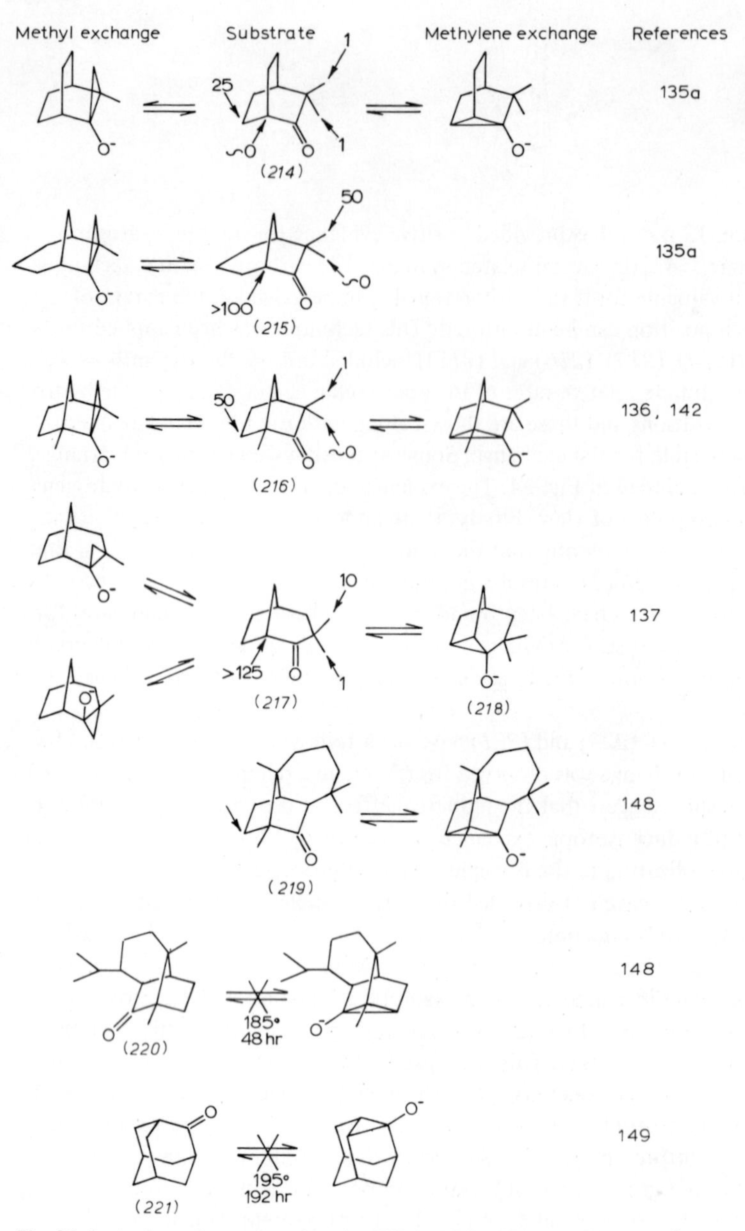

Fig. 26. Isotopic exchange reactions in $(CH_3)_3COD$–$(CH_3)_3COK$ at 185°. (Relative rates within each molecule are on a per hydrogen basis.)

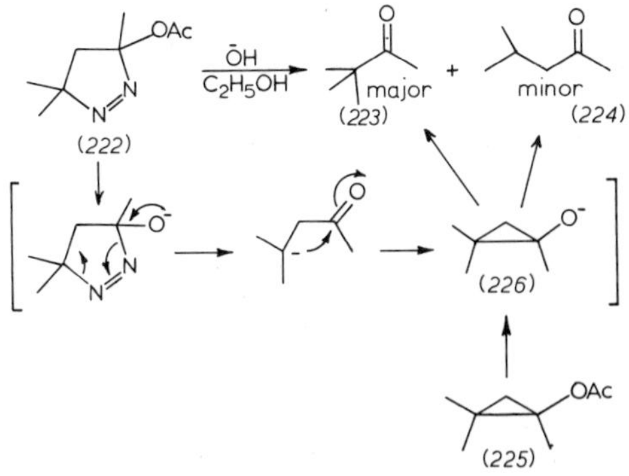

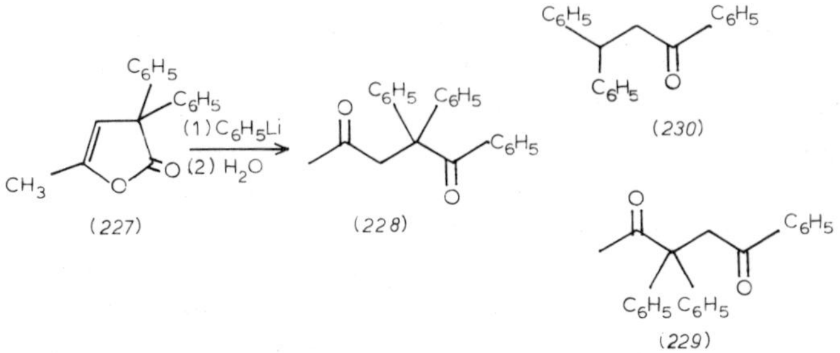

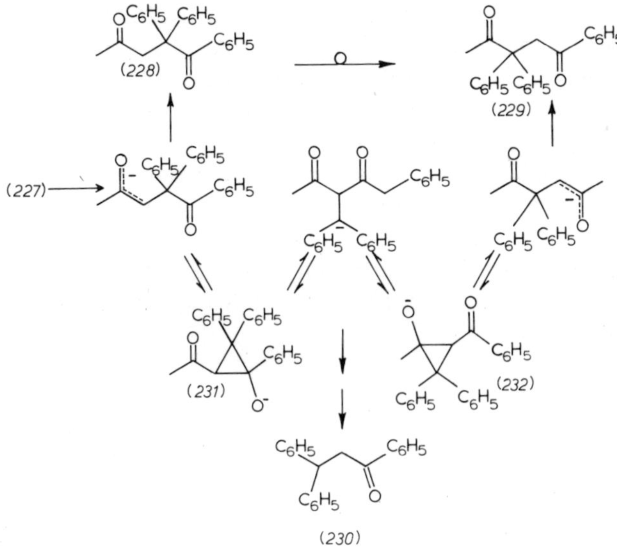

triphenyl-1,4-pentadione (*228*), 1,3,3-triphenyl-1,4-pentadione (*229*) and 3,3-diphenylpropiophenone (*230*). This complex reaction was simplified by noting that the ratio of (*229*) to (*228*) increases with time suggesting that (*228*) is the initial product and (*229*) is the result of a rearrangement. The rearrangement has been rationalized as occurring through two consecutive homoenolate intermediates ((*231*) and (*232*)). Two major pieces of evidence were obtained to substantiate the mechanistic proposal. First, consecutive 1,2-phenyl migrations to interconvert the enolates of (*228*) and (*229*) were ruled out by preparing the necessary intermediate (*233*) in this process and show-

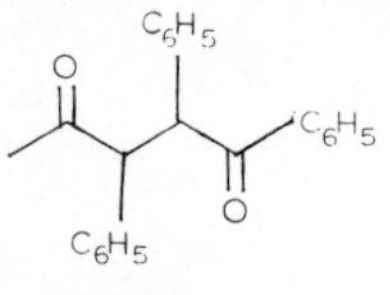

(*233*) 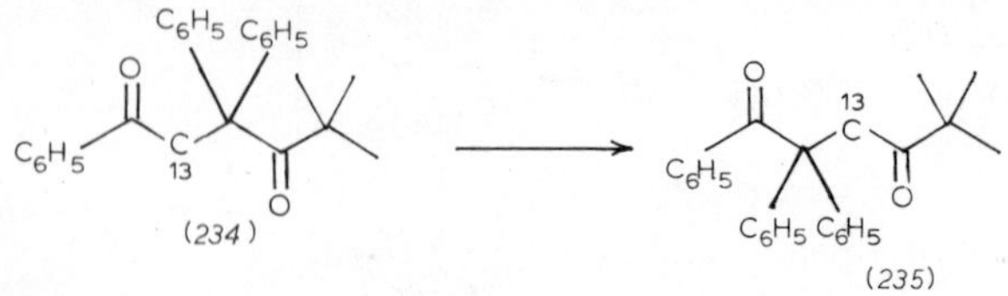 (*234*) (*235*)

ing it did not rearrange under the reaction conditions. The second was the effective use of carbon-13 labelling [152a,b]. It was demonstrated that (*234*), the *tert*-butyl analogue of (*228*) labelled with carbon-13, rearranged to (*235*) upon treatment with sodium methoxide in ether. In the rearranged material (*235*), the enriched carbon was still the methylene carbon showing that the rearrangement was indeed a skeletal rearrangement consistent with the proposed mechanism.

While most of the systems discussed here point to the intervention of homoenolate anions in both bicyclic and open chain systems, there is still the question of whether homoenolate formation provides assistance in carbanion formation. There are some pieces of evidence that point to assistance in the bicyclic systems. It is possible to estimate the approximate pK_a of the homoenolizable methylene hydrogens in the bicyclic systems (Figs. 24 and 26) to be about 35–36 by assuming a relationship between kinetic and thermodynamic acidities. The conditions required for isotopic exchange and rearrangement of many of the bicyclics (potassium *tert*-butoxide in *tert*-butyl alcohol at 185° for several hours) are similar but slightly milder than for isotopic exchange of the α and ring positions of 2-phenylbutane [153], of the methyl groups in 2,4,4-trimethyl-2-pentene [33], and the ring positions of various naphthalenes [57,142]. This places the pK_a of the homoenolizable protons of the bicyclic ketones near propene (35.5), benzene (37) or cumene (37) on the MSAD scale [1a–d]. This places their pK_a about midway between normal methylenes (e.g. 45 for cyclohexane) and enolizable protons (e.g. 21 for acetone) showing considerable assistance by the carbonyl group in carbanion formation. Since there is also significant stereoselectivity in carbanion formation (Fig. 26), it seems unlikely that an inductive (or field) effect of the carbonyl group is the only effect at work in enhancing kinetic acidity. On this basis it would appear that the mesomeric effect of the carbonyl group in the bicyclic ketones is of energetic importance in stabilizing the transition state for base abstraction of homoenolizable protons.

B. Homoallylic Anions

The effect of placing a double bond in direct conjugation with a carbanionic centre (e.g. allylic anion) would seem to stabilize the carbanion by about 7 pK_a units on the

MSAD [1a–d] scale (ethane, 42; propene, 35.5). This contrasts with the effect of the carbonyl group (enolates), as discussed in the earlier section, which produced a change of 22 pK_a units. On this basis the homoconjugative effect of a double bond would not be expected to be large. There appear to be no studies which have a direct bearing on the effect of a homoconjugated double bond on either kinetic or thermodynamic acidity. Nonetheless, there are a variety of studies revealing rearrangements between cyclopropylcarbinyl anions (*236*) and allylcarbinyl anions (*238*) which appear to pro-

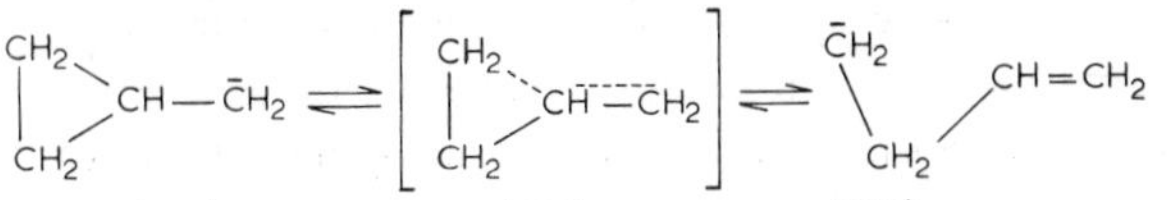

$$\text{cyclo-}(CH_2)_2CH{-}\bar{C}H_2 \;(236) \rightleftharpoons [\text{cyclo-}(CH_2)_2\dot{C}H{\cdots}CH_2]\;(237) \rightleftharpoons \bar{C}H_2{-}CH_2{-}CH{=}CH_2 \;(238)$$

ceed through homoallylic intermediates or transition states (*237*). All of the examples come from magnesium and lithium salts in ether solvents.

With the Grignard reagents the position of the equilibrium lies on the side of the allylcarbinyl anion (*238*) in all cases studied (Fig. 27). In fact, it was necessary to generate the Grignard reagent from cyclopropylcarbinyl bromide (*239*) at –24° in dimethyl ether to obtain evidence for the existence of cyclopropylcarbinyl magnesium bromide (*240*) before it rearranged [155]. However, several studies have revealed an interesting rearrangement of the more stable allylcarbinyl Grignard reagents (*241*) themselves (Fig. 28). In some cases ((*242*), (*243*) and (*244*)) the products of quenching of a la-

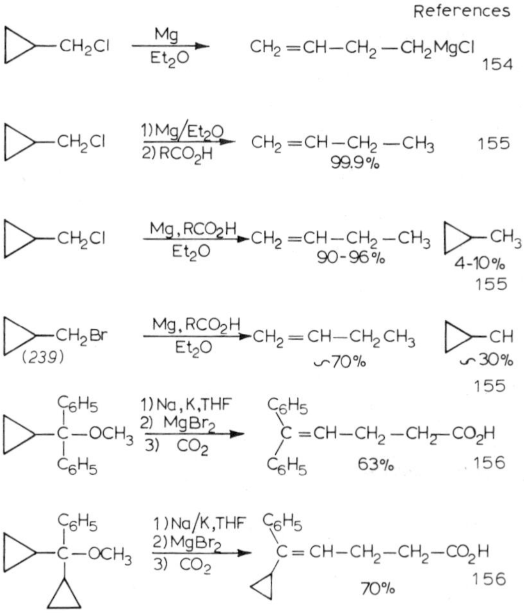

Reaction	Products	References
$\triangleright$–CH_2Cl $\xrightarrow[Et_2O]{Mg}$	$CH_2{=}CH{-}CH_2{-}CH_2MgCl$	154
$\triangleright$–CH_2Cl $\xrightarrow[2)\,RCO_2H]{1)\,Mg/Et_2O}$	$CH_2{=}CH{-}CH_2{-}CH_3$ 99.9%	155
$\triangleright$–CH_2Cl $\xrightarrow[Et_2O]{Mg,\,RCO_2H}$	$CH_2{=}CH{-}CH_2{-}CH_3$ 90-96%; $\triangleright$–CH_3 4-10%	155
$\triangleright$–CH_2Br (*239*) $\xrightarrow[Et_2O]{Mg,\,RCO_2H}$	$CH_2{=}CH{-}CH_2CH_3$ ~70%; $\triangleright$–CH ~30%	155
$\triangleright$–$C(C_6H_5)_2{-}OCH_3$ $\xrightarrow{1)\,Na,K,THF;\ 2)\,MgBr_2;\ 3)\,CO_2}$	$(C_6H_5)_2C{=}CH{-}CH_2{-}CH_2{-}CO_2H$ 63%	156
$\triangleright$–$C(C_6H_5)(\triangle){-}OCH_3$ $\xrightarrow{1)\,Na/K,THF;\ 2)\,MgBr_2;\ 3)\,CO_2}$	$(C_6H_5)(\triangleleft)C{=}CH{-}CH_2{-}CH_2{-}CO_2H$ 70%	156

Fig. 27. Examples of cyclopropylcarbinyl to allylcarbinyl Grignard rearrangements.

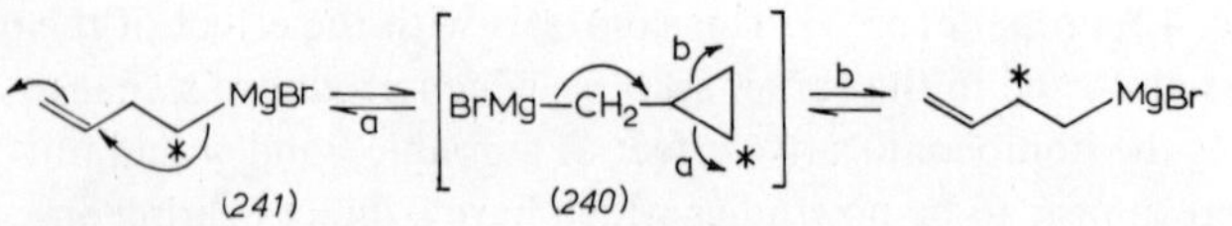

belled Grignard reagent revealed that the two methylene carbons had been scrambled and this rearrangement involved scrambling of the carbons rather than their substituents. In the other cases ((*245*), (*246*), (*247*) and (*248*)) the interconversions were followed by n.m.r. and the rearrangements were slow enough for rate measurements. A detailed kinetic study [159] has been made of the Grignard reagents from (*246*) and (*247*) and the sensitivity to solvent, concentration, temperature and metal have been

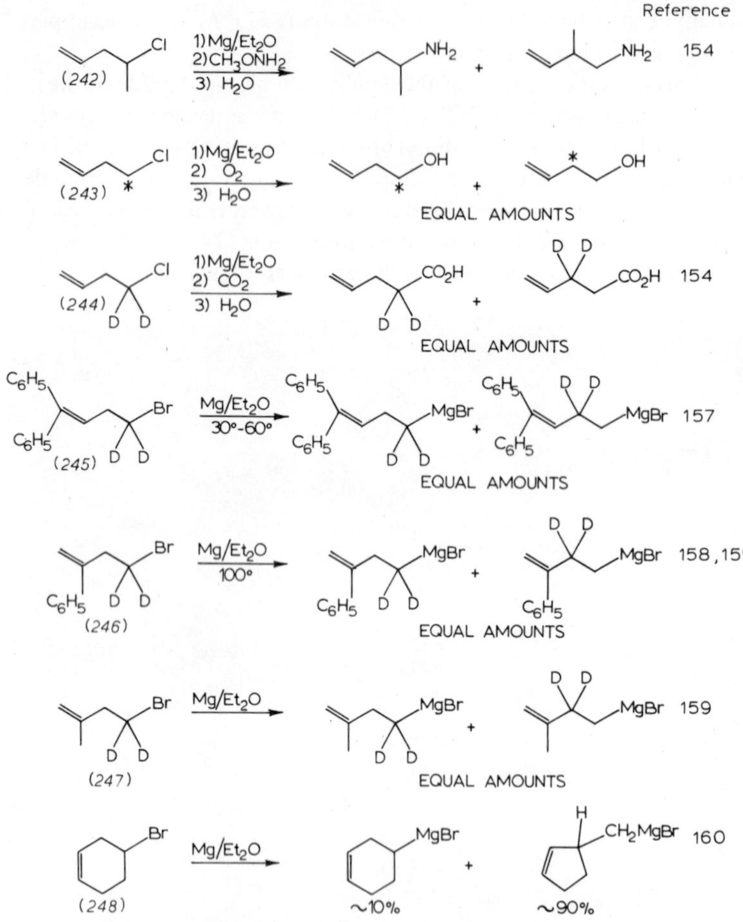

Reference

(242) 1) Mg/Et_2O 2) CH_3ONH_2 3) H_2O — 154

(243) 1) Mg/Et_2O 2) O_2 3) H_2O — EQUAL AMOUNTS

(244) 1) Mg/Et_2O 2) CO_2 3) H_2O — EQUAL AMOUNTS — 154

(245) Mg/Et_2O 30°-60° — EQUAL AMOUNTS — 157

(246) Mg/Et_2O 100° — EQUAL AMOUNTS — 158,159

(247) Mg/Et_2O — EQUAL AMOUNTS — 159

(248) Mg/Et_2O — ~10% + ~90% — 160

Fig. 28. Examples of skeletal rearrangements of allylcarbinyl Grignard rearrangements.

noted. All of these examples can be rationalized by postulating the intermediacy of a cyclopropylcarbinyl species (*240*) which opens either way resulting in an interchange of methylene groups. The fact that these rearrangements are not fast and that the cyclopropylcarbinyl species is not observed suggests that it is a high energy species but nonetheless accessible. In contrast, the isomeric cyclobutyl species does not appear accessible in spite of attempts to observe rearrangements through it [157,158,160,161]. Furthermore, 1-phenylcyclobutyl magnesium bromide (*249*) could not be induced to

MgBr C_6H_5 C_6H_5 MgBr (*249*) MgBr C_6H_5

rearrange even upon heating to 175°. The preference for homoallylic anions over cyclobutyl anions has been rationalized by LCAO molecular orbital calculations [162].

Besides Grignard reagents, analogous lithium, sodium and potassium compounds have also been studied. Cyclopropylcarbinyl lithium (*250*) could be prepared and

CH_2I $\xrightarrow[-70°]{s\text{-BuLi}}$ CH_2Li (*250*) ⇄ $LiCH_2$ (*251*)

C_6H_5CHO ↓ C_6H_5CHO ↓

$CH_2CH(OH)$-C_6H_5 C_6H_5-CH(OH)-$CH_2CH_2CH{=}CH_2$

trapped [161] at –70° in ether–petroleum ether but after 4 h it had been half-converted to allylcarbinyl lithium (*251*) thus behaving in a manner analogous to the Grignard reagents. The allylcarbinyl lithium compound (*252*) has also been shown [159] to

Li D D (*252*) ⇌ D D Li

rearrange with apparent scrambling of the methylene carbons and at a rate similar to the Grignard reagents. Thus, these lithium compounds are behaving in a manner quite similar to the magnesium compounds. One interesting difference is the observation [159] that on changing from ether to tetrahydrofuran solvent, the rate of methylene scrambling in the allylcarbinyl lithium (*252*) increases but decreases in the allylcarbinyl magnesium bromide (*241*).

$(C_6H_5)_2C^- Li^+$–cyclopropyl (253) red $\underset{THF}{\overset{Et_2O}{\rightleftharpoons}}$ $(C_6H_5)_2C{=}CH{-}CH_2CH_2Li$ (254) colorless

↓ CO_2 ↓ CO_2

$(C_6H_5)_2C(CO_2H)$–cyclopropyl (255) $(C_6H_5)_2C{=}CH{-}CH_2{-}CH_2{-}CO_2H$ (256)

In marked contrast to the results discussed above cyclopropyldiphenylcarbinyl lithium (*253*) shows a reversible rearrangement to the allylcarbinyl lithium (*254*) which depends upon the solvent [156]. In tetrahydrofuran the solution of the salt (*253*) is bright red in colour and on carbonation gives a 72% yield of the ring-closed carboxylic acid (*255*) and no open chain acid. In diethyl ether the solution is colourless (*254*) and carbonation yields 72% of open chain carboxylic acid (*256*) and no ring-closed acid. With a 2:1 ether–tetrahydrofuran mixture a yield of 46% of (*255*) and 3% of (*256*) was obtained on carbonation. This unique behaviour was interpreted in terms of a solvent dependent equilibrium between covalent and ionic species. In tetrahydrofuran the cyclic ionic species (*253*) is favoured while in ether the covalent open chain form (*254*) predominates. Results in the same study also indicated that the position of the equilibrium between cyclic and open chain forms showed a sensitivity to cation (Fig. 29). Thus the potsssium ((*257*), (*258*), (*259*) and (*260*)) and sodium (*261*) salts prefer to exist in the cyclic ionic form while the lithium and magnesium (*262*) salts prefer the open chain covalent form.

All of the results obtained for the cyclopropylcarbinyl–allylcarbinyl anions are interpretable in terms of an equilibrium between two distinct species ((*236*) and (*238*)) and do not point to the delocalized homoallylic anion (*237*) as a particularly stable intermediate. Rather, the delocalized homoallylic anion would seem to play more the role of a transition state interconnecting the cyclic and open chain species. However, a different picture may emerge from studies on more conformationally rigid systems (e.g. bicyclics). An indication of this comes from an investigation [163] of the norbornenyl anion (*263*)–nortricyclyl anion (*264*) rearrangements of short-lived anions. The anions were generated from the corresponding hydrazonium oxalates upon treatment with potassium periodate and potassium hydroxide in water. Under these conditions the same ratio of norbornene (*265*) to nortricyclene (*266*) was obtained (44% (*265*) and 56% (*266*)) from either hydrazine. Since the likely anionic intermediates should be quenched rapidly in this medium, the results are consistent with rapidly equilibrating

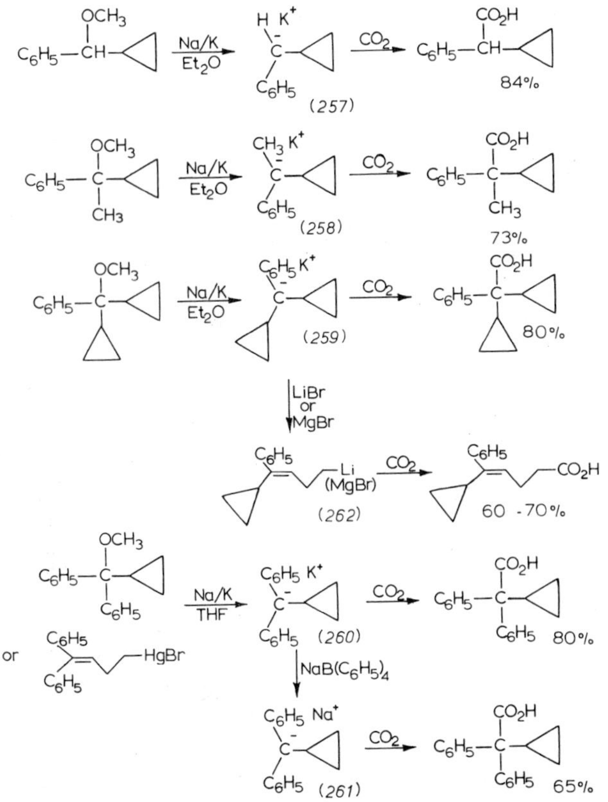

Fig. 29. Products of carbonation of organo-alkalies and Grignard reagents from cyclopropenylcarbinyl methyl ethers.

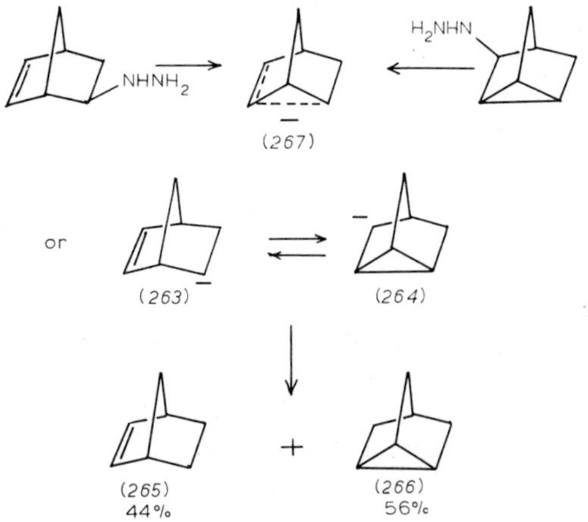

(267)

or

(263) (264)

(265) 44% (266) 56%

localized anions of similar energy ((*263*)⇌(*264*)) or one delocalized homoallylic anion (*267*).

Evidence for a further rearrangement in this system came from the use of a deuterium labelled substrate (*268*). Recovered norbornene was found to be a mixture of 50% *exo*-5-deuterionorbornene (*269*) and 50% *anti*-7-deuterionorbornene (*270*). These

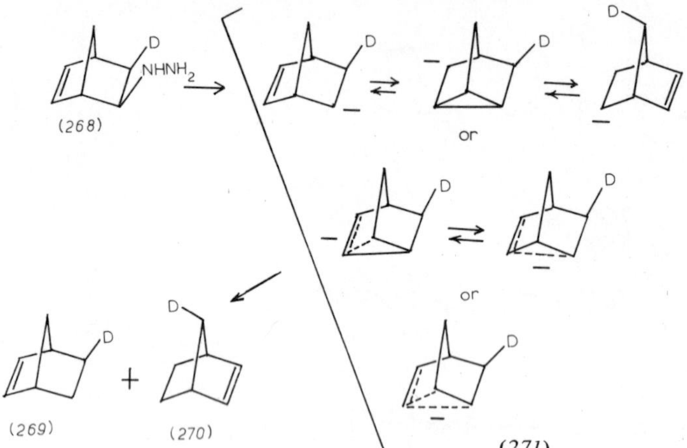

(268) (269) (270) (271)

results require either a thorough equilibration of localized anions, an equilibration of partially delocalized anions, the formation of a delocalized symmetrical anion (*271*) or some combination of these possibilities. Whatever the correct interpretation, it is clear that the bicyclic system is behaving in a manner different from the open chain systems.

C. *Bishomocyclopentadienyl Anions*

While a number of homoconjugated analogues of the cyclopentadienyl anion can be envisaged, the type which has received experimental attention is a bishomocyclopentadienyl anion (*272*). This particular anion can be considered as arising from the through

(272)

space interaction of an allylic anion and a double bond. There are ample theoretical arguments [164–167] to support the contention that this would be a stabilizing interaction. An impression of the possible maximum magnitude of the stabilization can be obtained by comparing the stability of the allylic anion with the cyclopentadienyl anion. Propene has a pK_a of about 36 on the MSAD scale [1a–d] and cyclopentadiene

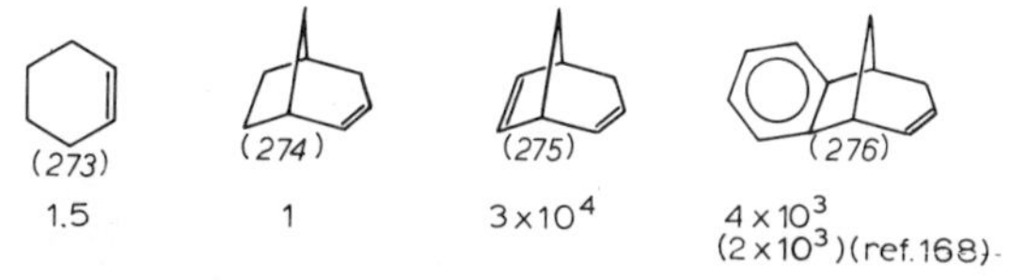

Fig. 30. Relative isotopic exchange rates in $(CH_3)_3COK-CD_3SOCD_3$ on a per-hydrogen basis [165,166,168,169].

of 15, thus suggesting an additional stabilization of 21 pK_a units. This should be related to similar comparisons made for the homoenolate (22 pK_a units) and homoallylic anion (7 pK_a units). The actual stabilization that might be experienced would naturally depend on the stereochemical details. Thus even the homoenolate interactions in the bicyclic systems were estimated to be worth about only 9 of a possible 24 pK_a units.

A number of studies [165,166,168,169] have been directed at estimating the stabilizing influence of the double bond on the allylic anion in the bicyclo[3.2.1] octane skeleton. The relative rates of isotopic exchange in dimethylsulfoxide-d_6 catalyzed by potassium *tert*-butoxide have been measured for cyclohexene (*273*), bicyclo[3.2.1] octa-2-ene (*274*), bicyclo [3.2.1] octa-2,6-diene (*275*) and benzo-6,7-bicyclo [3.2.1] octa-2,6-diene (*276*) (Fig. 30). The enhanced exchange rates of (*275*) and (*276*) relative to (*273*) and (*274*) were interpreted as indicating a stabilization of the intermediate carbanion by participation of the double bond in (*275*) or the benzene ring in (*276*). A further feature of interest was the stereochemistry of exchange for in both (*275*) and (*276*) the *exo*-hydrogen exchanged about six times faster than the *endo*-hydrogen. However, in this medium (potassium *tert*-butoxide in dimethyl sulfoxide-d_6), it is very likely that isotopic exchange measures only a small fraction of the actual rate of carbanion formation as evidenced by the very high intramolecularities in 1,3 proton transfer reactions in this medium (see p. 151). Unfortunately it is not certain that the ratio of isotopic exchange to actual carbanion formation will be constant from structure to structure. Thus ambiguity is added to the interpretation of the absolute or relative rates of isotopic exchange in terms of kinetic acidity for this medium.

A second approach [170–173] to this problem involved the direct preparation and n.m.r. analysis of the postulated bishomocyclopentadienyl anion (*278*). Treatment of

(277) Na-K THF → (278) → H^+ → (275)

(278) δ VALUES

3.5x10^{-5} 2.1x10^{-3} 1.4x10^{-2}

	(280)	(279)	(275)
25°	0.04 %	91.5 %	8.5 %
75°	0.7 %	86.4 %	12.9 %

Fig. 31. Equilibrium percentages of the isomers of bicyclo[3.2.1]octa-2,6-diene (*275*) and *pseudo*-first order rate constants for interconversion at 75° [167].

4-*exo*-methoxybicyclo[3.2.1]octa-2,6-diene (*277*) with sodium–potassium alloy in tetrahydrofuran produced an orange coloured solution. Quenching of this solution produced exclusively (*275*) and the n.m.r. of the orange solution was consistent with a species having the symmetry of (*278*). The upfield shift of the vinyl protons of 1.9 to 2.8 p.p.m. comparing (*278*) with (*275*) or (*277*) are in accord with delocalization of negative charge into the double bond but cannot be used in a quantitative manner to assign charge density. The quenching [173] of the orange solution provided interesting stereoselectivity with methanol-d_4 producing greater than 85% *endo*-deuteration and dimethylsulfoxide-d_6 providing >85% *exo*-deuteration. This result was rationalized by assuming a contact ion pair with the potassium on the *endo* face of (*278*) and acting to guide the methanol into the same face of the anion but not to guide in the dimethyl-sulfoxide.

The possibility of rearrangements occurring through anion (*278*) has been investigated using cesium cyclohexylamide in cyclohexylamine as catalyst [167] at 25° and 75°. It was found that the three isomers (*275*), (*279*) and (*280*) could be equilibrated and at equilibrium the tricyclic compound (*279*) predominates (Fig. 31). For all three hydrocarbons the isotopic exchange rates greatly exceed the isomerization rates and thus one anionic intermediate (e.g. (*278*)) is not sufficient to rationalize the exchange results, although the n.m.r. results are consistent with one predominant and symmetrical species. Ion pairing has been proposed as a possible source of other anionic intermediates as well as competition between *exo* and *endo* proton isotopic exchange. A related system has been studied [174]. Again the isomer analogous to (*279*) is the preferred, and in this case exclusive, product. Reaction of (*281*) as the methyl or *tert*-

RO_2C RO_2C CO_2R (281) RO^-, HOR → RO_2C RO_2C CO_2R (282)

exo and endo

butyl ester, with potassium methoxide in methanol at 90° for 4 h or with potassium *tert*-butoxide in *tert*-butyl alcohol at room temperature for 5 min, yielded similar amounts of the *exo* and *endo* form of (*282*) as the exclusive product. None of the product corresponding to (*275*) was found.

A reaction which had been described earlier under Section IV is also pertinent here since a bishomocyclopentadienyl anion seems to be involved [121]. Reaction of (*139*) with potassium *tert*-butoxide in dimethylsulfoxide at 70° for 20 h, produced a new tricyclic product (*283*) analogous to (*279*) and (*282*). In this case the bishomocyclopentadienyl anion (or anions) arises indirectly through a cyclopropyl anion ring-opening. The structure of the product (*283*) seems consistent with thermodynamic control rather than kinetic control and may well be the thermodynamically most stable form.

In contrast to the examples above, the reaction of either the tricyclic bromide (*284*) or the tetracyclic chloride (*285*) with metals yielded quite different product distributions [167] (Fig. 32). The reactions with magnesium and sodium can be envisaged as proceeding through carbanionic intermediates. Here the major product (*275*) is not the thermodynamically favoured species (*279*) and these results are analogous but not identical to the quenching result on the potassium salt of (*278*) discussed above.

The bicyclo[3.2.2] system has also received attention as a source of homoconju-

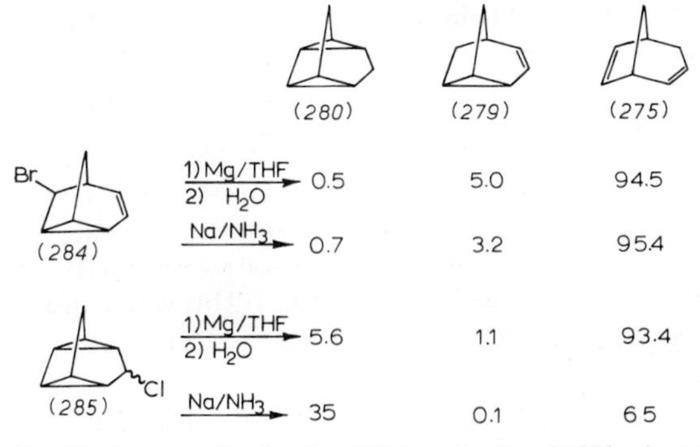

Fig. 32. Product distribution (%) in reduction [167] of a tricyclicbromide (*283*) and a tetracyclic-chloride (*284*).

gated anions and information has been gathered about n.m.r. spectra, isotopic exchange rates and rearrangements. The bicyclo[3.2.2]nona-2,6-dienyl anion [175] (*286*) and the bicyclo[3.2.2]nonatrienyl anion [175–177] (*287*) have both been pre-

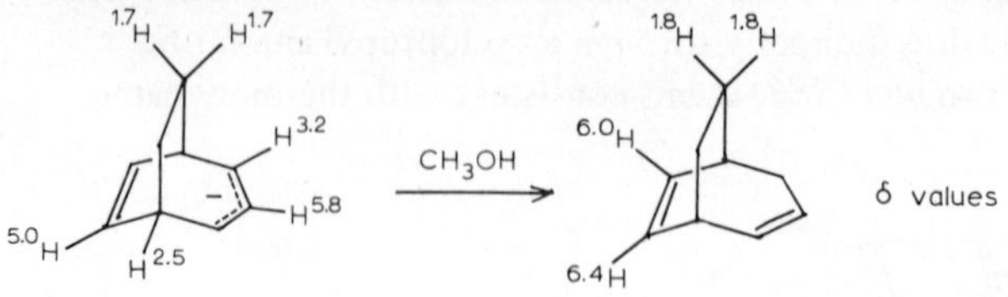

CH_3OH δ values

(*286*) (*288*)

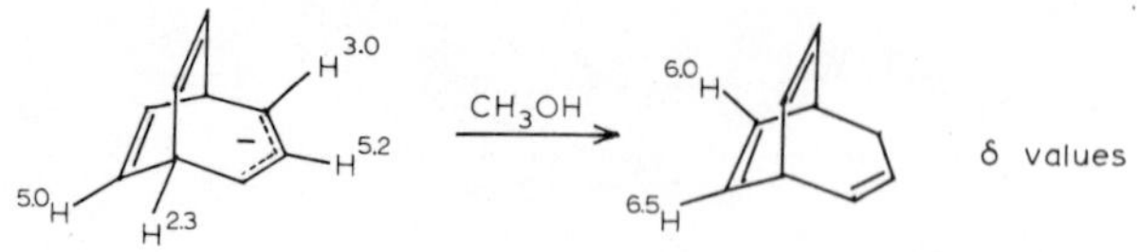

CH_3OH δ values

(*287*) (*289*)

pared by sodium–potassium alloy cleavage of the corresponding methoxy ether in THF and have been studied by proton n.m.r. and quenching. Quenching of (*286*) with methanol gave 68% yield of bicyclo[3.2.2]nona-2,6-diene (*288*) and quenching of the green solution of the quite insoluble potassium salt of (*287*) yielded mostly bicyclo[3.2.2]nonatriene (*289*). The n.m.r. spectra of both anions are consistent with symmetrical species like (*286*) and (*287*) and the spectrum of (*287*) was unchanged from –30° to 100°.

While it seems reasonably clear that the anions (*286*) and (*287*) have allylic character, the shift in the vinyl protons (1.0–1.5 p.p.m.) relative to (*288*) and (*289*) is only half that observed for the anion (*278*) leaving ambiguous the extent of delocalization into the double bonds of the bicyclo[3.2.2] system.

The similar chemical shifts in (*286*) and (*287*) suggest that the additional double bond in (*287*) does not lead to very different charge distributions. The anions (*286*) and (*287*) were found [175] to be deprotonating the 1,2-dimethoxyethane and tetrahydrofuran solvents on prolonged standing near or above room temperature (20°–90°). By following the *pseudo*-first order rates of deprotonation of solvent, it was found that (*286*) is about 10–30 times more reactive than (*287*). This would also seem to indicate that the extra double bond in (*287*) does not provide much more stabilization.

2-Methyl bicyclo[3.2.2]nonatriene (*290*) has been investigated as a source of a bicyclo[3.2.2]nonatrienyl anion [178a]. The rate of isotopic exchange in (*290*) was compared with bicyclo[3.2.2]nona-2,6-diene (*291*) and was found to exchange

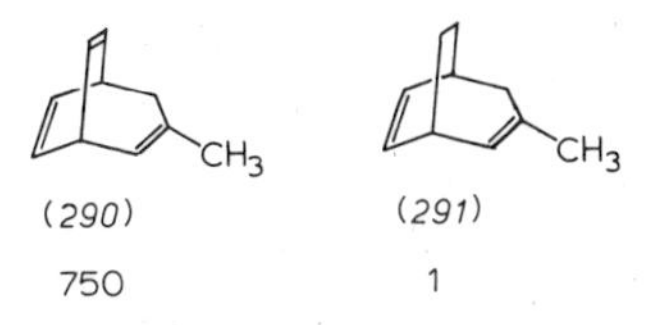

Fig. 33. Relative rates of isotopic exchange [178a] in $(CH_3)_3COK–CD_3SOCD_3$ at 25°.

750 times faster than (*291*) in dimethylsulfoxide-d_6 with potassium *tert*-butoxide. The parent hydrocarbons, (*288*) and (*289*), have also been studied [178b] in this medium and a rate ratio of 150 at 25° and 75 at 75.6° was observed. Unfortunately, as mentioned for similar experiments with the bicyclo[3.2.1]octanes (Fig. 33 and p. 151), the rate of isotopic exchange in this medium may not accurately reflect the actual rate of carbanion formation. Nonetheless, the enhanced rate of exchange of (*289*) to (*288*) and of (*290*) to (*291*) is permissive evidence for enhanced stability of the anion.

Rearrangements have also been observed in the bicyclo[3.2.2] anions. When anion (*287*) was prepared [177] with a deuterium label at the 2-position and was allowed to warm to 25°, n.m.r. revealed that the deuterium was not lost from the molecule but was being scrambled throughout the molecule. The details of the scrambling process are yet to be elucidated in this sytem but further information has been obtained [175] using the 2-deuterium labelled benzo analogue (*292*). Again on standing at 32° the deuterium in this anion appeared to be scrambling. Carbon-13 n.m.r. spectra of the benzo-6,7-bicyclo[3.2.2]nona-2,6,8-triene obtained after a methanol quench revealed that the deuterium had reached all the non-benzo positions in a nearly statistical manner. This extensive scrambling has been rationalized by invoking the anions (*293*) and (*294*) as intermediates. Anion (*293*) serves to scramble deuterium on positions 2,3 and 8 while (*294*) allows deuterium to be incorporated at the bridgehead positions.

A simpler rearrangement, consistent with part of the scheme suggested for (*292*), has been observed [178] for (*290*). When (*290*) was treated with potassium amide in

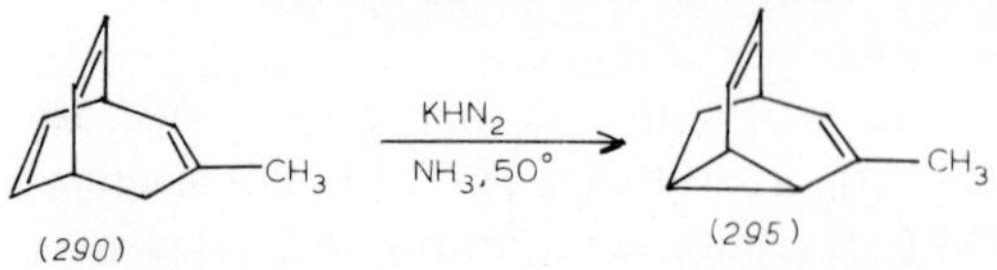

KHN_2, NH_3, 50°

(290) (295)

liquid ammonia at 50° for 3 h a rearrangement occurred to yield 3-methyltricyclo-[3.3.1.0^{2,8}]nona-3,6-diene (*295*) as the only product. This rearrangement is analogous to those shown in Fig. 31 and (*295*) perhaps represents the thermodynamic product.

An example of a rearrangement that could proceed through a second type of bishomocyclopentadienyl anion has been observed [179] during a Wolff–Kishner reduction of a bicyclononatrienone (*296*). Apparently the first formed bishomocyclopenta-

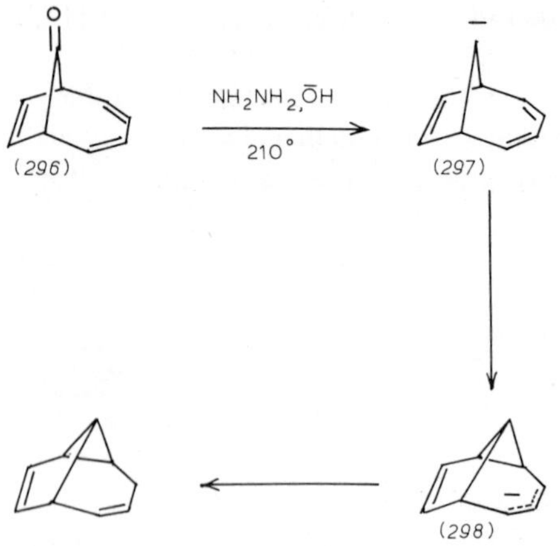

NH_2NH_2, $\bar{O}H$, 210°

(296) (297) (298)

dienyl anion (*297*) preferentially rearranges to the tricyclic allylic anion (*298*) which yields the product.

D. Bishomocyclopropenyl Anions

There have been attempts to identify the role of antiaromaticity [180] in homoconjugated systems. The relative rates of isotopic exchange of the 7-hydrogen in 7-cyanonorbornane (*299*) and *anti*-7-cyanonorbornene (*300*) have been measured [180] in potassium *tert*-butoxide in *tert*-butyl alcohol-*O-d*. It was found that (*300*) reacts only 1.4 times as fast as (*299*) thus providing no evidence for destabilization. Stereochemis-

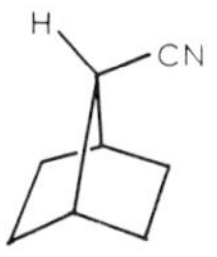

(299)  (300)

try has also been used as a probe for antiaromaticity in an analogous system [181]. The *syn* and *anti*-7-hydrazinonorbornenes ((*301*) and (*302*)) have been reacted under conditions expected to produce the anion; potassium periodate in basic deuterium oxide and *tert*-butyl alcohol-*O-d.* Both (*301*) and (*302*) produced in the *anti*-7-deuterio-norbornene (*303*) as the major isomer; 96% *anti* versus 4% *syn.* This selectivity can

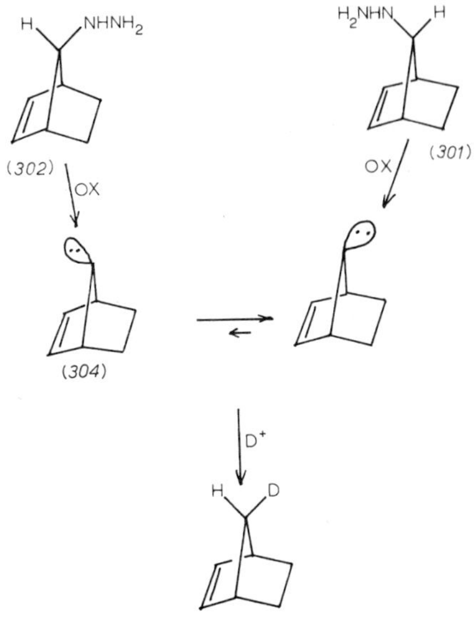

(302) (301) (304) (303)

be interpreted in terms of the electron–electron repulsion in the *anti*-aromatic *syn*-7-carbanion (*304*).

While there are also examples of homoconjugated radical anions and dianions, these have been arbitrarily omitted from this Chapter since the scope has been restricted to carbanions rather than carbanides.

VI. SIGMATROPIC REARRANGEMENTS

In contrast to carbonium ions, carbanions have shown a great reluctance to undergo rearrangements which involve migration of sigma-bonded groups. One such class of rearrangements where migration is across a π-conjugated system has been termed sigmatropic reactions [75] and orbital symmetry arguments have been developed to rationalize and predict the stereochemistry of such processes. Although examples involving carbon to carbon migration are rare there are a number of examples when other elements are present. Examples of the latter include the Wittig rearrangement [182a] (oxygen), the Stevens and Sommelet–Hauser rearrangements [182b,c] (nitrogen), as well as analogous sulfur [183a,b] systems. Since many of the latter reactions have been known for some time and have been reviewed periodically, the present discussion will be restricted to more recent examples involving carbon to carbon migration. These examples have been further subdivided into those cases involving migration of hydrogen and those involving migration of carbon.

A. Hydrogen Migration

Intramolecular migration of hydrogen in carbanions is readily amenable to analysis by orbital symmetry considerations. Figure 34 summarizes the stereochemistry of the allowed thermal migration for a (1,2), (1,4) and a (1,6) shift of hydrogen in successive-

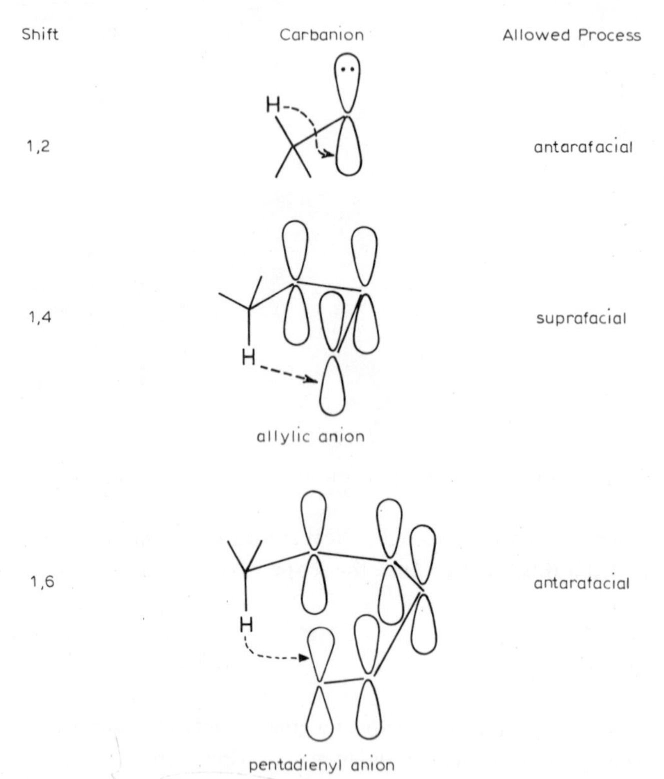

Fig. 34. Stereochemistry of thermally induced orbital symmetry allowed migrations of hydrogens in carbanions.

ly lengthening carbanion chains. This type of analysis points to the unlikelihood of a 1,2 migration of hydrogen occurring as a concerted or intramolecular process. The absence of an example of such a rearrangement in carbanions does not provide positive support for this type of analysis but certainly provides permissive support, particularly when carbanions are contrasted to the myriad of examples of 1,2 hydrogen migrations in carbonium ions and which are orbital symmetry allowed suprafacial processes.

The absence of examples of sigmatropic rearrangements of hydrogen also extends to 1,4 shifts which are allowed suprafacial processes. An example [184] of a case where a search for a 1,4 hydrogen shift produced negative results is provided by 3,4-diphenyl-1-butene (*305*). Treatment of (*305*) at 0° with either *n*-butyllithium in THF–

C_6H_5 (305) C_6H_5 ⇌ C_6H_5 (307) C_6H_5 → C_6H_5 (306) C_6H_5

C_6H_5 (308) C_6H_5

hexane or hexane–*N,N,N',N'*-tetramethylethylenediamine produced a deep red solution which, on quenching with water, yielded 17% of (*305*) and 78% of 1,2-diphenyl-*cis*-2-butene (*306*) and 5% of uncharacterized material. Both the geometry of (*306*) and the analogies discussed in Section III. A. of this Chapter point to the illustrated geometry (*307*) for the postulated allylic anion. Thus in spite of an apparently favourable geometry, conversion of (*307*) to (*308*) was not occurring. The possibility of a reversible 1,4 hydrogen migration but with (*307*) favoured was tested by employing 1,1-dideuterio-(*305*). Similar treatment but with an overnight reflux of the red solution followed by quenching provided no evidence for scrambling of the deuterium label.

Surprisingly the only examples of sigmatropic rearrangements of hydrogen are of 1,6 shifts which are predicted to be antarafacial. One of the first indications of such a possibility came from a study [185] of the reaction of linoleyl alcohol and its methyl ether (*309*) with *n*-butyl lithium in diethylether at room temperature. The products of quenching with carbon dioxide and methylation revealed a considerable mixture of esters which had two conjugated double bonds indicating carbonation at the termini of the pentadienyl anions. Mass spectrometric analysis of this mixture after hydrogenation suggested the migration of the postulated pentadienyl anion over a considerable portion of the carbon chain (*310*). A series of intramolecular 1,6 hydrogen shifts was postulated.

Further examples as well as evidence for the intramolecular nature of such rear-

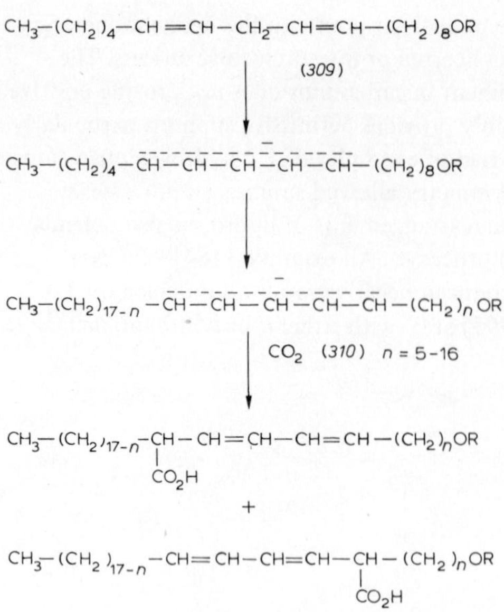

$CH_3{-}(CH_2)_4{-}CH{=}CH{-}CH_2{-}CH{=}CH{-}(CH_2)_8OR$

(309)

$CH_3{-}(CH_2)_4{-}\overline{CH{-}CH{-}CH{-}CH{-}CH}{-}(CH_2)_8OR$

$CH_3{-}(CH_2)_{17-n}{-}\overline{CH{-}CH{-}CH{-}CH{-}CH}{-}(CH_2)_nOR$

CO_2 (310) $n = 5\text{-}16$

$CH_3{-}(CH_2)_{17-n}{-}CH(CO_2H){-}CH{=}CH{-}CH{=}CH{-}(CH_2)_nOR$

+

$CH_3{-}(CH_2)_{17-n}{-}CH{=}CH{-}CH{=}CH{-}CH(CO_2H){-}(CH_2)_nOR$

rangements came from a study [186] of structurally simpler pentadienyl anions. It was discovered from n.m.r. that (*311*), prepared from 5-methyl-1,4-hexadiene and *n*-butyllithium in THF–hexane, partially rearranged at 40° to produce 52% of (*312*).

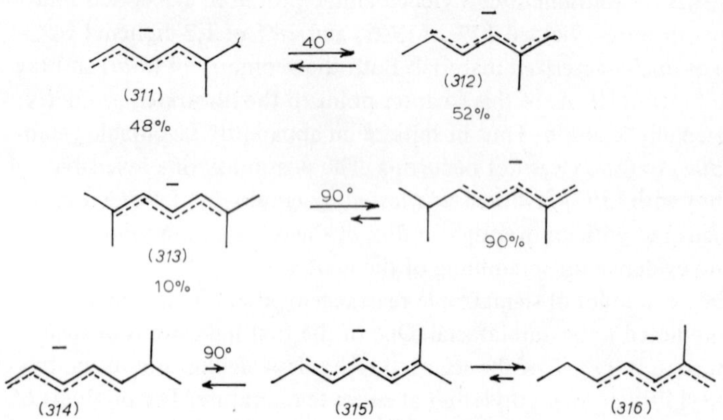

40°

(311) 48%

(312) 52%

90°

(313) 10%

90%

90°

(314) 10%

(315) 0%

(316) 90%

This was shown to be an equilibrium mixture by starting from (*312*). In an analogous manner anions (*313*), (*314*), (*315*) and (*316*) were also observed to undergo rearrangement apparently to equilibrium mixtures of the anions. The equilibrium percentages are shown beside the anions and the temperature used for equilibration is indicated.

The intramolecular nature of the rearrangement of (*315*) was demonstrated [186]

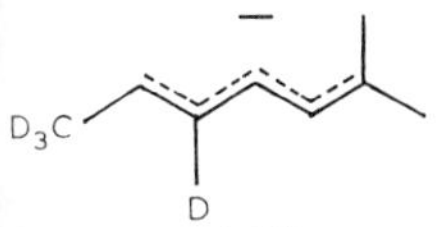

(*317*)

in two ways using the tetradeuterio analogue (*317*). First (*317*) was heated at 90° for 30 min and then protonated. Mass spectrometry on the product dienes showed only tetradeuterio-materials. Second, heating of a 1:1 mixture of (*315*) and (*317*) showed no interchange of deuteriums between the anions.

Permissive but negative evidence for the predicted antarafacial 1,6 shift in pentadienyl anions was obtained using the cycloheptadienyl anions (*318*) and (*319*) as models. In contrast to the open chain anions, which presumably can adopt a geometry

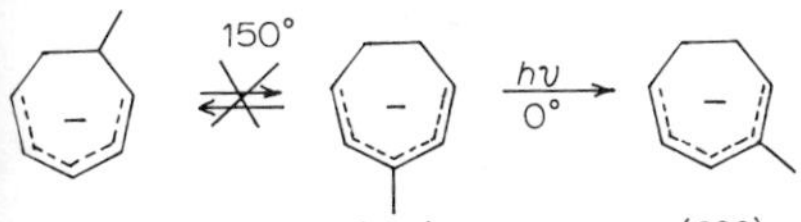

(*318*) (*319*) (*320*)

appropriate for antarafacial 1,6 hydrogen shift, neither (*318*) nor (*319*) could be persuaded to interconvert at 150° for 30 min. Interestingly interconversion of anion (*319*) to (*320*) could be achieved upon irradiation using a high pressure mercury lamp indicating that suprafacial shifts are possible, although (*318*) did not react under similar circumstances. The unique ability of the open chain pentadienyl anions to undergo 1,6 hydrogen shifts may reflect their ability to adopt a particularly favourable shape for reaction. Similar flexibility is not available for (1,2) or (1,4) hydrogen shifts in open chain anions or 1,6 shifts in cyclic anions.

More complex rearrangements which seem to involve 1,6 hydrogen shifts have been observed in other systems. Thus anion (*321*) was observed [186] to rearrange quanti-

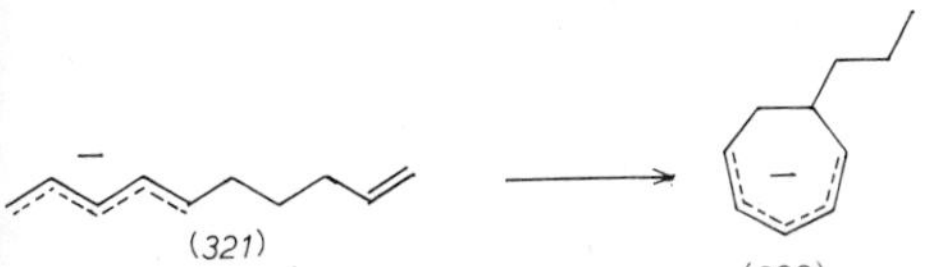

(*321*) (*322*)

tatively to the cycloheptadienyl isomer (*322*) on heating at 110°. This presumably is the result of 3 successive hydrogen shifts followed by electrocyclization (see Section IV. C.). In an analogous system anion (*323*), prepared from 3-vinyl-1,4-heptadiene, rearranged [91] at 33° to the cycloheptadienyl anion (*324*). Again this can be viewed as the result of two sigmatropic 1,6 hydrogen shifts followed by electrocyclization.

Likewise, the reaction [187] of linolenyl alcohol or its methyl ether (*325*) with butyl lithium in ether yielded products of carbonation, methylation and hydrogena-

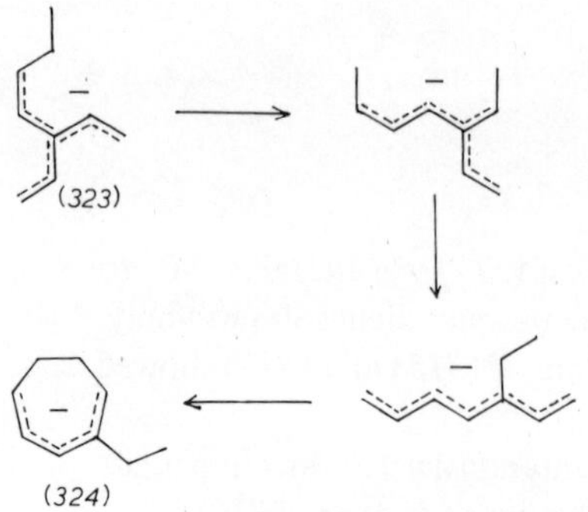

(323)

(324)

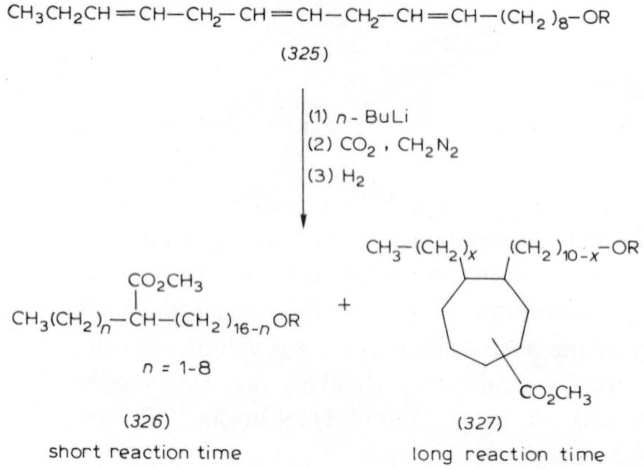

$CH_3CH_2CH{=}CH{-}CH_2{-}CH{=}CH{-}CH_2{-}CH{=}CH{-}(CH_2)_8{-}OR$

(325)

(1) n-BuLi
(2) CO_2, CH_2N_2
(3) H_2

$CH_3(CH_2)_n{-}CH(CO_2CH_3){-}(CH_2)_{16-n}OR$

n = 1-8

(326)
short reaction time

+

$CH_3{-}(CH_2)_x$ $(CH_2)_{10-x}{-}OR$ CO_2CH_3

(327)
long reaction time

tion (*326*) consistent with heptatrienyl anions at short reaction time and products (*327*) consistent with cycloheptadienyl anions on further standing. Protonation and deuteration as well as ozonolysis were employed in product identification.

B. Carbon Migration

The migration of carbon can be conveniently subdivided into two cases, saturated carbon and unsaturated carbon. Orbital symmetry requirements can be applied to the migration of saturated carbon and the consequence of this on 1,2 shifts is illustrated in Fig. 35. There are two orbital symmetry allowed pathways for 1,2 migration of saturated carbon but both require unfavourable transition state geometries. Thus it is not too surprising that no examples of this type of concerted rearrangement are known. Also examples of 1,4 or 1,6 migrations have not been discovered. There is one example [188], however, of an apparent [3,2]-sigmatropic rearrangement which has been interpreted as occurring with a concerted component. When 9-(3-methylbut-2-enyl)-9-(lithiomethyl)-fluorene (*328*), prepared from the corresponding bromo compound and

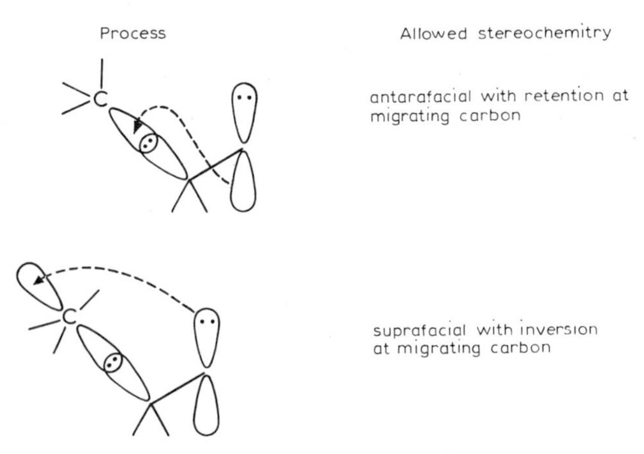

Fig. 35. Stereochemistry of thermally induced orbital symmetry allowed 1,2 migration of saturated carbon in carbanions.

lithium in THF at $-70°$, was warmed to $-20°$, three products were obtained ((*329*), (*330*), (*331*)) after quenching. The rearranged product (*329*) was interpreted as arising through a concerted route with (*330*) and (*331*) being cleavage products. The tempera-

Li
(328)
(1) $-20°$, THF
(2) H_2O

(329) 70%
(330) 10%
(331) 20%

ture dependence of the product distribution was presented as supportive evidence for this interpretation.

The sigmatropic migration of an unsaturated carbon must necessarily occur with retention and as a result in an antarafacial manner along the C–C chain in a 1,2 shift. However a non-sigmatropic and allowed pathway is also available and this can be viewed as a nucleophilic addition of the carbanion to the unsaturated carbon followed by elimination. Within this category fall rearrangements of the carbonyl group (*332*) the olefinic double bond (*333*) and the phenyl ring (*334*). There are examples of all

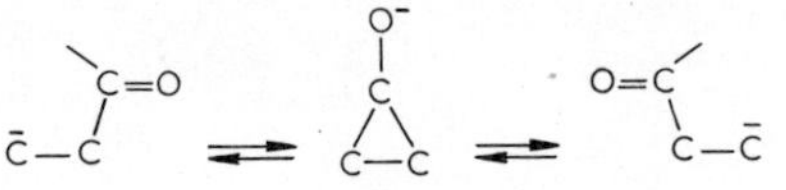

(*332*)

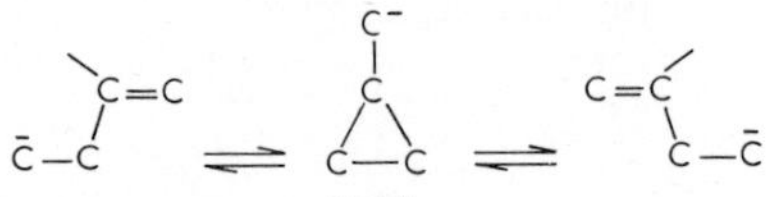

(*333*)

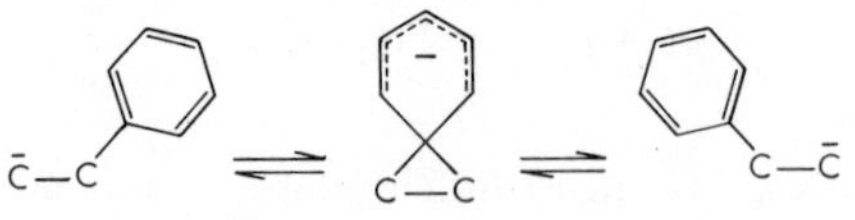

(*334*)

three cases and the rearrangement of the carbonyl group and the olefinic double bond have been discussed in Section V as examples of homoconjugative rearrangements. The rearrangement of the phenyl ring, known as the Grovenstein–Zimmerman rearrangement, will not be discussed in detail here since it has been reviewed periodically [1e]. It is worth noting, however, that firm evidence has been obtained [189] to support the intermediacy of a species like that shown in (*334*) in this type of a rearrangement.

REFERENCES

1 a D.J. Cram, Fundamentals of Carbanion Chemistry, Academic Press, New York, 1965, p. 47. b D.J. Cram, Fundamentals of Carbanion Chemistry, Academic Press, New York, 1965, pp. 85–105. c D.J. Cram, Fundamentals of Carbanion Chemistry, Academic Press, New York, 1965, Chap. V. d D.J. Cram, Fundamentals of Carbanion Chemistry, Academic Press, New York, 1965, pp. 204–210. e D.J. Cram, Fundamentals of Carbanion Chemistry, Academic Press, New York, 1965, p. 237.
2 R.J. Gillespie, J. Chem. Educ., 47 (1970) 18.
3 R.C. Fort and P. von R. Schleyer, Advan. Alicyclic Chem., Academic Press, New York, 1966, Vol. 1, pp. 287–292.
4 R.E. Kari and I.G. Csizmadia, J. Chem. Phys., 50 (1969) 1443; 56 (1972) 4337.
5 D.H. Owens and A. Streitwieser, Jr., Tetrahedron, 27 (1971) 4471.
6 E. Weiss and G. Hencken, J. Organometal. Chem., 21 (1970) 265.
7 E. Weiss and G. Sauermann, J. Organometal. Chem., 21 (1970) 1.
8 E. Weiss and G. Sauermann, Chem. Ber., 103 (1970) 265.
9 H. Dietrich, Acta Crystallogr., 16 (1963) 681.
10 R.P. Zerger and G.D. Stucky, Chem. Commun., (1973) 44.
11 S.P. Patterman, I.L. Karle and G.D. Stucky, J. Amer. Chem. Soc., 92 (1970) 1150.
12 J.J. Brooks and G.D. Stucky, J. Amer. Chem. Soc., 94 (1972) 7333.
13 J.J. Brooks, W. Rhine and G.D. Stucky, J. Amer. Chem. Soc., 94 (1972) 7339.

14 J.F. Coetzee and C.D. Ritchie, Solute–Solvent Interactions, Marcel Dekker, New York, 1969.
15 M. Szwarc, Carbanions, Living Polymers and Electron-Transfer Processes, Interscience, New York, 1968.
16 M. Szwarc, Ions and Ion Pairs in Organic Reactions, Wiley-Interscience, New York, 1972.
17 a T.L. Brown, Pure Appl. Chem., 23 (1970) 447. b T.L. Brown, Accounts Chem. Res., 1 (1968) 23. c T.L. Brown, Advan. Organometal. Chem., 3 (1965) 365.
18 P. West, R. Waack and J.I. Purmort, J. Amer. Chem. Soc., 92 (1970) 840.
19 a J.M. Motes and H.M. Walborsky, J. Amer. Chem. Soc., 92 (1970) 3697. b H.M. Walborsky and J.M. Motes, J. Amer. Chem. Soc., 92 (1970) 2445.
20 a S.J. Miller and W.G. Lee, J. Amer. Chem. Soc., 81 (1959) 6316. b D.H. Hunter and D.J. Cram, J. Amer. Chem. Soc., 86 (1964) 5478. c D.H. Hunter and D.J. Cram, J. Amer. Chem. Soc., 88 (1966) 5765. d H.M. Walborsky and L.M. Turner, J. Amer. Chem. Soc., 94 (1972) 2273.
21 Lead References include a E.J. Corey and T.H. Lowry, Tetrahedron Lett., (1965) 803. b D.J. Cram and T.A. Whitney, J. Amer. Chem. Soc., 89 (1967) 4651. c F.G. Bordwell, E. Doomes and P.W.R. Corfield, J. Amer. Chem. Soc., 92 (1970) 2581. d L.A. Paquette, J.P. Freeman and M.J. Wyvratt, J. Amer. Chem. Soc., 93 (1971) 3216.
22 F.G. Bordwell, Accounts Chem. Res., 3 (1970) 281.
23 H.E. Zimmerman, in P. de Mayo, (Ed.), Molecular Rearrangements, Interscience, Vol. 1, New York, 1963, Chapter 6.
24 C.D. Broaddus, Accounts Chem. Res., 1 (1968) 231.
25 A.J. Hubert and H. Reimlinger, Synthesis, (1969) 97.
26 D.J. Cram and R.T. Uyeda, J. Amer. Chem. Soc., 84 (1962) 4358.
27 D.J. Cram and R.T. Uyeda, J. Amer. Chem. Soc., 86 (1964) 5466.
28 D.H. Hunter and D.J. Cram, J. Amer. Chem. Soc., 86 (1964) 5478.
29 S.W. Ela and D.J. Cram, J. Amer. Chem. Soc., 88 (1966) 5791.
30 J. Klein and S. Brenner, Chem. Commun., (1969) 1020.
31 J.M. Figuera, J.M. Gamboa and I. Santos, J. Chem. Soc. Perkin II, (1972) 1434.
32 S. Bank, C.A. Rowe and A. Schriesheim, J. Amer. Chem. Soc., 85 (1963) 2115.
33 D.H. Hunter and R.W. Mair, Can. J. Chem., 47 (1969) 2361.
34 R.B. Bates, R.H. Carnighan and C.E. Staples, J. Amer. Chem. Soc., 85 (1963) 3032.
35 W. von E. Doering and P.P. Gaspar, J. Amer. Chem. Soc., 85 (1963) 3043.
36 C.D. Broaddus, Accounts Chem. Res., 1 (1968) 231.
37 a D.J. Cram, F. Willey, H.P. Fischer and D.A. Scott, J. Amer. Chem. Soc., 86 (1964) 5370. b D.J. Cram, F. Willey, H.P. Fischer, H.M. Relles and D.A. Scott, J. Amer. Chem. Soc., 88 (1966) 2759.
38 R.D. Guthrie, W. Meister and D.J. Cram, J. Amer. Chem. Soc., 89 (1967) 5288.
39 R.D. Guthrie, D.A. Jaeger, W. Meister and D.J. Cram, J. Amer. Chem. Soc., 93 (1971) 5137.
40 D.J. Cram and R.D. Guthrie, J. Amer. Chem. Soc., 88 (1966) 5760.
41 R.D. Guthrie and J.L. Hedrick, J. Amer. Chem. Soc., 95 (1973) 2971.
42 D.H. Hunter and S.K. Sim, Can. J. Chem., 50 (1972) 678.
43 D.H. Hunter and R.P. Steiner, Can. J. Chem., (1975) in press.
44 R.D. Guthrie and G.R. Weisman, Chem. Commun., (1969) 1316.
45 G. Bergson and L. Ohlsson, Acta Chem. Scand., 21 (1967) 1393.
46 G. Bergson and A-M. Wiedler, Acta Chem. Scand., 17 (1963) 862; 18 (1964) 1483.
47 G. Bergson, Acta Chem. Scand., 17 (1963) 2691.
48 a J. Almy and D.J. Cram, J. Amer. Chem. Soc., 91 (1969) 4459. b J. Almy, R.T. Uyeda and D.J. Cram, J. Amer. Chem. Soc., 89 (1967) 6768.
49 J. Almy, D.C. Garwood and D.J. Cram, J. Amer. Chem. Soc., 92 (1970) 4321

50 a C.K. Ingold, Structure and Mechanism in Organic Chemistry, 2nd Edn., Cornell University Press, Ithaca, New York, 1969, pp. 834–837. b C.K. Ingold, Structure and Mechanism in Organic Chemistry, 2nd Edn., Cornell University Press, Ithaca, New York, 1969, p. 826.
51 H.O. House, Modern Synthetic Reactions, 2nd Edn., W.A. Benjamin, Inc., Menlo Park, California, 1972, Chap. 9.
52 D. Turnbull and S.H. Maron, J. Amer. Chem. Soc., 65 (1943) 212.
53 W.E. Hugh and C.A.R. Kon, J. Chem. Soc., (1930) 775.
54 S.W. Ela and D.J. Cram, J. Amer. Chem. Soc., 88 (1966) 5777.
55 A. Gero, J. Org. Chem., 19 (1954) 469, 1960.
56 D.A. Jaeger and D.J. Cram, J. Amer. Chem. Soc., 93 (1971) 5153.
57 D.H. Hunter and J.B. Stothers, Can. J. Chem., 51 (1973) 2884.
58 A. Streitwieser, R.A. Caldwell, R.G. Lawler and G.R. Ziegler, J. Amer. Chem. Soc., 87 (1965) 5399.
59 C.D. Broaddus, J. Amer. Chem. Soc., 90 (1968) 5504.
60 M. Saunders and E.H. Gold, J. Amer. Chem. Soc., 88 (1966) 3376.
61 R. Kuhn and D. Rewicki, Tetrahedron Lett., (1965) 3513.
62 A.J. Birch, and G.S. Rao, Advances in Organic Chemistry, Vol. 8, Wiley-Interscience, New York, 1972, p. 1.
63 A.J. Birch, Quart. Rev. Chem. Soc., 4 (1950) 69.
64 R.B. Bates, S. Brenner, C.M. Cole, E.W. Davidson, G.D. Forsythe, D.A. McCombs and A.S. Roth, J. Amer. Chem. Soc., 95 (1973) 926.
65 R.B. Bates, D.W. Gosselink and J.A. Kaczynski, Tetrahedron Lett., (1967) 199.
66 R.B. Bates, W.H. Dienes, D.A. McCombs and D.E. Potter, J. Amer. Chem. Soc., 91 (1969) 4608.
67 R.B. Bates and D.A. McCombs, Tetrahedron Lett., (1969) 977.
68 J. Hine, J. Org. Chem., 31 (1966) 1236.
69 For a recent review and useful comments see J.B. Stothers, Carbon-13 NMR Spectroscopy, Academic Press, New York, 1972, Chap. 6 (I).
70 R.J. Bushby and G.L. Ferber, Chem. Commun., (1973) 407.
71 G. Bergson and A-M. Wiedler, Acta Chem. Scand., 17 (1963) 1798.
72 A-M. Wiedler and G. Bergson, Acta Chem. Scand., 18 (1964) 1487.
73 G. Sorlin and G. Bergson, Ark. Kemi, 29 (1968) 593.
74 J. Almy, D.H. Hoffman, K.C. Chu and D.J. Cram, J. Amer. Chem. Soc., 95 (1973) 1185.
75 R.B. Woodward and R. Hoffmann, The Conservation of Orbital Symmetry, Academic Press, Weinheim, Germany, 1971.
76 A.Schriesheim and C.A. Rowe, Jr., Tetrahedron Lett., (1962) 405.
77 C.C. Price and W.H. Snyder, Tetrahedron Lett., (1962) 69.
78 S. Bank, A. Schriesheim and C.A. Rowe, J. Amer. Chem. Soc., 87 (1965) 3244.
79 S. Bank, J. Amer. Chem. Soc., 87 (1965) 3245.
80 a W.O. Haag and H. Pines, J. Amer. Chem. Soc., 82 (1960) 387. b G.J. Heiszwolf, J.A.A. Van Drunen and H. Kloosterziel, Rec. Trav. Chim. Pays-Bas, 88 (1969) 1377.
81 R. Hoffmann and R.A. Olofson, J. Amer. Chem. Soc., 88 (1966) 943.
82 R.B. Bates and W.A. Beaver, J. Amer. Chem. Soc., 96 (1974) 5001.
83 D. Seyferth and T.F. Jula, J. Organometal. Chem., 8 (1967) P13.
84 E.R. Dolinskaya, I.Y. Poddubnyi and I.V. Tsereteli, Dokl. Phys. Chem., 191 (1970) 279.
85 a W.H. Glaze and P.C. Jones, Chem. Commun., (1969) 1434. b W.H. Glaze, J.E. Hanicak, M.L. Moore and J. Chaudhuri, J. Organometal. Chem., 44 (1972) 39. c W.H. Glaze, J. Hanicak, D.J. Berry and D.P. Duncan, J. Organometal. Chem., 44 (1972) 49. d W.H. Glaze, J.E. Hanicak, J. Chaudhuri, M.L. Moore and D.P. Duncan, J. Organometal. Chem., 51 (1973) 13.
86 M.D. Carr, J.R.P. Clark and M.C. Whiting, Proc. Chem. Soc., London, (1963) 333.
87 R.B. Bates, D.W. Gosselink and J.A. Kaczynski, Tetrahedron Lett., (1967) 205.

88 G.J. Heiszwolf and H. Kloosterziel, Rec. Trav. Chim. Pays-Bas, 86 (1967) 807.
89 H. Kloosterziel and J.A.A. Van Drunen, Rec. Trav. Chim. Pays-Bas, 88 (1969) 1084.
90 R.B. Bates, S. Brenner and B.I. Mayall, J. Amer. Chem. Soc., 94 (1972) 4765.
91 R.B. Bates, S. Brenner and C.M. Cole, J. Amer. Chem. Soc., 94 (1972) 2130.
92 H. Kloosterziel and J.A.A. Van Drunen, Rec. Trav. Chim. Pays-Bas, 89 (1970) 270.
93 W.T. Ford and M. Newcomb, J. Amer. Chem. Soc., 96 (1974) 309.
94 H. Kloosterziel and J.A.A. Van Drunen, Rec. Trav. Chim. Pays-Bas, 88 (1969) 1471.
95 V.R. Sandel, S.V. McKinley and H.H. Freedman, J. Amer. Chem. Soc., 90 (1968) 495.
96 G.J. Heiszwolf and H. Kloosterziel, Rec. Trav. Chim. Pays-Bas, 86 (1967) 1345.
97 a H.H. Freedman, V.R. Sandel and B.P. Thill, J. Amer. Chem. Soc., 89 (1967) 1762. b J.W. Burley and R.N. Young, J. Chem. Soc. Perkin II, (1972) 1843 and references therein.
98 H. Kloosterziel and J.A.A. Van Drunen, Rec. Trav. Chim. Pays-Bas, 89 (1970) 37.
99 H. Kloosterziel and J.A.A. Van Drunen, Rec. Trav. Chim. Pays-Bas, 87 (1968) 1025.
100 S. Brenner and J. Klein, Isr. J. Chem., 7 (1969) 735.
101 H. Kloosterziel, Rec. Trav. Chim. Pays-Bas, 89 (1970) 300.
102 H. Kloosterziel and J.A.A. Van Drunen, Rec. Trav. Chim. Pays-Bas, 89 (1970) 667.
103 H. Kloosterziel and J.A.A. Van Drunen, Rec. Trav. Chim. Pays-Bas, 89 (1970) 32.
104 a T.J. Prosser, J. Amer. Chem. Soc., 83 (1961) 1701. b P. Canbere and M.F. Hochu, Bull. Soc. Chim. Fr., (1968) 459. c G. Courtois and L. Miginiac, Tetrahedron Lett., (1972) 2411. d A.J. Hubert, A. Georis, R. Warin and P. Teyssie, J. Chem. Soc. Perkin II, (1972) 366.
105 P. West, J.I. Purmont and S.V. McKinley, J. Amer. Chem. Soc., 90 (1968) 797.
106 F.J. Kronzer and V.R. Sandel, J. Amer. Chem. Soc., 94 (1972) 5751.
107 J.J. Lagowski and G.A. Moczygemba, The Chemistry of Non-aqueous Solvents, Volume II, Academic Press, New York, 1967, Chap. 7-C.
108 R. Huisgen and P. Eberhard, J. Amer. Chem. Soc., 94 (1972) 1346.
109 P. Eberhard and R. Huisgen, J. Amer. Chem. Soc., 94 (1972) 1345.
110 G. Boche and D. Martins, Angew. Chem. Int. Ed. Engl., 11 (1972) 724.
111 M. Newcomb and W.T. Ford, J. Amer. Chem. Soc., 95 (1973) 7186.
112 J.E. Mulvaney and D. Savage, J. Org. Chem., 36 (1971) 2592.
113 W.T. Ford and M. Newcomb, J. Amer. Chem. Soc., 95 (1973) 6277.
114 D.E. Applequist and A.H. Peterson, J. Amer. Chem. Soc., 83 (1961) 862.
115 H.M. Walborsky and F.M. Hornyak, J. Amer. Chem. Soc., 77 (1955) 6026.
116 a J.B. Pierce and H.M. Walborsky, J. Org. Chem., 33 (1968) 1962. b H.M. Walborsky, F.J. Impasto and A.E. Young, J. Amer. Chem. Soc., 86 (1964) 3283.
117 a S.W. Benson, Thermochemical Kinetics, Wiley, New York, 1968, p. 170. b M.J.S. Dewar and S. Kirschner, J. Amer. Chem. Soc., 93 (1971) 4290 and references therein.
118 T. Kauffmann, K. Habersaat and E. Koppelmann, Angew. Chem. Int. Ed. Engl., 11 (1972) 291.
119 G. Boche, D. Martens and W. Danzer, Angew. Chem. Int. Ed. Engl., 8 (1969) 984.
120 G. Wittig, V. Rautenstrauch and F. Wingler, Tetrahedron Suppl., 7 (1966) 189.
121 M.E. Londrigan and J.E. Mulvaney, J. Org. Chem., 37 (1972) 2823.
122 P.R. Stapp and R.F. Kleinschmidt, J. Org. Chem., 30 (1965) 3006.
123 L.H. Slaugh, J. Org. Chem., 32 (1967) 108.
124 H. Kloosterziel and J.A.A. Van Drunen, Rec. Trav. Chim. Pays-Bas, 89 (1970) 368.
125 P.J. Garratt and K.A. Knapp, Chem. Commun., (1970) 1215.
126 D.J. Atkinson, M.J. Perkins and P.Ward, Chem. Commun., (1969) 1390.
127 S.W. Staley and J.P. Erdman, J. Amer. Chem. Soc., 92 (1970) 3832.
128 a D.H. Hunter and S.K. Sim, J. Amer. Chem. Soc., 91 (1969) 6202. b D.H. Hunter and S.K. Sim, Can. J. Chem., 50 (1972) 669, 678.
129 E.A. Zeuch, D.L. Crain and R.F. Kleinschmidt, J. Org. Chem., 33 (1968) 771.

130 H. Kloosterziel and G.M. Gorter-la Roy, Chem. Commun., (1972) 352.
131 H. Kloosterziel and E. Zwanenburg, Rec. Trav. Chim. Pays-Bas, 88 (1969) 1373.
132 H. Kloosterziel, J.A.A. Van Drunen and P. Galama, Chem. Commun., (1969) 885.
133 R.B. Bates, L.M. Kroposki and D.E. Potter, J. Org. Chem., 37 (1972) 560.
134 a A. Nickon and J.L. Lambert, J. Amer. Chem. Soc., 84 (1962) 1604. b A. Nickon and J.L. Lambert, J. Amer. Chem. Soc., 88 (1966) 1905.
135 a D.M. Hudyma, J.B. Stothers and C.T. Tan, Org. Magn. Res., in press. b J.B. Stothers and C.T. Tan, Chem. Commun., (1974) 738.
136 A.L. Johnson, J.B. Stothers and C.T. Tan, Can. J. Chem., in press.
137 A.L. Johnson, N.O. Petersen, M.B. Rampersad and J.B. Stothers, Can. J. Chem., in press.
138 A. Nickon, H. Kwasnik, T. Swartz, R.O. Williams and J.B. DiGiorgio, J. Amer. Chem. Soc., 87 (1965) 1615.
139 R.M. Coates and J.P. Chen, Chem. Commun., (1970) 1481.
140 R. Howe and S. Winstein, J. Amer. Chem. Soc., 87 (1965) 915.
141 T. Fukunaga, J. Amer. Chem. Soc., 87 (1965) 916.
142 D.H. Hunter, A.L. Johnson, J.B. Stothers, A. Nickon, J.L. Lambert and D.F. Covey, J. Amer. Chem. Soc., 94 (1972) 8582.
143 a A. Nickon, J.H. Hammons, J.L. Lambert and R.O. Williams, J. Amer. Chem. Soc., 85 (1963) 3713. b A. Nickon, J.L. Lambert, R.O. Williams and N.H. Werstiuk, J. Amer. Chem. Soc., 88 (1966) 3354.
144 G.C. Joshi, W.D. Chambers and E.W. Warnhoff, Tetrahedron Lett., (1967) 3613.
145 A.J.H. Klunder and E. Zwanenburg, Tetrahedron Lett., (1971) 1721.
146 W.T. Borden, V. Varma, M. Cabell and T. Ravindranathan, J. Amer. Chem. Soc., 93 (1971) 3800.
147 A. Nickon, J.L. Lambert and J.E. Oliver, J. Amer. Chem. Soc., 88 (1966) 2787.
148 K.W. Turnbull, S.J. Gould and D. Arigoni, Chem. Commun., (1972) 597.
149 J.E. Nordlander, S.P. Jindal and D.J. Kitko, Chem. Commun., (1969) 1136.
150 J.P. Freeman and J.H. Plonka, J. Amer. Chem. Soc., 88 (1966) 3662.
151 a P. Yates, G.D. Abrams and S. Goldstein, J. Amer. Chem. Soc., 91 (1969) 6868. b P. Yates, G.D. Abrams, M.J. Betts and S. Goldstein, Can. J. Chem., 49 (1971) 2850.
152 a M.J. Betts and P. Yates, J. Amer. Chem. Soc., 92 (1970) 6982. b P. Yates and M.J. Betts, J. Amer. Chem. Soc., 94 (1972) 1965.
153 D.J. Cram, C.A. Kingsbury and B. Rickborn, J. Amer. Chem. Soc., 83 (1961) 3688.
154 M.S. Silver, P.R. Shafer, J.E. Nordlander, C. Ruchardt and J.D. Roberts, J. Amer. Chem. Soc., 82 (1960) 2646.
155 D.J. Patel, C.L. Hamilton and J.D. Roberts, J. Amer. Chem. Soc., 87 (1965) 5144.
156 A. Maercker and J.D. Roberts, J. Amer. Chem. Soc., 88 (1966) 1742.
157 M.E.H. Howden, A. Maercker, J. Burdon and J.D. Roberts, J. Amer. Chem. Soc., 88 (1966) 1732.
158 A. Maercker and K. Weber, Justus Liebigs Ann. Chem., 756 (1972) 20.
159 A. Maercker and K. Weber, Justus Liebigs Ann. Chem., 756 (1972) 43.
160 A. Maercker and R. Geuss, Angew. Chem. Int. Ed. Engl., 9 (1970) 909.
161 P.T. Lansbury, V.A. Pattison, W.A. Clement and J.D. Sidler, J. Amer. Chem. Soc., 86 (1964) 2247.
162 M.E.H. Howden and J.D. Roberts, Tetrahedron Suppl. 2, 19 (1963) 403.
163 J.K. Stille and K.N. Sannes, J. Amer. Chem. Soc., 94 (1972) 8494.
164 a M.J. Goldstein, J. Amer. Chem. Soc., 89 (1967) 6357. b M.J. Goldstein and R. Hoffmann, J. Amer. Chem. Soc., 93 (1971) 6193.
165 J.M. Brown and J.L. Occolowitz, J. Chem. Soc., B (1968) 411.
166 J.M. Brown, E.N. Coin and M.C. McIvor, J. Chem. Soc., B (1971) 730.
167 S. Winstein, Aromaticity, Special Publication No. 21, The Chemical Society, London, 1967, p. 34.

168 J.M. Brown and J.L. Occolowitz, Chem. Commun., (1965) 376.
169 J.W. Rosenthal and S. Winstein, Tetrahedron Lett., (1970) 2683.
170 J.M. Brown, Chem. Commun., (1967) 638.
171 S. Winstein, M. Ogliaruso, M. Sakai and J.M. Nicholson, J. Amer. Chem. Soc., 89 (1967) 3656.
172 S. Winstein, Quart. Rev., 23 (1969) 141.
173 J.M. Brown and E.N. Coin, J. Amer. Chem. Soc., 92 (1970) 3821.
174 W. Eberbach and H. Prinzback, Chem. Ber., 102 (1969) 4164.
175 M.V. Moncur and J.B. Grutzner, J. Amer. Chem. Soc., 95 (1973) 6449.
176 J.B. Grutzner and S. Winstein, J. Amer. Chem. Soc., 90 (1968) 6562.
177 J.B. Grutzner and S. Winstein, J. Amer. Chem. Soc., 94 (1972) 2200.
178 a S.W. Staley and D.W. Reichard, J. Amer. Chem. Soc., 91 (1969) 3998. b M.J. Goldstein and S. Natowsky, J. Amer. Chem. Soc., 95 (1973) 6451.
179 M. Sakai, D.L. Harris and S. Winstein, Chem. Commun., (1972) 861.
180 R. Breslow, R. Pagni and W.N. Washburn, Tetrahedron Lett., (1970) 547.
181 J.K. Stille and K.N. Sannes, J. Amer. Chem. Soc., 94 (1972) 8489.
182 A recent review a A. Jefferson and F. Scheinmann, Quart. Rev. Chem. Soc., 22 (1968) 391. b S.H. Pine, Organic Reactions, Vol. 18, Wiley, New York, 1970, Chap. 4. c A.R. Lepley and A.G. Giumanini, in B.S. Thyagarajan (Ed.), Mechanisms of Molecular Rearrangements, Wiley-Interscience, New York, 1971, Vol. 3, p. 297.
183 a V. Rautenstrauch, Chem. Commun., (1970) 4 and references therein. b A. Wright, D. Ling, P. Boudjouk and R. West, J. Amer. Chem. Soc., 94 (1972) 4784.
184 R.M. Magid and S.E. Wilson, Tetrahedron Lett., (1971) 19.
185 J. Klein, S. Glily and D. Kost, J. Org. Chem., 35 (1970) 1281.
186 R.B. Bates, S. Brenner, W.H. Dienes, D.A. McCombs and D.E. Potter, J. Amer. Chem. Soc., 92 (1970) 6345.
187 J. Klein and S. Gilly, Tetrahedron, 27 (1971) 3477.
188 J.E. Baldwin and F.J. Urban, Chem. Commun., (1970) 165.
189 E. Grovenstein and G. Wentworth, J. Amer. Chem. Soc., 89 (1967) 2348.

APPENDIX

As a guide to the reader, the specific compounds that have been referenced in this chapter are compiled here according to type. They are referenced either according to section (e.g. IIB), figure (e.g. f 11) or table (e.g. t 15). The compounds are organized under the general headings of Organometallic and Organic.

A. Organometallic

Organolithium
–, allyl, n.m.r., IIIB, f 19
–, allyl, rotational barrier, f 20
–, benzyl, aggregation, t 1
–, benzyl, X-ray crystal structure, IB
–, 1-benzyl-1-phenylallyl, sigmatropic reaction, VIA
–, 1-bicyclo[1.1.0]butyl, X-ray crystal structure, IB
–, *cis*-bicyclo[3.3.0]octenyl, from electrocyclization, IVB
–, 3-butenyl, rearrangement, VB
–, *tert*-butyl, aggregation, t 1
–, 1-carboxylate-2,3-diphenylallyl, from ring opening, IVA
–, 2-carboxylate-1,3-diphenylallyl, from ring opening, IVA
–, 1-cyano-2,2-diphenylcyclopropyl, ring opening, IVA
–, 2-cyano-1,3-diphenylallyl, from ring opening, IVA
–, cycloheptadienyl, f 11
–, cycloheptadienyl, from electrocyclization, IVC
–, cyclononadienyl, f 11
–, cyclooctadienyl, electrocyclization, IVB
–, *exo*- and *endo*-7-cycloprop(a)acenaphthylenyl, ring opening, IVA
–, cyclopropylcarbinyl, rearrangement, VB
–, cyclopropyldiphenylcarbinyl, rearrangement, VB
–, 4-cyclopropyl-4-phenyl-3-butenyl, f 29
–, 1,2:3,4-dibenzocyclononatetraenyl, n.m.r. and cyclization, IVB
–, 1,1- and -2,2-dideuterio-3-methyl-3-butenyl, rearrangement, VB
–, 1,3-dimethylcycloheptadienyl, from electrocyclization, IVC
–, 2,6-dimethylheptatrienyl, electrocyclization, IVC
–, dimethylcyclohexadienyl, f 11
–, 1,3-dimethylidenecyclohexyl, f 11
–, 1,1-dimethylpentadienyl, n.m.r., t 15
–, 1,1-dimethylpentadienyl, sigmatropic reaction, VIA
–, 1,4-dimethylpentadienyl, sigmatropic reaction, VIA

B. Organic

Chapter 5

UTILIZATION OF DEUTERIUM LABELING IN ORGANIC PHOTOCHEMICAL REARRANGEMENTS

JOHN S. SWENTON

Department of Chemistry, The Ohio State University, Columbus, Ohio 43210 (U.S.A.)

I. INTRODUCTION

The excited state processes of simple organic molecules have been studied by gas phase photochemists and spectroscopists for many years. However, it was only recently that a significant segment of organic chemists became interested in the mechanism of photochemical rearrangements. Thus, since the late 1950's an area of investigation termed "organic photochemistry" has developed. It differs from classical physical photochemical studies in that the work is done primarily in solution, the molecules involved are often quite complex, and the synthetic utility as well as the mechanism of the photoreaction is of interest. In this area many detailed studies have been directed toward understanding the basic reactive processes of excited organic compounds. It is from the total results of these kinetic, quantum efficiency, substituent effect, and labeling studies that some of the structure–reactivity relationships of excited organic molecules have been elucidated. The intent of this survey is to emphasize those contributions made by virtue of isotopic labeling, bearing in mind that conclusions drawn from these studies are in many instances strongly enforced by other non-isotopic labeling investigations.

Photochemical studies employing deuterium labeled substrates may be conveniently classified into two groups. In the first of these, the major emphasis is on the isotope effect of the reaction concerned. Such studies have been systematically applied to only a few systems, and for this survey discussion will be limited to: (1) interactions of excited carbonyl compounds with ground state olefins, and (2) hydrogen abstraction reactions of carbonyl compounds. The second group comprises investigations wherein deuterium is used as a label to establish the structural features of a photochemical rearrangement. This latter class is by far the more common type of organic photochemical study. In contrast to the first classification an inherent assumption in these studies is that isotopic substitution does not alter the basic processes of concern, but serves only as a labeling substituent. As will be noted later, this assumption is not always valid. In this review the information which may be obtained from the results of deuterium labeling studies will be illustrated by selected examples in several areas of organic chemistry.

II. EFFECT OF DEUTERIUM SUBSTITUTION ON SINGLET LIFETIMES, INTERSYSTEM CROSSINGS, AND TRIPLET LIFETIMES

While the interpretation of isotope effects in ground state reactions is by no means always straightforward, the understanding of an isotope effect for an organic photochemical system is even more complex. In addition to the usual isotope effects due to the breaking of bonds and hydridization changes, there is also the potential isotope effect on the non-reactive processes of the molecule (i.e., k_2, k_3, k_4). Consider the processes available to the lowest excited singlet and triplet state of a molecule shown below. Certainly one of the easiest and most commonly measured properties of a

$M \xrightarrow{h\nu} M^{*1}$	excitation
$M^{*1} \xrightarrow{k_1} M + h\nu'$	fluorescence
$M^{*1} \xrightarrow{k_2} M$	radiationless transition to the ground state
$M^{*1} \xrightarrow{k_3} M_2^{*3}$	intersystem crossing to a second triplet state
$M^{*1} \xrightarrow{k_4} M_1^{*3}$	intersystem crossing to the lowest triplet
$M^{*1} \xrightarrow{k_5}$ Product	formation of photochemical products
$M_1^{*3} \xrightarrow{k_6} M + h\nu''$	phosphorescence
$M_1^{*3} \xrightarrow{k_7} M$	radiationless transition to the ground state
$M_1^{*3} \xrightarrow{k_8}$ Products	formation of photochemical products

photochemical reaction is the quantum yield. However, the interpretation of a change in quantum efficiency upon deuteration may not be readily discernible since the quantum yield is actually a ratio of the rate for the process of interest divided by the rates of other processes which destroy the excited state (bimolecular processes will not be

$$\Phi_n = \frac{k_n M^{*1}}{\sum_{n=1}^{n} k_n M^{*1}} = \frac{k_n}{\sum_{n=1}^{n} k_n}$$

considered here). Thus, for a full understanding of the effect of deuterium substitution on a quantum yield, knowledge must be available on the effect of isotopic substitution on the important pathways for excited state decay. Unfortunately, such a detailed understanding of the effect of deuterium substitution is generally not available except for aromatic compounds and some carbonyl compounds, and here most of the work has been concerned with the triplet excited state. While this review will not attempt a detailed coverage of deuteration effects on excited state decay, the experimental organic chemist should be aware of the profound influence deuterium substitution may exert on excited state processes, even if his interest is only in labeling studies. Thus, some of the important experimental consequences of deuteration on singlet and triplet lifetimes will be noted first.

A. Triplet Lifetimes

The effect of deuterium substitution on the radiative and radiationless processes of the triplet state has been of interest to both spectroscopists and theoreticians since the initial report of Hutchison and Mangum [1a] on the lifetime of naphthalene and naphthalene-d_8 in a durene crystal. These workers found that while the phosphorescence lifetime of naphthalene in durene was 2.1 sec, the lifetime of naphthalene-d_8 was 16.9 sec. The phosphorescence lifetime of ordinary naphthalene in deuterated durene did not differ appreciably from that in durene. This remarkable enhancement of phosphorescence lifetime invoked many more experimental studies [1b–d], and a number of theories were proposed to account for this deuteration effect. As numerous review articles have discussed the theoretical ramifications of the deuterium enhancement of triplet lifetimes relative to the theories of radiationless transitions [2–6], this review will primarily deal with the experimental results of these studies.

The major effect of deuterium substitution on the triplet excited state is generally agreed to involve a reduction in the rate constant for radiationless conversion to the ground state, $T_1 \rightarrow S_0$*† Franck–Condon factors account for a major portion of this decrease in the radiationless transition rate in going from protium to deuterium systems [10]. Simply stated, the Franck–Condon factor depends upon the overlap of the vibrational wave function for the lowest vibrational level of the initial electronic state with an isoenergetic high vibrational level of the ground state. For isoenergetic vibrational levels in the two states, this overlap generally decreases with increasing vibrational quantum number. Since the high energy C–H stretching vibrations are dominant

* The mean radiative lifetime of the excited state, $\tau_0 = 1/k_{\text{radiative}}$, is the lifetime observed if the quantum yield for emission is 1 (i.e., no radiationless processes from the excited state). For most molecules radiationless processes do account for excited state decay; thus, the lifetime measured from the phosphorescence or fluorescence emission is less than the radiative lifetime and $\Phi_{\text{emission}} = k_{\text{radiative}}/(k_{\text{radiative}} + k_{\text{radiationless}})$. Under these conditions, the measured lifetime is the reciprocal of the sum of the rates for radiative and radiationless decay, $\tau = 1/(k_{\text{radiative}} + k_{\text{radiationless}})$. A test that the radiationless process is most affected upon deuteration is to establish that the ratio of the quantum yields for emission is proportional to the ratio of the lifetimes, $\Phi_H/\Phi_D = \tau_H/\tau_D$. However, reliable measurements of these quantities are often difficult and thus this relationship is often only approximately satisfied.

† Several recent studies have proposed that, for benzene and naphthalene, deuteration also alters the radiative rate constant from the triplet to the ground state [7–9]. However, interpretation of these results is complicated since the effect on the radiative rate may depend on the media (i.e., EPA versus Argon matrix). Thus, for this elementary discussion it will be assumed that deuteration primarily affects the radiationless process.

TABLE 1

POSITIONAL EFFECT OF PHOSPHORESCENCE LIFETIME

Compound	Lifetime (sec)
Naphthalene[a]	2.51
Naphthalene-1-d_1	2.94
Naphthalene-2-d_1	2.75
Naphthalene-2, 3-d_2	3.08
Naphthalene-1, 4-d_2	3.61
Naphthalene-2, 3, 4, 5, 6, 7, 8-d_7	11.08
Naphthalene-1, 3, 4, 5, 6, 7, 8-d_7	13.51
Naphthalene-d_8	21.73
Toluene[a]	8.02
Toluene-α, α, α-d_3	8.03
Toluene-2, 3, 4, 5, 6-d_5	10.24
Toluene-d_8	10.54
2-Naphthaldehyde[b]	0.37
2-Naphthaldehyde-α-d_1	1.35
2-Naphthaldehyde-1,3,4,5,6,7,8-d_7	0.46
2-Naphthaldehyde-d_8	4.80

[a] Measurements at 77°K in 3-methylpentane.
[b] Measurements at 77°K in ether–pentane alcohol.

in acquiring the vibrational energy in the radiationless transition, for a given energy gap, fewer quanta of modes primarily involving C–H stretching motion would be required than for modes which primarily involve C–D stretches. As a consequence, the deuterated molecule has a slower rate of radiationless decay to the ground state*. While the deuterium effect has been most studied in aromatic triplets, limited studies in carbonyl systems such as acetone [12a,b], acetophenone [13], biacetyl [14a–c], and 2-naphthaldehyde [15] indicate increased phosphorescence lifetimes upon deuteration in these systems also.

Investigations have also appeared on the effect of partial deuteration on excited state phosphorescence lifetimes of acetophenone [13], benzene [16], toluene [17], biphenyl [18], naphthalene [19], phenylacetylene [20a], and 2-naphthaldehyde [15], and theoretical interpretations of the effect have been presented [21a–e]. Table 1 summarizes the data for naphthalene, toluene, and 2-naphthaldehyde from which some generalizations arising from these studies may be illustrated. Firstly, while the lifetimes increase with increasing deuteration, they do not do so in a linear fashion.

* It should be noted that other vibrational modes may also accept energy in the radiationless process. Thus, for benzene it is calculated that 66% of the energy goes into a C–H breathing mode, the other major energy sinks being ring vibrations [11].

Secondly, the position of the deuterium is important in its effect on the lifetime. For naphthalene, substitution of deuterium at the α-position has a larger effect than for β-substitution. The positional effect is especially dramatic for toluene, phenylacetylene, and 2-naphthaldehyde. Substitution of deuterium at the methyl group of toluene has very little effect on lifetime whereas ring substitution is effective. For 1-naphthaldehyde it is estimated that the contribution by the aldehydic C–H vibration in the non-radiative process is approximately five times that of all seven aromatic hydrogens. For phenylacetylene, deuterium substitution at the acetylenic hydrogen has nearly the same effect as deuterium substitution for all five aromatic hydrogens. A very recent study involving stilbene has led to the conclusion that deuteration of the vinyl position increases the stilbene triplet lifetime by 30% while ring deuteration has no effect [20b].

The effect of solvent and environment on the decay of the lowest triplet state is not fully understood. While changes in the lifetime of the triplet state upon alteration of media may be large, an interpretation of the effect is complicated since in general external atoms apparently may affect both the radiative and non-radiative $T_1 \rightarrow S_0$ transitions [3,21]. An extensive survey of the effect of deuterated media on triplet lifetimes will not be attempted here, although two examples should be noted which illustrate that such effects can be important. Thus, it was observed that in going from benzene to benzene-d_6 as solvent the phosphorescence of both biacetyl and biacetyl-d_6 was enhanced [14a–c]. In a study of the effect of complexation between benzene and chloroform at 77°K, it was noted that substitution of chloroform-*d* for chloroform greatly increases the phosphorescence efficiency and the decay time for benzene triplet [22]. For this system it was proposed that vibronic coupling of the triplet benzene with the C–H vibration in chloroform accelerated the radiationless decay of benzene. This effect was much less important in chloroform-*d* where a C–D vibration would be involved.

B. Singlet Lifetimes and Intersystem Crossing Rates

The question of a deuterium isotope effect on excited singlet state lifetimes has been much less extensively studied than that for the triplet state. An important aspect in discussing an isotope effect on the singlet state lifetime concerns the important pathways for excited singlet decay. Singlet excited states can decay by fluorescence, intersystem crossing to a triplet state, or radiationless conversion to a lower singlet state. Assuming that deuterium substitution will not strongly alter the radiative rate*, the

---

* The influence of deuterium substitution in altering the absorption spectrum of simple organic molecules is known [23]. However, the effect of deuterium substitution on the radiative lifetime of the singlet state does not appear to have been systematically studied. It seems reasonable to suppose that its effect on radiative properties will be small judging from the results noted for the triplet state.

principal effect of isotopic substitution on the singlet state lifetime will reside in its effect on the radiationless processes. Such factors as the electronic matrix element, the distortion of molecular geometry, the anharmonicity of molecular vibrations, and the energy gap for the process all contribute to the overall radiationless rate [3]. For many organic molecules the energy gap between S_1 and S_0 is quite large; thus, the $S_1 \rightarrow S_0$ radiationless conversion is not very important as a decay mechanism for the excited singlet state. Intersystem crossing on the other hand is an important route for excited singlet decay. For the case of intersystem crossing, an additional factor is the relative importance of $S_1 \rightarrow T_2$ versus $S_1 \rightarrow T_1$ crossing. For an $S_1 \rightarrow T_1$ process the energy gap in some molecules is large enough for an isotope effect, analogous to that observed for the $T_1 \rightarrow S_0$ process previously discussed, to be expected. On the other hand the much smaller energy gap for $S_1 \rightarrow T_2$ process might imply the lack of an appreciable effect. Scharf has presented a theoretical treatment which indicates that a reverse deuteration effect could be exhibited by the $S_1 \rightarrow T_1$ and $S_1 \rightarrow T_2$ non-radiative decays [24a,b]; that is, the deuterated compound would undergo these processes faster than the compound containing only hydrogen. Thus, theory indicates both normal and inverse deuterium isotope effects can be expected for $S_1 \rightarrow T_n$ intersystem crossings depending upon the particular system. Indeed, examples of both effects are known.

While early results on several aromatic systems in condensed media suggested that fluorescence lifetimes for perprotonated and perdeuterated compounds were identical within experimental error [25–28], more detailed studies indicate that in general this is not true [9]. Azulene provided the first example of enhanced fluorescence upon deuteration [29a], and recently the deuterium isotope effect on its fluorescence lifetime has been reported in both the vapor state and in solution [29b]. In the vapor state at 296°K the ratio of fluorescence lifetimes τ_{h_8}/τ_{d_8} was 1.4–1.6 while in 10^{-4} *M* methylcyclohexane the value was 1.3. Since azulene is exceptional in emitting from its second excited singlet state ($S_2 \rightarrow S_0$), the deuterium effect could arise from a decrease in the radiationless rate of the $S_2 \rightarrow S_1$ or $S_2 \rightarrow T_n$ process; the authors favor the former alternative. Several other investigations have focused attention on the effect of deuterium substitution on intersystem crossing. In a temperature dependent study of intersystem crossing in pyrene and pyrene-d_{10}, it was concluded that two mechanisms operate [30]. One mechanism involves intersystem crossing via an intermediate triplet state, while the second is a direct conversion of $S_1 \rightarrow T_1$. The rate for this latter process is temperature independent and decreases upon deuteration. For benzene vapor [31,32] it was observed that the fluorescence lifetime is enhanced upon deuteration (τ_{h_6} = 72 nsec versus τ_{d_6} = 92 nsec, 50 torr, 266–247 nm excitation) [32]. In a study of excitation to several different vibronic levels of the $^1B_{2u}$ state the lifetime of fluorescence varied with the vibrational level acting as the origin for the emission, but in each case the deuterium isotope effect was noted. Using literature values for the quantum yields of intersystem crossing and fluorescence in benzene and benzene-d_6, Breuer and Lee [31] have concluded that intersystem crossing in benzene was twice as rapid as in benzene-d_6. The benzene B_{2u} excited singlet ($E_s = \sim 110$ kcal mol^{-1}) could conceivably undergo crossing to two triplet levels, the $^3B_{1u}$ ($E_T = \sim 84$ kcal mol^{-1}) or the $^3E_{1u}$ ($E_T = \sim 106$ kcal mol^{-1}). It was considered that the magnitude of the isotope effect was more consistent with the intersystem crossing $^1B_{2u} \rightarrow {}^3B_{1u}$ being more important [31].

In addition to those examples in which deuterium substitution slows decay processes in the excited singlet state, the decay from certain vibrational levels in excited naphthalene vapor [33] and for anthracene [34] is enhanced upon deuteration. In naphthalene and naphthalene-d_8 vapor, the fluorescence quantum yields from different vibrational levels of the lowest singlet state were studied. For emission from several of these vibrational levels, the quantum yield was higher for naphthalene than for the d_8 compound. The higher efficiency of naphthalene fluorescence from several vibrational levels of the lowest singlet is consistent with Scharf's theoretical treatment if crossing from the naphthalene singlet to a nearby higher triplet state is occurring. However, since theory now indicates that the $S_1 \rightarrow T_1$ process can also exhibit an inverse isotope effect, these conclusions may need modification [24a,b]. Interestingly, the fluorescence yield of the d_8 compound was greater than that of naphthalene if excitation into the second and third absorption bands was performed. In the latter case the ratio of $\Phi_{F(D)}/\Phi_{F(H)}$ increases smoothly with increasing excitation energy. The results from excitation into the second or third absorption band seemingly show enhancement of singlet lifetime on deuteration due to an appreciable energy gap between S_2 (or S_3) and T_n.

A second system in which the inverse effect of deuterium substitution has been noted is in anthracene. Anthracene is a particularly good case for evaluating the reverse deuteration effect when crossing of S_1 to a nearby higher triplet is involved in a radiationless process; since several lines of evidence suggest intersystem crossing in S_1 anthracene occurs via a second triplet state ($S_1 \rightarrow T_2$). When the fluorescence lifetimes and quantum efficiencies of anthracene and anthracene-d_{10} were studied in nujol and cyclohexane at room temperature, it was found that the fluorescence lifetime was longer for the protio compound ($\tau_H = 6.1$ nsec versus $\tau_D = 5.4$ nsec) and the quantum yield for fluorescence higher ($\Phi_H = 0.32$, $\Phi_D = 0.29$ in nujol). Thus, fluorescence from certain vibrational levels in naphthalene and anthracene indicates that aside from the usual isotope effect on fluorescence lifetimes, a reverse deuteration effect in certain molecules having S_1 and T_n states close in energy may be observed.

While most deuterium isotope effects on fluorescence have been largely limited to aromatic compounds, some results have been reported for other systems. Thus, the fluorescence intensity of biacetyl-d_6 is increased relative to biacetyl by 50% in EPA at $-198°$ and by 20% at 25° in benzene [14a,b]. In studying the vapor phase fluorescence of ketones and aldehydes it was noted that deuteration of cyclopentanone increases the fluorescence lifetime by a factor of ~ 1.7 [35]. A slightly smaller effect was observed for deuteration of acetone.

Detailed studies on the effect of deuterated solvents on singlet state decay processes have been mainly concerned with water versus deuterium oxide. The results from several studies noted below will serve to illustrate some of the general points. For molecules which undergo ionization in the excited state (i.e., $Ar{-}NH_3^+$ or Ar–OH) changing the solvent from water to deuterium oxide may cause an alteration in both the quantum yield and spectral distribution of the fluorescence, the effect being especially dramatic if one of the components in the acid–base equilibrium is weakly or non-fluorescent. It is known that the acidity of 2-naphthol is much greater in the excited singlet state than in the ground state and that the fluorescence spectrum of this compound in aqueous solution consists of two components [36]. The major component (λ_{max} = 352 nm) arises from unionized 2-naphthol while the minor emission (λ_{max} = 416 nm) derives from the 2-naphtholate anion, the anion being formed by proton transfer to the solvent during the lifetime of the fluorescing excited state. In changing from water to deuterium oxide, the 416 nm emission decreases markedly and the 352 nm component increases in intensity. This change in spectral distribution is accounted for by the slower deuteron transfer rate in excited 2-naphthol to solvent (deuterium oxide); thus, the major proportion of the fluorescence in deuterated media is from the unionized 2-naphthol. In support of this isotope effect on the rate of proton transfer during the excited state lifetime, it was noted that in 1 *M* hydrochloric acid, where virtually all the emission arises from the neutral molecule, or 1 *M* sodium hydroxide, where the emission arises from the anion, no change in either spectral distribution or quantum yield between water and deuterium oxide was observed. From the emission data, the ratio of the forward and reverse rate constants in the excited state equilibrium could be calculated

$$\text{2-naphthol}^* + H_2O \underset{k_2}{\overset{k_1}{\rightleftharpoons}} \text{2-naphtholate}^* + H_3O^+$$

$$(k_1/k_2)_{H_2O} = 9.6 \times 10^{-4},\ (k_1/k_2)_{D_2O} = 3.7 \times 10^{-4}$$

A second illustration of the effect of deuterium oxide versus water on fluorescence is the case of 2-naphthylamine [37]. In its lowest singlet state 2-naphthylamine is a much weaker base than in the ground state. Even in aqueous sulfuric acid solutions up to 2 *M*, excitation of the cation produces mainly the fluorescence of 2-naphthylamine due to proton transfer to the solvent. In studying the efficiency of molecular to cationic fluorescence of 2-naphthylamine in sulfuric acid–water mixtures, it was noted that with increasing acidity the fluorescence efficiency of the molecular species decreased while that of the cationic molecule increased. However, in the turnover region from molecular to cationic fluorescence, strong quenching of both fluorescence components was observed. Past the turnover region, the total quantum yield for fluorescence again increased. While the total fluorescence efficiency was greater in deuterium oxide than water throughout the acidity range, the largest deuteration effect on fluorescence was observed in the turnover region (i.e., $\Phi_D = 0.019$ versus $\Phi_H = 0.006$).

The strong quenching of the total fluorescence in the turnover region together with the maximum solvent isotope effect observed suggested that in this system proton transfer participates in the internal conversion process responsible for quenching. Thus, proton transfer from the excited cation to solute appears to open radiationless pathways not available to either the cationic or neutral molecule itself. The reader is referred to the original paper for a discussion of mechanisms to account for this radiationless decay [37].

Aside from effects involving specific proton (deuteron) transfer in the excited singlet state, results have been reported in which solute–solvent interaction is directly responsible for radiationless internal conversion. In 5-amino-1-naphthalenesulfonate and 8-amino-1,3,6-naphthalenetrisulfonate the fluorescence quantum yields ratios in water versus deuterium oxide were 3.05 and 3.70 respectively [36]. While this was originally attributed to ionization processes of the amino hydrogen in the excited singlet state, subsequent work has yielded an alternate explanation [38]. Thus, 1-*N,N*-dimethyl-1-amino-5-naphthalenesulfonate, which has no amino hydrogens, also shows high fluorescence in deuterated media, $\Phi_{D_2O}/\Phi_{H_2O} = 1.75$. The isotope effect in this system was attributed to specific solute–solvent interaction, perhaps involving coupling between the electronic motion within the dissolved molecule and the nuclear motions of one or more solvent molecules. A similar solvation effect apparently is also important in tryptophan [39]; however, the results are somewhat complicated as intramolecular fluorescent quenching by proton transfer may also be occurring [40].

III. SECONDARY DEUTERIUM ISOTOPE EFFECTS IN EXCITED KETONE–OLEFIN INTERACTIONS

While the preceding section has illustrated the complex manner in which deuteration can affect the basic processes of the excited state, it should not be judged that isotope effect studies are too complex to be useful in studying various processes of excited molecules. Consider the several mechanisms that have been proposed for the interaction of an excited carbonyl compound with a ground state olefin. The first and simplest of these involves triplet energy transfer from the excited carbonyl to the olefin, producing the excited triplet state of the olefin. This well-known process would

Triplet energy transfer

$$\rangle C{=}O^{*t} + \rangle{=}\langle \longrightarrow \rangle C{=}O + \left[\rangle{=}\langle\right]^{*t}$$

occur in a collision complex and at rates approaching diffusion control if the overall exothermicity of the reaction is > 3 kcal mol^{-1} [i.e., $E_T(\rangle C{=}O) > E_T(\rangle C{=}C\langle)$ by 3 kcal mol^{-1}]. The net photochemical reaction for the olefin triplet state is *cis–trans* isomerization (of course, in some systems this process is not observable due to the symmetry of the olefin). A second mechanism, which was first proposed by Schenck [41,42], involves interaction of the excited sensitizer with olefin to give a biradical

Schenck mechanism for olefin isomerization

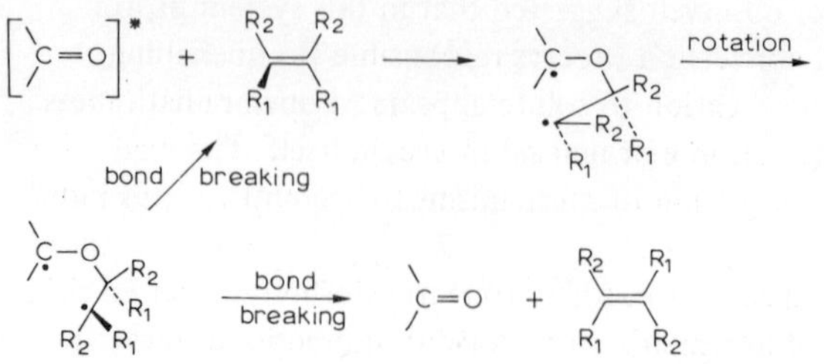

intermediate. For the case of an excited ketone the biradical intermediate would be reasonably pictured as above. Rotation in the biradical intermediate followed by bond cleavage would result in overall *cis–trans* isomerization of the olefin. Yang had modified the Schenck mechanism for the case of $n{-}\pi^*$ triplet carbonyl compounds so that energy transfer as well as oxetane formation could occur via the biradical [43]. As illustrated below, the biradical intermediate could cyclize to give oxetane or dissociate to give back the carbonyl compound and the triplet state of the olefin in a non-planar configuration. Since the triplet energy of an olefin decreases as the departure from planarity increases, the latter alternative was especially attractive in explaining an energy transfer process which would be endothermic for vertical excitation of an olefin (i.e.,

Schenck mechanism as modified by Yang

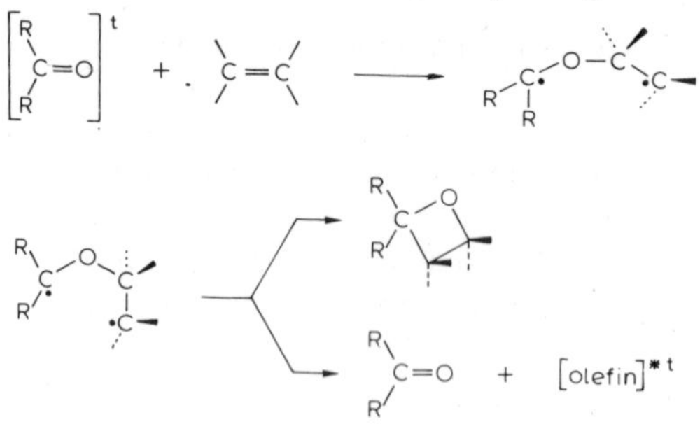

excitation of planar olefin to the planar triplet state) [44]. A fourth possible scheme for excited ketone–olefin interaction involved formation of an excited state complex, exiplex, between ketone and olefin which subsequently decayed to a biradical species [45]. The formation of excited state complexes had received direct support in many singlet state systems and their importance has been inferred in several triplet state reactions [45].

Exiplex mechanism

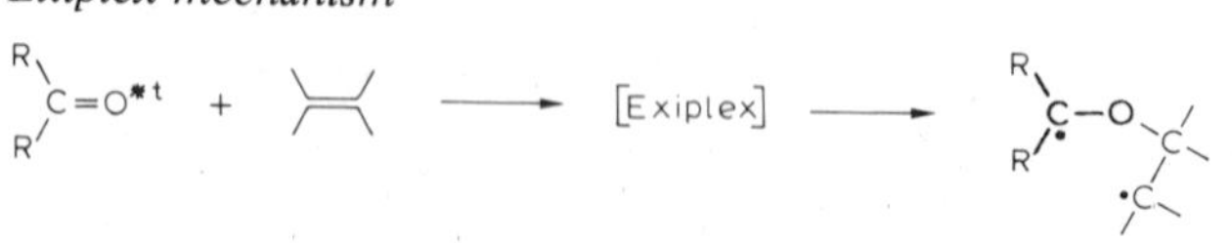

A. Vertical Excitation Transfer

While the precise mechanism for the interaction of excited carbonyl compounds with ground state olefins will certainly depend on the electronic features of the carbonyl system and the olefin, secondary deuterium isotope effect studies have been informative in several systems for deciding on the most reasonable model for the interaction [45–50]. The most straightforward case is when the triplet transfer process is exothermic by 3 kcal mol^{-1} or greater. Thus, for the isomerization of *trans*-β-methylstyrene (E_T = 59.8 kcal mol^{-1}) to *cis*-β-methylstyrene by propiophenone (E_T = 74.6 kcal mol^{-1}) and benzophenone (E_T = 68.5 kcal mol^{-1}), no isotope effect is observed for energy transfer to *trans*-β-methylstyrene deuterated at the β, α, *p*, or *m* position [46,47]. Similarly, in a non-carbonyl system such as triplet benzene (E_T = 84.4 kcal mol^{-1}) transferring energy to butadiene (E_T = 59.6 kcal mol^{-1}), no isotope effect on the rate of transfer to butadiene and butadiene-d_6 was noted [51]. These observations are as expected since in these highly exothermic transfers any differences in triplet energy between the protium and deuterium compounds would be unimportant. Furthermore, there is no reason to expect geometric change in the course of the excitation (non-vertical excitation), as ample energy is available in the sensitizer for efficient vertical excitation.

An especially interesting instance of excited carbonyl–olefin interaction is one in which there may be insufficient energy in the triplet state of the carbonyl compound to populate the planar triplet state of the olefin. It is in these instances that reactions proceeding via a Schenck mechanism appeared most attractive. Caldwell and coworkers [45–50] have studied this problem extensively by the use of secondary deuterium isotope effects. The basic premise of their studies is that if bi-radical formation is the initial step in excited carbonyl–olefin interaction (Schenck mechanism), then an inverse secondary isotope effect should be exhibited by the system since an sp^2 center is being converted to an sp^3 center. This premise has much analogy in ground state chemistry where conversion of an sp^2 carbon bearing deuterium to the sp^3 state is accompanied by such an inverse isotope effect. For example [52], addition of methyl radical to styrene versus styrene-β,β-d_2 has an isotope effect k_D/k_H = 1.1. In the potassium thiocyanate catalyzed isomerization of maleic acid-2,-3-d_2, a reaction involving conversion of an $sp^2 \rightarrow sp^3$ center in the transition state, an inverse isotope effect of k_D/k_H = 1.18–1.13 was noted [53]. Perhaps the most pertinent analogy to the following discussion is the iodine-catalyzed isomerization of *cis*-β-methylstyrene [45] which shows an inverse isotope effect for a β-deuterium of k_D/k_H = 1.04.

In investigating the isomerization of *trans*-β-methylstyrene (E_T = 59.8 kcal mol^{-1}) sensitized by 2-acetonaphthone, chrysene, biacetyl, and fluorenone the results shown in Table 2 were obtained [47]. These values are the apparent product isotope effects calculated from the ratio

$$\frac{\text{Unlabeled product}}{\text{Labeled product}} : \frac{\text{Unlabeled reactant}}{\text{Labeled reactant}}$$

TABLE 2

ISOTOPE EFFECTS ON THE ISOMERIZATION OF *TRANS*-β-METHYLSTYRENE TO *CIS*-β-METHYLSTYRENE

	Position of monodeuteration			
Sensitizer (E_T kcal mol^{-1})	β	α	*p*	*m*
2-Acetonaphthone (59.3)	1.13			
Chrysene (56.6)	1.13	1.05	1.04	
Biacetyl (54.9)	1.14	1.05	1.05	1.01
Fluorenone (53.3)	1.15	1.05	1.06	1.02

The observation of direct, rather than inverse, isotope effects was interpreted as ruling out the direct formation of biradical intermediates in these sensitizations. Furthermore, the isotope effect for a given position of deuteration appears quite independent of the nature of the sensitizer. The triplet states of 2-acetonaphthone, chrysene, and fluorenone are π–π*, while that for biacetyl is *n*–π*. If a specific bonding interaction between sensitizer and olefin were involved, one would certainly not have expected the positional effect to be so insensitive to sensitizer. The positional isotope effect showed a good correlation with the change in free valence index between the planar ground and planar first excited state suggesting that a vertical excitation was occurring in this stystem. Thus, the results from isotope effects complemented conclusions based on kinetic and photostationary state studies; namely, that vertical excitation was involved in these sensitizations. A similar study with *cis*-β-methylstyrene was also reported; however, here the conclusions as to the type of sensitizer–olefin interaction were less certain.

The observed direct isotope effects in the sensitizer–*trans*-β-methylstyrene system are comparable to those noted by Schmidt and Lee in the quenching of benzene, acetone, and biacetyl triplets by deuterated olefins [51], the isotope effects on the benzene–ethylene, acetone–ethylene, and biacetyl–butadiene systems being 1.16, 1.18, and 1.25 per deuterium, respectively. Both Lee and Caldwell have discussed in detail the possible origins for the direct isotope effects noted in these systems. Among the more attractive possibilities is the slightly higher triplet energy of the deuterated olefin, which would cause an increased activation energy for an already endothermic process.

B. The Exiplex Mechanism

For the *cis*–*trans* isomerizations of alkyl ethylenes by *n*–π* ketone triplets, Yang et al. [43] and Saltiel et al. [42] both concluded that collisional energy transfer did not fully account for the experimental data. Thus, Yang noted that the higher energy benzaldehyde triplet (E_T = 71.5 kcal mol^{-1}) was a poorer donor of triplet energy to

TABLE 3

RATE CONSTANTS FOR QUENCHING AROMATIC KETONE TRIPLETS

Sensitizer (E_T, kcal mol^{-1})	Quenching rate constant (1 mol^{-1} sec^{-1})		
	cis-2-butene	*trans*- -dichloroethylene	isobutylene
4,4-Dimethoxybenzophenone (69.4)	4×10^6	2.4×10^7	1.1×10^7
Benzophenone (68.6)	7×10^7	1.3×10^7	3×10^7
4-Trifluoromethyl benzophenone (67.6)	2.2×10^8	6×10^6	7×10^7
4-Benzoyl pyridine (67.1)	3.3×10^8		

3-methyl-2-pentene than was the benzophenone triplet (E_T = 68.0 kcal mol^{-1}). In a study of the acetone and acetophenone sensitized isomerization of 2-pentenes, Saltiel noted that the decay ratio of triplet olefin to ground state *cis*- and *trans*-isomers depended on the sensitizer. Since a common decay ratio did not result, Saltiel concluded that different intermediates were produced with different sensitizers. Both groups of workers suggested the involvement of Schenck type intermediates in these isomerizations. Caldwell has recently reported extensive studies involving quantum yields, rate constants, and substituent effects on the quenching of benzophenone triplet by *cis*-2-butene, isobutylene, and *trans*-dichloroethylene [45,48,49]. For *trans*-dichloroethylene, for which a direct excitation mechanism for quenching had been proposed [49], decreasing quenching efficiencies occur with decreasing energy of the ketone sensitizer (Table 3). This result would certainly be expected for an endothermic energy transfer process. However, *cis*-2-butene and isobutylene quenching efficiencies increase with decreasing sensitizer energy (Table 3). Such behavior is not consistent with an excitation transfer process and this conclusion is in agreement with the earlier results of Yang and Saltiel. Furthermore, such a dependence on quenching rate with sensitizer would be consistent with an exiplex mechanism if some charge contribution to the complex was important.

Having excluded excitation transfer, there remain the Schenck and the exiplex mechanisms. While the non-isotopic experiments seem more in agreement with the exiplex process, no rigorous distinctions between the two could be made. Caldwell et al. [45] have extensively applied secondary isotope effects to this problem by studying the quenching of benzophenone triplet by *cis*-2-butene-2,3-d_2, *trans*-2-butene-2-d_1, *cis*-3-methyl-2-pentene-2-d_1, and *trans*-3-methyl-2-pentene-2-d_1. The product isotope effects in these systems were very small, the deviation from unity no greater than 0.015 per deuterium. These results reconfirmed the absence of a direct excitation mechanism as this would have been expected to show a direct large product isotope effect*. Furthermore, since the Schenck mechanism would be expected to show an inverse isotope effect, these results render this mechanism less attractive also.

To obtain further insight into the steps in the benzophenone–alkyl ethylene system, the mechanism of oxetane formation was studied and compared with the *cis–trans* isomerization results. The product isotope effects were reported for the oxetanes from benzophenone with *cis*-2-butene-2,3,-d_2 (*1d*) and *trans*-2-butene-2-*d* (*6d*). Irradiation of benzophenone and (*1d*) yielded oxetanes (*2d*) and (*3d*). When the intermolecular competition was performed with *cis*-2-butene (*1*) and *cis*-2-butene-d_2, (*1d*)

(1d) (2d) 6 parts (3d) 1 part

the product isotope effect for discrimination between (*1*) and (*1d*) in oxetane formation was 1.03 ± 0.01, (*1*) being the slightly more reactive. This should be compared with a product isotope of 1.02 ± 0.01 for isomerization of the same reaction mixture to *trans*-2-butene. These results clearly demonstrate that there is no isotopic discrimination in bi-radical (*4*) (formed via the exiplex) between fragmentation and ring closure. Since the ring closure reaction might have been expected to show an inverse

(4d) ring closure → (2d) + (3d)
fragmentation → (1d) + (5d)

isotope effect ($sp^2 \rightarrow sp^3$ conversion) and the fragmentation route a direct isotope effect ($sp^3 \rightarrow sp^2$ process), the authors interpret this as indicative of an extremely reactive bi-radical undergoing reactions with little or no energy barrier. Such a proposal would certainly not be unreasonable based on calculations of the potential energy surface of tetramethylene [54].

A second study designed to obtain evidence on the selectivity of bi-radical formation from the exiplex intermediate employed the oxetane formed from benzophenone and *trans*-2-butene-2-d_1. In this system the analysis of the ratio of the four oxetane products showed that (*7d*) and (*8d*) were favored by a factor of 1.10 over (*7d'*) and (*8d'*). Since the results with the *cis*-butene-2,3-d_2 indicated no isotopic discrimination at the bi-radical stage between closure and fragmentation, these results indicate that

* For the benzophenone sensitized *cis–trans* isomerization of the 1,2-dichloroethylenes, which occurs via excitation transfer, product isotope effects of 1.15 (*cis–trans*) and 1.18 (*trans–cis*) were recorded.

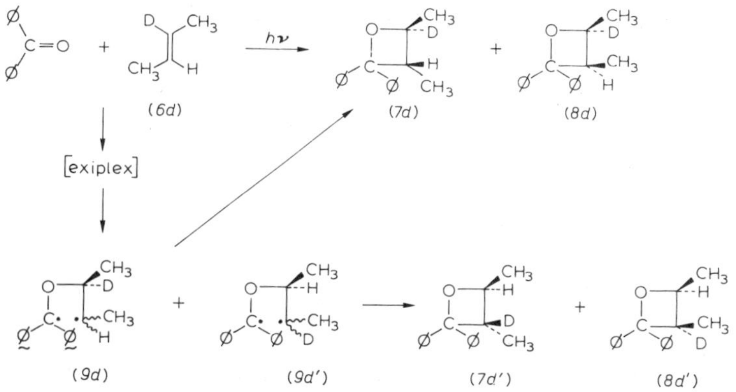

the preference for *(9d)* to *(9d')* is also 1.10. Thus, while the inverse isotope effect is not exhibited in the initial phase of the benzophenone–alkene interaction, it is observed in the biradical formation step. These results then seem most compatible with the exiplex mechanism.

In summary, the secondary isotope effect studies allow a basis for distinction between excitation transfer, the direct biradical mechanism, and exiplex formation. For energy transfer processes which are exothermic by 3 kcal mol^{-1}, no isotope effects on energy transfer to deuterated olefins are observed. In systems in which a slightly endothermic excitation transfer process has been proposed (i.e., benzophenone-*trans*-dichloroethylene), a moderate positive isotope effect has been noted. Finally, in the benzophenone–alkene system the negligible isotope effects as well as kinetic data support a mechanism involving complex formation followed by decay to the biradical. The direction of collapse of the complex is not determined solely by biradical stability as charge distributions and polarizabilities in the complex might also be involved. This latter statement seems appropriate in that the product distributions in oxetane formation with unsymmetrical olefins do not always derive from the "most stable biradical" [55]. Finally, while such studies as this have afforded mechanistic information on ketone–olefin interactions, the future is sure to yield additional insights into the mechanism of these processes.

IV. HYDROGEN ABSTRACTION PROCESSES OF EXCITED CARBONYL COMPOUNDS

A. Hydrogen Abstraction from C–H and O–H Bonds of Alcohols

Hydrogen abstraction reactions of excited carbonyl compounds are a second area in which hydrogen–deuterium isotope studies have been extensively employed. For many aldehydes and ketones irradiation in primary or secondary alcohols results in hydrogen abstraction by the oxygen to produce a corresponding α-hydroxy radical. Thus, when benzophenone is irradiated in benzene with benzhydrol as the hydrogen source, benzpinacol is produced in high efficiency. In one of the early reports of the isotope effect on excited state hydrogen abstraction [56], the rate for proton abstraction versus deuterium abstraction of triplet benzophenone from benzhydrol was determined as $k_H/k_D = 2.7$. Especially in isopropyl alcohol, initial hydrogen abstraction

$$\phi-\overset{O}{\overset{\|}{C}}-\phi \quad + \quad \phi_2CD-OH \xrightarrow[h\nu]{\text{benzene}} \phi-\underset{\phi}{\overset{OD}{C}}-\underset{\phi}{\overset{OH}{C}}-\phi$$

(10) (11d) (12)

may be followed by radical transfer processes as well as dimerization. Presumably such a radical transfer reaction complicates the photochemistry of benzophenone in isopropyl alcohol where benzpinacol is the major product. The isotope effect here for triplet benzophenone was determined as $k_H/k_D = 2.8$ [57]. That these results were not complicated by hydrogen abstraction of the hydroxyl hydrogen by triplet benzophenone was established by studying the rate of the reaction in deuterium isopropoxide. Several determinations of the rate of photoreduction in this media yielded results with no pronounced deviation from that of the parent system. In a recent study [58]

$$\phi\overset{O}{\overset{\|}{C}}\phi + CH_3\underset{H}{\overset{OH}{C}}-CH_3 \xrightarrow{h\nu} \phi-\overset{OH}{\dot{C}}-\phi + CH_3\overset{OH}{\dot{C}}-CH_3$$

$$CH_3-\overset{OH}{\dot{C}}-CH_3 + \phi\overset{O}{\overset{\|}{C}}\phi \rightarrow CH_3\overset{O}{\overset{\|}{C}}CH_3 + \phi\overset{OH}{\dot{C}}-\phi$$

$$2\phi\overset{OH}{\dot{C}}-\phi \rightarrow \phi-\underset{\phi}{\overset{OH}{C}}-\underset{\phi}{\overset{OH}{C}}-\phi$$

of the photochemistry of isobutyrophenone (*13*), a mixture of pinacols and carbinol was formed. Irradiation of (*13*) in deuterium isopropoxide gave a 73:27 mixture of the

$$\phi-\overset{O}{\overset{\|}{C}}-i\text{-}Bu + (CH_3)_2\underset{H}{C}-OH \xrightarrow{h\nu} \phi-\underset{i\text{-}Bu}{\overset{OH}{C}}-\underset{i\text{-}Bu}{\overset{OH}{C}}-\phi + \phi-\underset{H}{\overset{OH}{C}}-i\text{-}Bu + CH_3\overset{O}{\overset{\|}{C}}CH_3$$

(*13*) (*14*) (*d,l + meso*) (*15*)

undeuterated and deuterated carbinols, (*15*) and (*15d*). This result was interpreted as indicating an isotope effect of k_H/k_D = 2.7 for ketyl radical disproportionation [58].

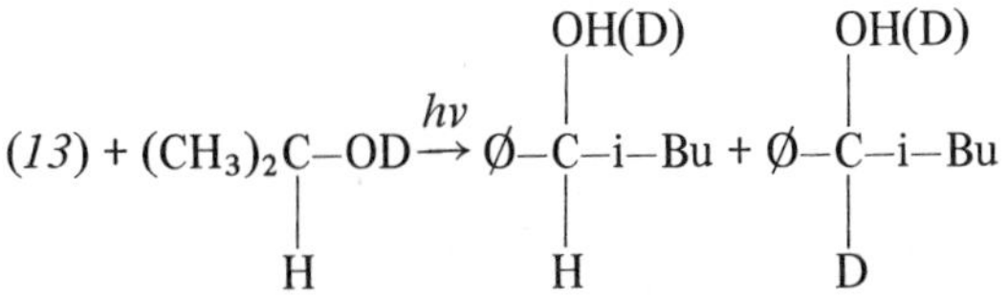

$$(13) + (CH_3)_2\underset{\underset{H}{|}}{C}-OD \xrightarrow{h\nu} \text{Ø}-\overset{\overset{OH(D)}{|}}{\underset{\underset{H}{|}}{C}}-i-Bu + \text{Ø}-\overset{\overset{OH(D)}{|}}{\underset{\underset{D}{|}}{C}}-i-Bu$$

(*15*) (*15d*)

In photoreductions of excited ketones and aldehydes from primary and secondary alcohols, there is a high preference for the abstraction of α-hydrogen to produce an α-hydroxy radical as opposed to abstraction of the hydroxyl hydrogen to give an alkoxy radical. However, in the case of photoreductions in *tert*-butyl alcohol [59a–c], hydrogen abstraction from the hydroxyl group has been proposed to account for the production of solvent incorporation products at the α-position [59c] (see p. 268). In a recent study of hydrogen abstraction by phenyl radicals from methanol [60], a relative rate ratio of 9:1 (a C–H bond being three times more reactive than an O–H bond) for methyl versus hydroxyl abstraction was observed. Thus, in this system only a small part of the differences in bond energies is effective in the transition state of the two competing reactions [60]. These results suggest that for excited carbonyl systems which have slow rates of intramolecular rearrangements, hydrogen abstraction from the hydroxyl group of *tert*-alcohols may assume importance.

In the case of two rather special systems, isotope effects strongly indicate hydroxyl hydrogen abstraction by the excited state. Thus, the original conclusion of Halwer [61] from acid catalysis studies, that the 2′-hydroxyl group of (*16*) and (*17*) was involved in photobleaching of these compounds, has been supported by the isotope effect studies of Moore and Baylor [62]. These workers found that irradiation of (*17*)

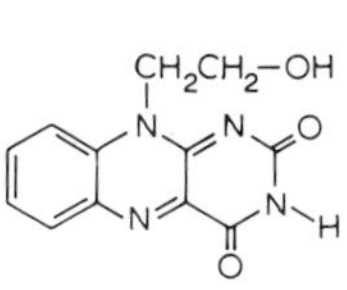

CH_2CH_2-OH (*16*)

$CH_2-C(CH_3)_2-OH$ (*17*)

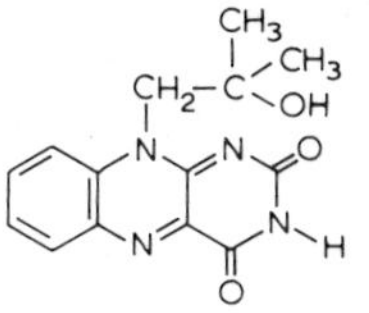

produced (*19*), and presumably (*20*), in addition to a product tentatively assigned as (*21*) ((*19*):(*21*) ratio = 55:45). Their proposed mechanism involved hydrogen abstraction of the hydroxyl hydrogen to produce biradical (*18*) which subsequently yielded the observed products. When the irradiation of (*17*) was performed simultaneously in water and deuterium oxide, the difference in the ratio of the rate constants was k_{H_2O}/k_{D_2O} = 4.9. Since no exchange at the aliphatic positions of (*17*) was noted, these results in the absence of any large solvent isotope effect indicate excited state abstraction of the hydroxyl hydrogen.

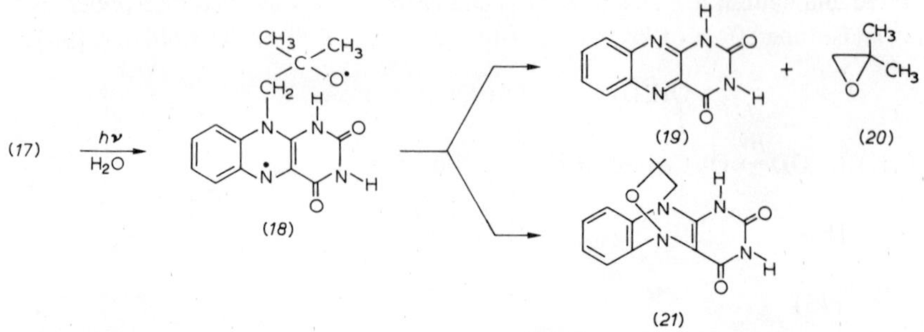

(17) $\xrightarrow[H_2O]{h\nu}$ (18) → (19) + (20); (21)

In a recent study the photoreduction of a series of ketones by cyclopropanols in benzene has been conducted [63]. For example, photoreduction of benzophenone by cyclopropanol (*22*) gives (*24*) and (*25*) in addition to a high yield of benzpinacol. For some cyclopropanol systems the quantum yield of benzophenone photoreductions was as high as 0.6 indicating that cyclopropanols are among the more reactive substrates in photoreduction processes. To establish that hydroxyl hydrogen abstraction by benzo-

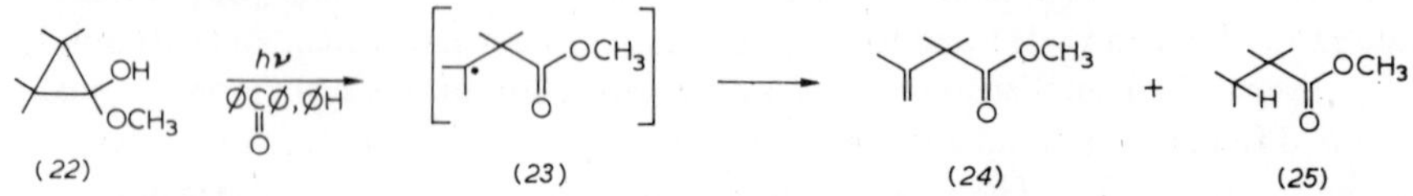

(22) $\xrightarrow[\phi C(O)\phi,\ \phi H]{h\nu}$ [(23)] → (24) + (25)

phenone was involved in these reactions, comparative irradiations of 1,2,2-trimethylcyclopropanol and 1,2,2-trimethylcyclopropanol-*O-d* were studied. The kinetic isotope effect for photoreduction of this system was between 3.6 and 6.6, supporting initial abstraction of the oxygen bound group. The unique ability of cyclopropanols to function as hydrogen donors is perhaps explained by synchronous ring opening of the 3-membered ring with abstraction of the hydroxyl hydrogen. Interestingly even a ketone such as fluorenone, which is unreactive toward photoreduction in ordinary alcohols, is reduced in the presence of cyclopropanols.

B. Isotope Effect on the Intramolecular Hydrogen Abstraction Reactions of Ketones

The effect of isotopic substitution on the hydrogen transfer from carbon has been most extensively studied in the case of intramolecular hydrogen abstraction of ketones, notably the Norrish type II reaction. From studies by several groups the essential steps in the intramolecular hydrogen abstraction of ketones have been established as shown below. For aliphatic ketones, initial hydrogen abstraction may occur from either the singlet or triplet excited state to afford biradical (*27*) [64,65]. Much evidence indicates that (*27*) can undergo reverse hydrogen transfer to afford (*26*), fragmentation to yield the enol (*28*) and olefin (*29*) or ring closure to afford cyclobutanol (*30*) [66]. The partitioning of the biradical among these three pathways is dependent upon the solvent and structural features of the particular ketone.

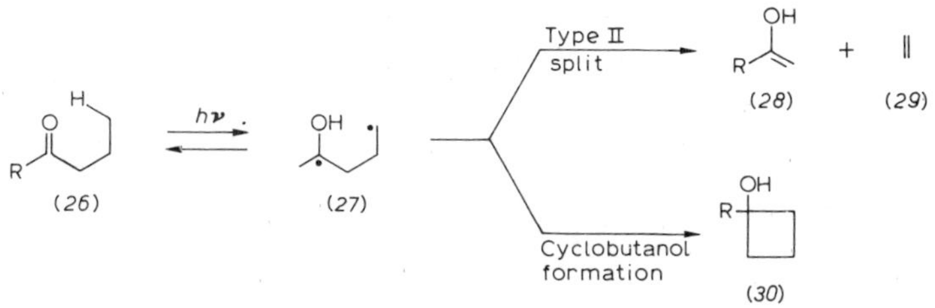

Type II split

Cyclobutanol formation

(26) (27) (28) (29) (30)

Early work on the photolysis of deuterated ketones was performed by Srinivasan who showed that vapor phase irradiation of 2-hexanone-5,5-d_2 yielded mainly *(33d)* and acetone which was 45% mono-deuterated [67]. These results were explained by

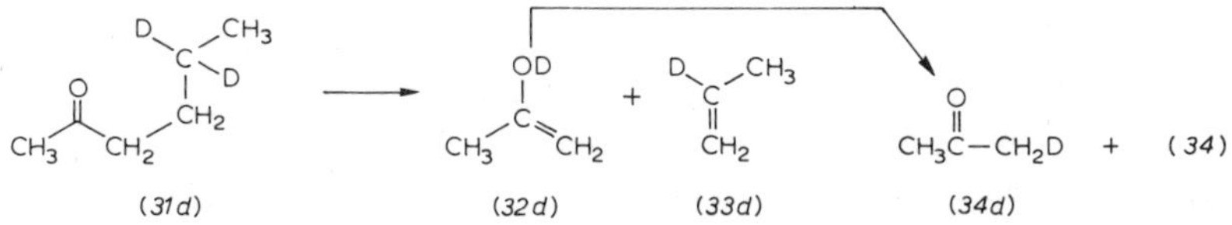

(31d) (32d) (33d) (34d) + (34)

the cleavage of *(31d)* to the enol *(32d)* and *(33d)*. The loss of deuterium in going from *(32d)* to *(34d)* was attributed to a ketonization of the enol on the walls of the reaction vessel accompanied by some deuterium hydrogen exchange. Subsequently, the enol *(32)* has been detected in the photolysis of *(31)* by long path infrared absorption and thus its intermediacy is rigorously established [68].

In a detailed study of the gas phase photochemistry of 2-pentanone-4,5,5-d_3 the effect of variables on the proportion of *(36d)*:*(37d)* was reported [69].

$$CH_3{-}\overset{O}{\overset{\|}{C}}CH_2CHDCD_2H \xrightarrow{h\nu} CH_3\overset{O}{\overset{\|}{C}}CH_3 + CHD{=}CD_2 + CHD{=}CHD$$

(35d) *(34)* *(36d)* *(37d)*

The ratio of *(36d)*:*(37d)* in the gas phase is independent of intensity, but increases with increasing wavelength and ketone concentration, and decreases with increasing temperature and oxygen pressure. From the temperature dependence of the *(36d)* to *(37d)* ratio in the liquid phase, a difference in activation energy for deuterium versus hydrogen transfer was calculated as about 1 kcal mol^{-1}. While no studies were reported which established the multiplicity of the reacting state in solution, subsequent work [64] suggests that these solution phase reactions involved primarily the triplet state of 2-pentanone.

It was in the photochemistry of 2-hexanone that the dissection of the isotope effect for hydrogen abstraction into its singlet and triplet contributions was first reported. Thus, Coulson and Yang [70] studied the quantum yields of the type II cleavage and cyclobutanol formation for 2-hexanone, 2-hexanone-5-d_1, 2-hexanone-5-d_2, and

TABLE 4

QUANTUM YIELDS FOR 2-HEXANONE IRRADIATION

Compound	Quantum Yields					
	Type II			Cyclobutanols		
	Singlet	Triplet	Total	Singlet	Triplet	Total
2-Hexanone	0.097	0.155	0.252	0.007	0.068	0.075
2-Hexanone-5-d	0.072	0.201	0.273	0.006	0.090	0.096
2-Hexanone-5, 5-d_2	0.047	0.251	0.298	0.005	0.108	0.113
2-Hexanone-d_{12}	0.043	0.251	0.294	0.004	0.111	0.115

2-hexanone-d_{12}. The quantum yields for the singlet reaction were measured at high concentrations of *cis*-1,2-dichloroethylene, and the quantum yields for the triplet process were calculated by subtracting that of the singlet reaction from the total quantum yield. The data presented in Table 4 summarize the results of this study. The relative ratio of propene-2-d_1 to propene in the photolysis of 2-hexanone-5-d_1 in the absence of quencher was 5.7, while a ratio of 2.7 was observed in the presence of quencher. When the former value was corrected for the 26% of singlet state reaction, the ratio of the deuterated to undeuterated propene for the triplet state was 6.7. Thus, these values indicate a substantial preference for abstraction of hydrogen versus deuterium in both the singlet and triplet ketone excited state. Inspection of the quantum yield data of Table 4 reveals an interesting aspect of the effect of deuterium substitution. Both the quantum yields for the type II reaction and cyclobutanol formation decrease with increasing deuterium substitution in the singlet state. However, in the case of the triplet state, which shows a higher selectivity for hydrogen over deuterium abstraction, the quantum yields increase with deuterium substitution for the d_1 and d_2 systems. Interestingly, estimates of the lifetimes of these deuterated molecules from the slopes of the Stern–Volmer plots indicate that the triplet lifetimes of 2-hexanone-5-d_1 and 2-hexanone-5-d_2 are about 2.3 and 7.0 times longer than for 2-hexanone itself.

Yang and Coulson considered three factors in interpreting these results. First, since deuterium substitution lowers the efficiency of the singlet state process, more efficient intersystem crossing could result, giving a higher steady-state concentration of triplets. This effect alone does not seem of sufficient magnitude to entirely explain the increased quantum efficiencies for the triplet reactions of the deuterated ketones. Second, deuterium substitution at the 5-position should markedly alter the reversal of the bi-radical to the ground state ketone, since it will require the breaking of an O–D bond instead of an O–H bond. Since deuterium substitution should not markedly affect either the fragmentation or ring closure reactions of the biradical, this effect would increase the quantum efficiency of product formation by retarding reversal of the bi-radical to starting ketone. The third factor is the possibility that the γ-C–D and

γ-C–H bonds may interact with the excited carbonyl group in a manner analogous to hydrogen bonding. Since deuterium would be less effective than hydrogen in aiding energy removal to the ground state, due to Franck–Condon factors, the molecule substituted with deuterium would have a longer triplet lifetime. Indeed, the mono- and di-deuterated 2-hexanones do have longer triplet lifetimes; however, it was not established whether this is due to a slower rate constant for hydrogen abstraction or the operation of a Franck–Condon factor. It seems most probable that all three factors are operating simultaneously, and their individual importance will vary with the particular system of interest.

Several other studies [71a,b] on the Norrish type II reaction in which deuterium isotope effects were observed, have been reported. Utilizing methods similar to those of Yang, the effect of deuterium substitution on the singlet- and triplet-state chemis-

(38d) $\xrightarrow{h\nu}$ (39d) CH_3 + CH_3–C(=O)– (40d) + Olefins

try on 5-decanone was reported [71a]. The quantum yield data of Table 5 established that, as in Yang's system, the total quantum efficiency for the singlet state decreased upon deuterium substitution while that for the triplet state increased slightly. In contrast to the 2-hexanone case, the total quantum yield of the reaction decreased. The isotope effect for abstraction in the singlet state was calculated as about 3.6 while for the triplet state a value of about 5.2 was reported.

While this discussion has emphasized the magnitude of the isotope effect in the γ-hydrogen abstraction reaction, it should be noted that the isotope effect may have synthetic consequences arising from alteration of product ratios in the photolysis of labeled compounds. One striking example [72] of this effect is (*41*) which gives (*42*) and (*43*) in yields of 45% and 28%, while (*41d*) affords (*42d*) and (*43d*) in yields of 55% and $<$ 4%. Since deuterium substitution retards the efficiency of the singlet-state reaction but often enhances the efficiency of the triplet-state reaction, this change in product ratio would result if the ratio of (*43*):(*42*) from the singlet state were higher than from the triplet state. Inspection of Yang's data on the ratio of type II versus cyclobutanol formation for the singlet and triplet state of 2-hexanone suggests this

TABLE 5

QUANTUM YIELDS FOR APPEARANCE OF 2-HEXANONE AND 2-HEPTANONE

Starting ketone	2-Hexanone			2-Heptanone		
	Singlet	Triplet	Total	Singlet	Triplet	Total
5-Decanone (*38*)	0.049	0.030	0.079	0.055	0.033	0.088
5-Decanone-2,2-d_2 (*38d*)	0.050	0.054	0.104	0.014	0.010	0.024

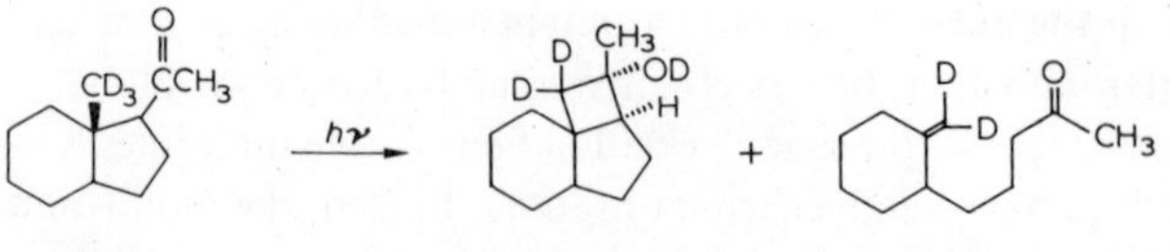

(41d) $\xrightarrow{h\nu}$ (42d) + (43d)

rationale is quite plausible. A further contributing factor to the change in product ratio could be an inverse isotope effect on bi-radical closure favoring formation of *(42d)*.

In the case of alkylphenone photolysis, no correction is needed for singlet reaction, as the type II processes for aromatic ketones arise entirely from the triplet state. In comparing the photochemistry of nonanophenone *(44)* versus nonanophenone-γ,γ-d_2 *(44d)* substitution of hydrogen for deuterium produced no change in the quantum

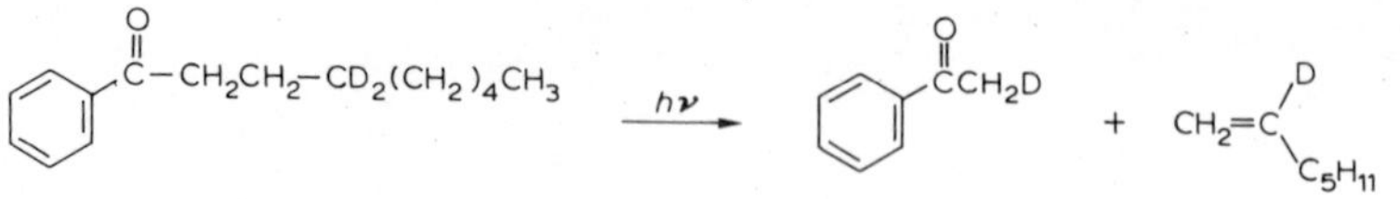

(44d) $\xrightarrow{h\nu}$ (45d) + (46d)

yield of the reaction but increased the triplet lifetime of *(44d)* by a factor of three [66]. The calculated value of the isotope effect for hydrogen versus deuterium abstraction was 4.8. In the case of the photochemistry of γ-hydroxy-γ-phenylbutyrophenone, a deuterium isotope effect on hydrogen abstraction of 1.7 was noted [73].

In a more recent study [74], the photochemistry of α-cyclopropoxyacetophenones was reported. For the dimethyl compound *(47)* irradiation resulted in inefficient formation of *(48)* and *(49)* as major products, the production of *(48)* being somewhat unusual. For the deuterated derivative *(47d)*, both the lifetime of the triplet state ($\tau_{(47d)}/\tau_{(47)} = 2.4$) and the efficiency of *(48)* formation increased ($\Phi_{(48d)} = 7.8 \times 10^{-3}$, $\Phi_{(48)} = 4.3 \times 10^{-3}$). To account for the low efficiency of hydrogen abstrac-

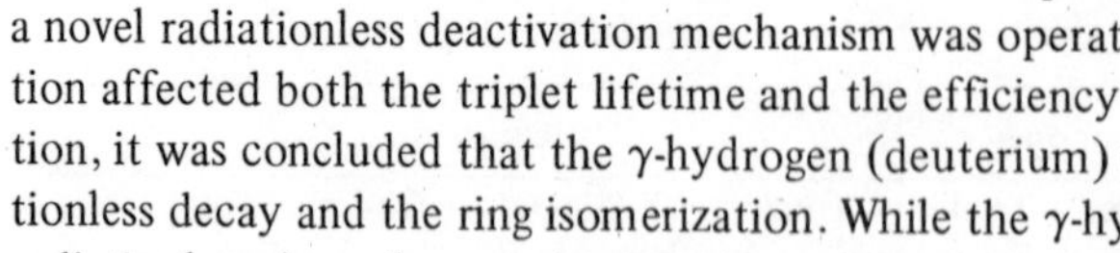

(47d) $\longrightarrow$ (48d) + (49d)

tion in this system, and the *cis–trans* isomerization process, the authors proposed that a novel radiationless deactivation mechanism was operating. Since deuterium substitution affected both the triplet lifetime and the efficiency of the *(47)* → *(48)* transformation, it was concluded that the γ-hydrogen (deuterium) was involved in both the radiationless decay and the ring isomerization. While the γ-hydrogen involvement in the radiationless decay has ample precedent, the increased efficiency of the *(47)* → *(48)* conversion in *(47d)* does not necessarily implicate the γ-hydrogen in the isomerization reaction. Certainly further work is needed before any firm conclusions may be drawn.

Photochemical processes derived from intramolecular hydrogen abstraction to produce 1,3-biradicals have also received mechanistic scrutiny through isotopic labeling. Two instructive examples in this area involve the photochemistry of aroylazetidines and aroylaziridines. Irradiation of azetidine (*50*) affords (*51*) while (*52*) yields a mixture of (*51*) and (*53*) [75]. Multiplicity studies on the rearrangements indicated that the photoreactions of (*50*) and (*52*) originated from their lowest triplet states. From the absence of significant quenching of the rearrangement it was proposed that the mechanism involved electron transfer from nitrogen to the excited carbonyl group

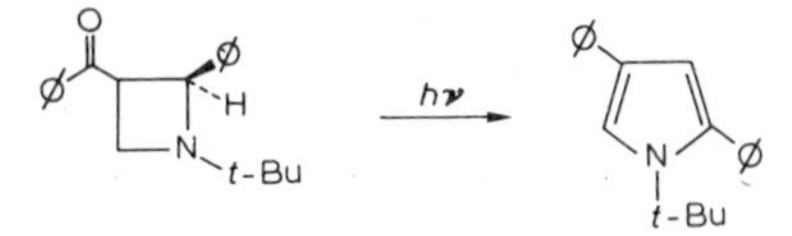

(50) → hν → (51)

(52) → hν → (53) + (51)

followed by proton transfer and electron reorganization to yield biradical (*56*). Ring

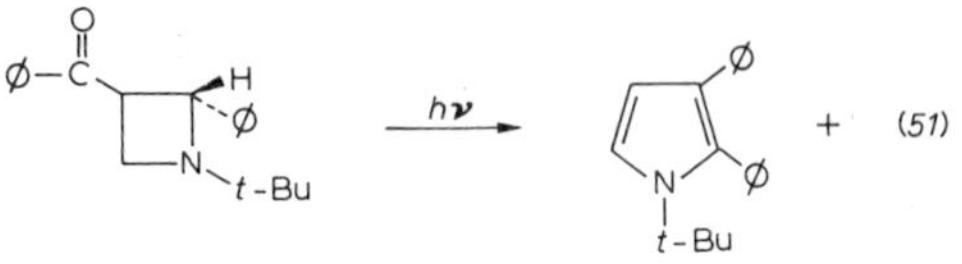

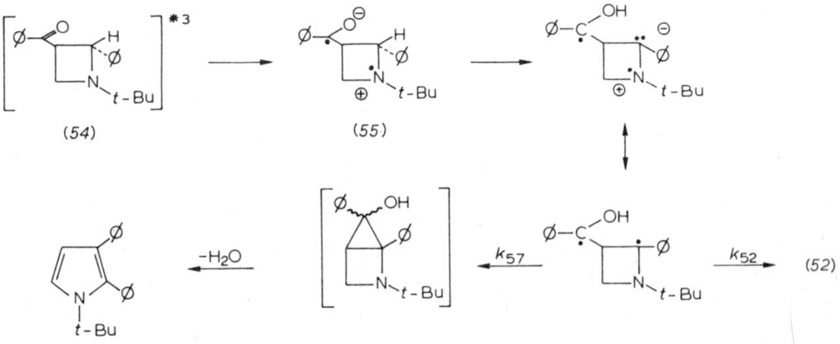

(54) → (55) → ... ↕ (56); (56) → k_{57} → (57) → $-H_2O$ → (53); (56) → k_{52} → (52)

closure of the biradical followed by dehydration yielded the observed products. The production of (*51*) would be explained by an analogous process involving hydrogen abstraction at the secondary position. In a detailed study of the effect of deuterium substitution on the course of the rearrangement, the photochemistry of the deuterated derivatives shown below was examined. The quantum yield data for the deuterated azetidines are given in Table 6. Interestingly, replacement of hydrogen by deuterium resulted in a moderate increase in overall quantum yields in both the *cis* and *trans* series of azetidines. The most significant result is that replacement of hydrogen by deuterium enhances the amount of ring closure at that position (compare the ratio of

TABLE 6

IRRADIATION OF DEUTERATED AZETIDINES

Compound	2, 3 : 2, 4 pyrrole	Φ pyrrole
(*50*)		0.046
(*50d*)		0.053
(*50d'*)		0.051
(*52*)	2.0	0.043
(*52d*)	1.4	0.087
(*52d'*)	2.3	0.065

2,3:2,4-pyrrole from (*52*), (*52d*), and (*52d'*)). It was proposed that in this system the low quantum yield and the failure to quench the triplet indicated that reverse transfer

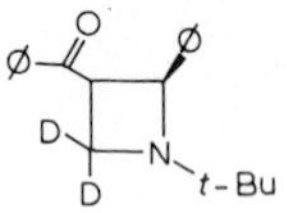

(*50d*) 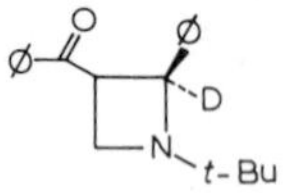 (*50d'*) 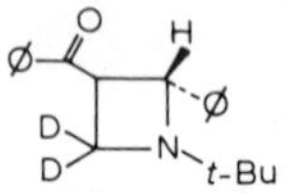 (*52d*) 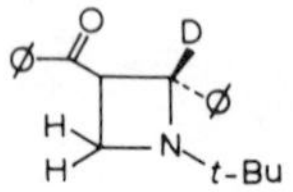 (*52d'*)

of hydrogen in the biradical back to carbon is faster than spin inversion and ring closure (i.e. $k_{52} > k_{57}$). Since substitution of deuterium for hydrogen would retard the reverse transfer process, ring closure would be enhanced at a position where deuterium had originally been abstracted. Further support for this rationale came from the actual calculation of the ratio of pyrroles for (*52*) versus (*52d*) [2,3:2,4 (*52*)/2,3:2,4 (*52d*)], the calculated value being 1.54 and the observed value being 1.43.

The effect of deuterium labeling in aroylaziridine photochemistry where the isotope effect is opposite to that just noted has also been reported. Irradiation of the *trans*-aziridine (*58*) affords the oxazole (*59*) (38%) and the conjugated ketone (*60*) (41%) in addition to minor amounts of (*61*) and (*62*) [76]. The oxazole (*59*) was proposed to arise by further oxidation of the dihydrooxazole (*64*) formed from dipolar

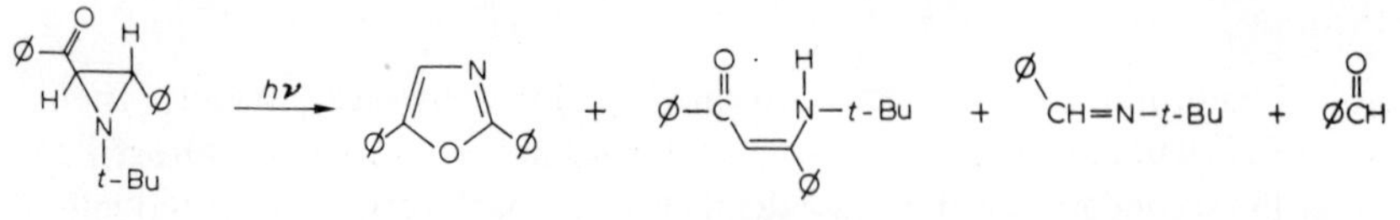

(*58*) (*59*) (*60*) (*61*) (*62*)

opening of excited (*58*). In a deuterium labeling study of the mechanism of the formation of (*60*), the photochemistry of (*58d*) and (*58d'*) was studied. The labeling results were consistent with the 1,4-hydrogen transfer mechanism shown below. Interestingly, the quantum yield for (*60*) was markedly decreased in (*58d*) (Φ = 0.006) versus (*58*) (Φ = 0.018) and (*58d'*) (Φ = 0.018). Thus, in contrast to the results observed in the

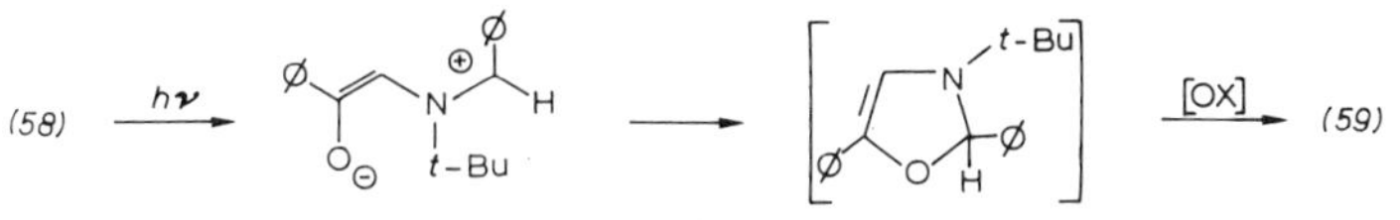

(58) $\xrightarrow{h\nu}$ (63) ⟶ [(64)] $\xrightarrow{[OX]}$ (59)

azetidine system, replacement of the abstracted hydrogen by deuterium leads to a much less efficient reaction. The authors propose that the lower efficiency for (*58d*) versus (*58*) indicates the hydrogen transfer is not reversible in the biradical (*65*) possibly because ring cleavage of (*65*) → (*66*) should be a rapid process. Thus, in the absence of a reverse transfer step, substitution of hydrogen for deuterium results in decreased efficiency for the reaction.

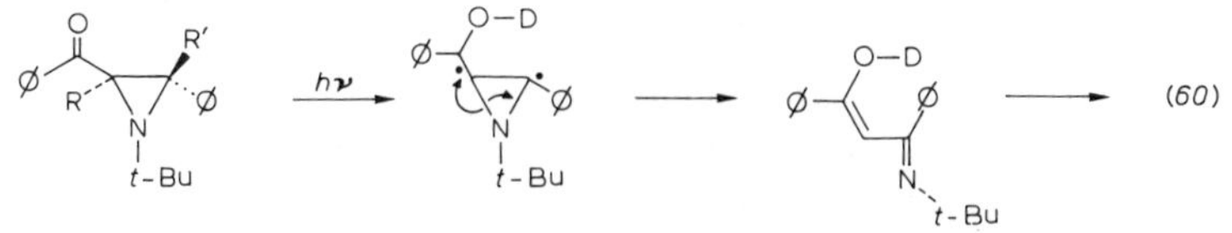

(*58d*) (R′ = D, R = H) $\xrightarrow{h\nu}$ (*65d*) ⟶ (*66d*) ⟶ (*60*)

(*58d′*) (R′ = H, R = D)

C. Hydrogen Abstraction Reactions of α,β-unsaturated Ketones and Esters

While the hydrogen abstraction reactions of dialkyl-, alkylaryl-, and diarylketones involve nearly exclusively the carbonyl oxygen, in excited α,β-unsaturated ketones deuterium labeling studies have aided in establishing that hydrogen abstraction may occur at either the carbonyl oxygen or the β-carbon [77a–c]. One of the consequences of carbonyl intramolecular hydrogen abstraction in certain α,β-unsaturated ketones is production of the β,γ-unsaturated isomer. For example, photolysis of 5-methyl-3-hexen-2-one (*67*) in methanol-*O-d* affords monodeuterated (*70d*) in agreement with a scheme involving *cis–trans* isomerization, intramolecular hydrogen abstraction to yield the dienol (*69*) and ketonization. More recently, 1-acetylcyclooctene

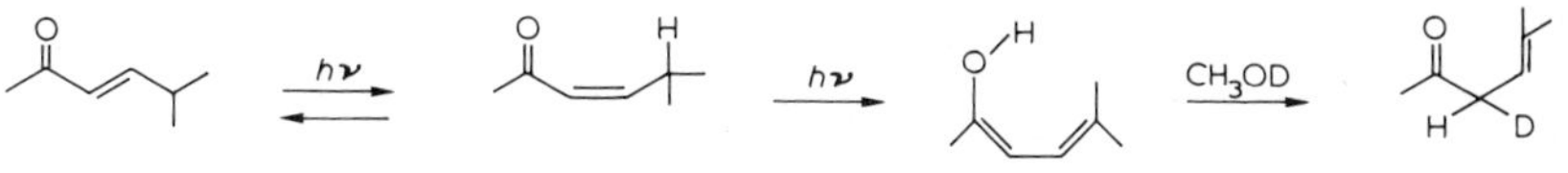

(*67*) $\underset{}{\overset{h\nu}{\rightleftharpoons}}$ (*68*) $\xrightarrow{h\nu}$ (*69*) $\xrightarrow{CH_3OD}$ (*70d*)

has been noted to undergo the same rearrangement; however, in this instance the intermediate dienols, (*73*) and (*74*), were fairly stable in the dark and could be readily detected by n.m.r. [79]. The mechanistic scheme shown below was proposed to account for the double bond isomerization.

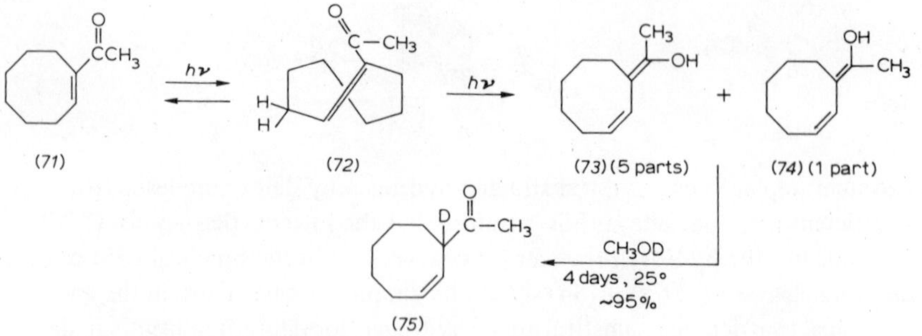

(71) (72) (73) (5 parts) (74) (1 part) (75)

$h\nu$; CH_3OD, 4 days, 25°, ~95%

Of a more unusual nature are the photoisomerizations of α,β- to β,γ-unsaturated ketones in systems where the intramolecular hydrogen abstraction seems quite unlikely [80–83]. Thus, in certain cyclic systems it appears unreasonable that intramolecular hydrogen abstraction of the γ-hydrogen by the carbonyl oxygen could occur. Schaffner and co-workers [81b,c] have extensively studied the kinetics, solvent effects, and wavelength effects for several such systems. For the octalone, (*76*), the concentration dependence for formation of (*77*) was indicative of a bimolecular process [81c]. This proposal was further enforced by simultaneously irradiating a mixture of *O*-acetyl-testosterone (*78*) and deuterated testosterone (*79d*). The isomerized products

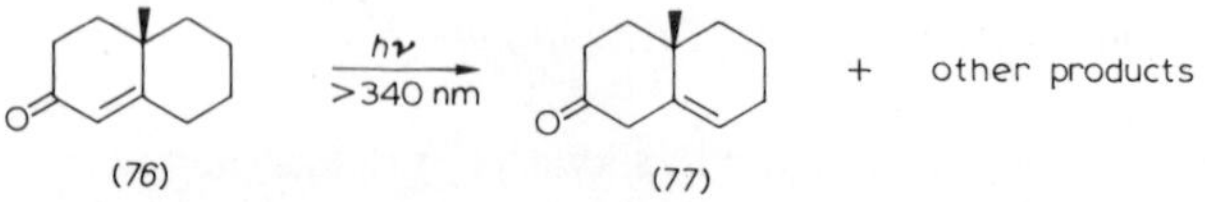

$h\nu$, >340 nm; + other products

(76) (77)

from this reaction showed loss of deuterium in (*81d*) and incorporation of deuterium in (*80d*) again supporting a bimolecular process for the double bond isomerization.

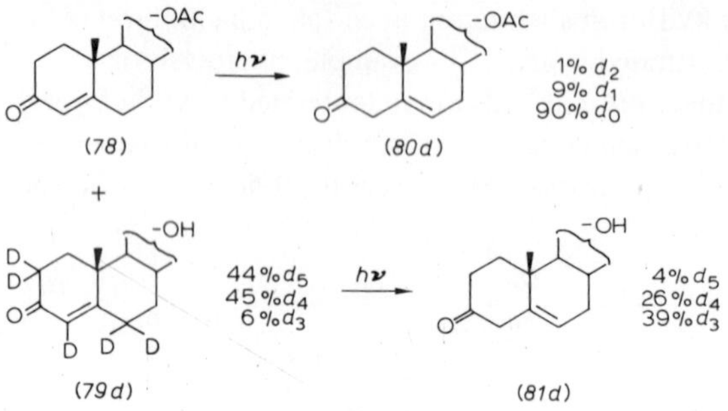

(78) + (79d): 44% d_5, 45% d_4, 6% d_3 —$h\nu$→ (80d): 1% d_2, 9% d_1, 90% d_0; (81d): 4% d_5, 26% d_4, 39% d_3

Furthermore, when the isomerization of (*78*) was performed in benzene containing *tert*-butyl alcohol-*O*-*d*, (*80d*) was formed containing 60% d_1 (position not specified but presumably at C-4). The workers interpreted the latter result as indicating the involvement of a hydroxyl bearing intermediate in the photochemical deconjugation [81c]. The partial reaction scheme below in which a n–π^* excited triplet state abstracts hydrogen from a ground state ketone is consistent with the kinetics and deuterium scrambling results. However, all the intermediates in this process have not as yet been

(76)* + (76) ⟶ (82) + (83)

(82) + (83) ⟶ (76) + (77)

identified. Interestingly, a second mechanism for double bond isomerization involving $\pi-\pi^*$ excitation occurs in *O*-acetyl-10α-testosterone [81b]. Since the proposed mechanism appears analogous to polar additions involving cycloolefins (see p. 278), this work will not be presented here.

Aside from hydrogen abstraction processes giving rise to photoreduction, interesting photochemical transformations can often result from such reactions as illustrated below. Thus, while irradiation of (*84*) in the $n-\pi^*$ region afforded mainly deconjugation to yield (*85*), photolysis at 254 nm afforded (*86*) as a major product [84]. Under preparative conditions in which (*85*) is continually isomerized to (*84*) by base, the

(85) ⟵ ($h\nu$, 340 nm; ØH) (84) ⟶ ($h\nu$, 253.7 nm) (86) (6 parts) + (85) (8 parts) + Unknown compound (1 part)

interesting oxapropellane can be obtained in 40% yield. Deuterium labeling studies on this novel $\pi-\pi^*$ excitation process strongly support the intramolecular nature of the reaction and the involvement of an enol intermediate. Furthermore, from simultaneous irradiation of (*84*) and (*84d*), the production of (*86*) was faster than that of

(84d) ⟶ ($h\nu$, 254 nm) (86d) — in iso-octane, 100% d_6; in *t*-BuOH, 66% d_6

(84d′) ⟶ ($h\nu$, 254 nm) (86d′) — in *t*-BuOH, 100% d_5

(*86d*) by a factor of 2.7, indicative of an expected isotope effect for the process. The reaction is conveniently viewed as hydrogen abstraction of a methoxy hydrogen in the non-planar excited state of (*84*) to produce the biradical (*87*) which after five-membered ring formation and ketonization affords (*86*).

(84)* → (87) → (88) → (86)

While hydrogen abstraction by the carbonyl oxygen is certainly the most well-known process, several literature examples strongly imply initial hydrogen abstraction may also occur at the β-carbon of α,β-unsaturated ketones [81c,85–87c], and deuterium labeling results in one system rigorously demonstrate such a process. Thus, from irradiation of (*76*) and (*90*) in toluene and *tert*-butyl alcohol respectively, products arising from addition of benzyl and *tert*-butoxy fragments to the α-positions were noted (see ref. 81c for a discussion of possible oxygen alkylated products from (*76*)).

(76) $\xrightarrow[\text{ØCH}_3]{h\nu}$ (89) + other products

(90) $\xrightarrow[t\text{-BuOH}]{h\nu}$ (91) + other products

This latter process would have involved hydrogen abstraction of the hydroxyl hydrogen of *tert*-butyl alcohol. Unfortunately, an attempt to verify hydroxyl hydrogen abstraction by (*90*) was unsuccessful since the formation of (*91*) from irradiation in *tert*-butyl alcohol-*O*-*d* was completely suppressed [87c]. However, for the cyclopentenone, (*92*), initial hydrogen (deuterium) abstraction by the β-carbon could be supported by deuterium labeling. This ketone was particularly suited for observing abstraction by the β-carbon as the type I and type II processes should not be particularly favorable, and the tertiary hydrogen five atoms removed from the β-carbon was ideally situated relative to the β-position. Indeed, irradiation of (*92*) produces (*93*), (*94*) and (*95*), which are the products expected from disproportionation and closure reactions of biradical

(92) R = H; (92d) R = D $\xrightarrow[\text{ØH}]{h\nu}$ (93) 59% + (94) 24% + (95) 10%

(*96*). When the deuterated compound (*92d*) was irradiated, the same distribution of products was formed although the reaction was much slower, suggesting an appreciable isotope effect in the process. Thus, for excited α,β-unsaturated ketones where the

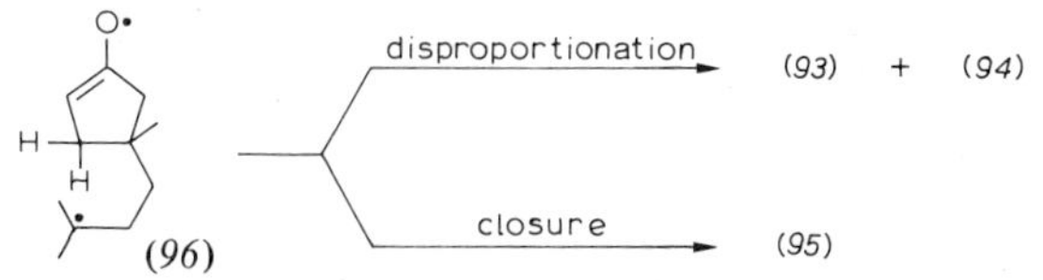

(96) — disproportionation → (93) + (94); closure → (95)

rates of unimolecular isomerizations and hydrogen abstraction processes by the carbonyl oxygen are low, hydrogen abstraction by the β-carbon may become quite important.

Solvent isotope effects have generally not been extensively studied in organic photochemical systems. A report by Jorgenson involving the irradiation of α,β-unsaturated esters in methanol and methanol-*O-d* is particularly striking and is noted here [88]. Irradiation of (*97*) in methanol primarily affords (*98*) while irradiation in methanol-*O-d* yields comparable amounts of (*98*) and (*99*). Simple kinetic studies established

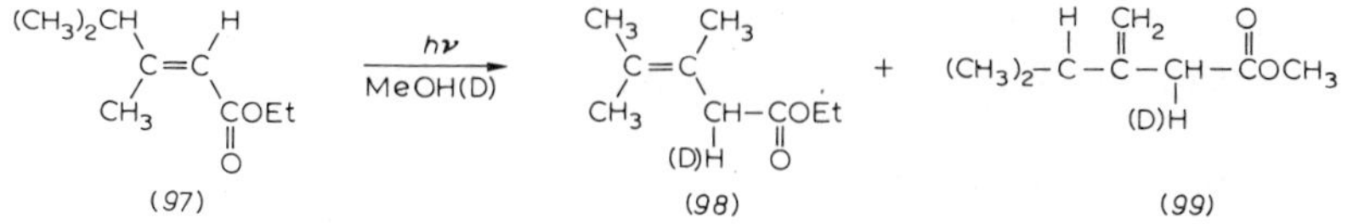

(97) —hν, MeOH(D)→ (98) + (99)

that the deuterated solvent did not affect the rate of formation of (*98*) but markedly

Reaction scheme for the irradiation of (97)

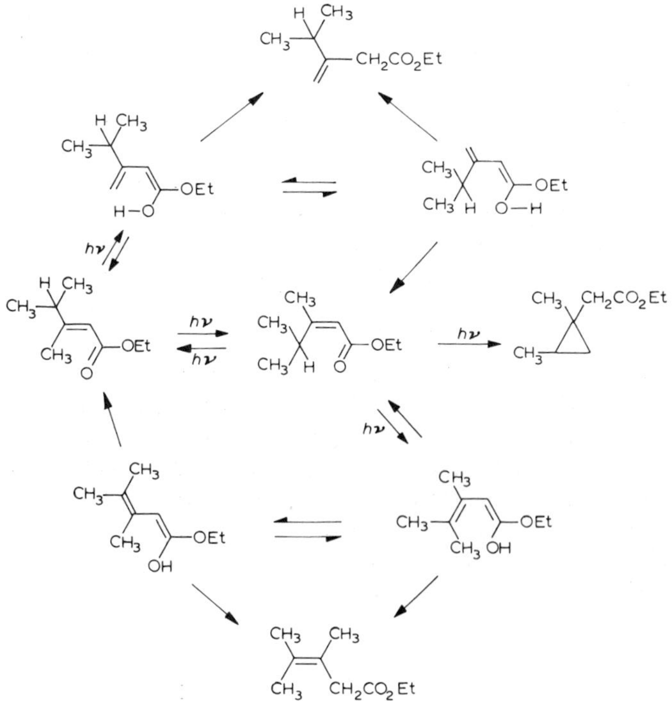

accelerated the production of (*99*). The solvent isotope effect (k_{CH_3OD}/k_{CH_3OH}) on the formation of (*99*) was estimated to be on the order of 15–50! In view of the complex scheme of reactions involved in the irradiation of (*97*) it is not surprising that a complete explanation for this large solvent isotope effect is lacking. These solvent isotope effects were also noted for ester (*100*). Irradiation of (*100*) in methanol yields the ester (*101*) as the major product. However, in methanol-*O*-d_1 ester (*102*) forms thirteen times faster than (*101*) and the final product mixture is comprised of 90–95% of (*102*). These large effects observed between methanol and methanol-*O*-*d* as solvent

C(CH3)3 CH3 OEt —hν, MeOH→ CH3 CH3 CH3 OEt + C(CH3)3 OEt

(*100*) (*101*) (*102*)

certainly merit more detailed study and may have synthetic as well as mechanistic interest in other systems in which intramolecular hydrogen abstractions are involved.

V. 1,7- AND 1,5-INTRAMOLECULAR HYDROGEN AND ALKYL SHIFTS IN OLEFIN SYSTEMS

In addition to the hydrogen abstraction processes discussed in the previous section, a second class of photochemical reactions in which hydrogen (carbon) is transferred from one atom to a second is conveniently designated as hydrogen (carbon) shift reactions. The important distinction between these two classes is that in the former radical intermediates are formed, while in the shift processes the migration is a concerted reaction leading to a non-radical product. Deuterium labeling studies have been especially useful in studying the mechanism of many of these concerted shifts. Not only have they established the occurrence of such processes where the reaction is degenerate, but these studies have also inferred the presence of photolabile intermediates in reactions where isolation of these intermediates was quite difficult.

A. 1,7-Shifts

The first reports of intramolecular photochemical hydrogen migrations in the early 1960's involved studies in cycloheptatriene systems. Since for cycloheptatriene itself a 1,7-hydrogen shift is a degenerate rearrangement, the reaction was first noted in the deuterated cycloheptatrienes. In a study of the cycloheptatriene produced from photolysis of diazomethane-d_2 in benzene, Doering and Gaspar [89] reported that the initially produced cycloheptatriene-7,7-d_2 (*104d*) underwent further reaction to afford the deuterium migrated product (*104d'*). In an independent study Roth [90]

+ CD_2N_2 —hν→ D D —hν→ D H D

(*103*) (*104d*) (*104d'*)

also showed the 1,7-shift process as occurring upon irradiation of cycloheptatriene-7-d_1; in addition 1,7-shifts in several substituted cycloheptatrienes were reported. Aside from demonstrating the occurrence of the 1,7-hydrogen shift in cycloheptatrienes, deuterium labeling has also been used to establish the relative efficiency of 1,7-hydrogen shift versus electrocyclic ring closure. Thus, ter Borg and Kloosterziel [91] reported that the photochemical 1,7-hydrogen shift ((*105d*) → (*105d'*)) occurs 500 times faster than cyclobutene formation ((*105d*) → (*106d*)). Since the isotope

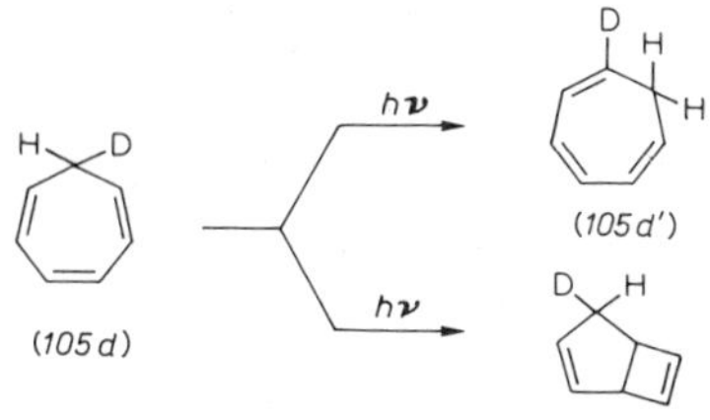

(105d) (105d') (106d)

effect for hydrogen versus deuterium migration is not known, the actual ratio of hydrogen migration to cyclobutene formation is undoubtedly greater than 500. In recent years several investigations of substituent effects on the directionality of 1,7-hydrogen shifts in unsymmetrical cycloheptatrienes [92] and the effects of substituents on the partitioning between 1,7-shift processes and bicyclo[3.2.0]hepta-2,5-diene formation have also been reported [93a,b].

A process formally analogous to the 1,7-shifts observed in the simple cycloheptatrienes occurs with high quantum efficiency and in good synthetic yield for 3,4-benzotropilidenes (i.e. (*107d*) → (*109d*)) [94,95]. While the exact mechanism of the formal

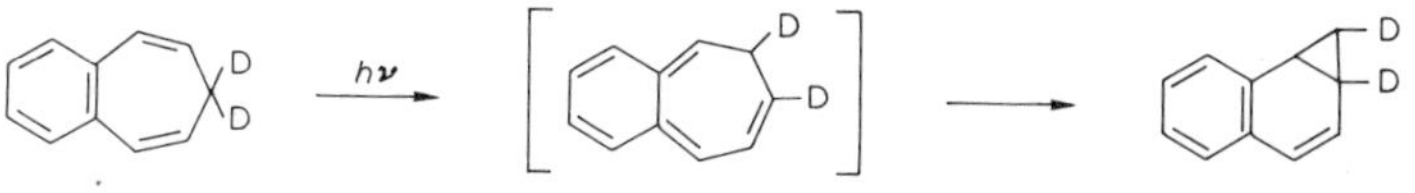

(107 d) (108d) (109d)

1,7-shift process has not been rigorously established, deuterium labeling and other data are consistent with the pathway shown above. Notice that while this reaction sequence involves disruption of the aromatic system, the quantum efficiencies of various 3,4-benzotropilidenes to benzonorcaradienes are uniformly high (0.6–1.0) [95]. Deuterium labeling studies were also used to probe the simultaneous operation of a 1,3-shift process in the photochemistry of (*107d*) (i.e. (*107d*) → (*110d*)). It was known that (*110*), which would be the product of a 1,3-hydrogen shift from (*107*), is readily photochemically transformed to (*109*) and thus might be difficult to detect in the irradiation of (*107*). If, however, this sequence were operative for (*107d*), some deuterium would have been expected at the cyclopropyl–benzylic position (i.e. (*109d'*)). Since within experimental error none of (*109d'*) was detected, and in addition the

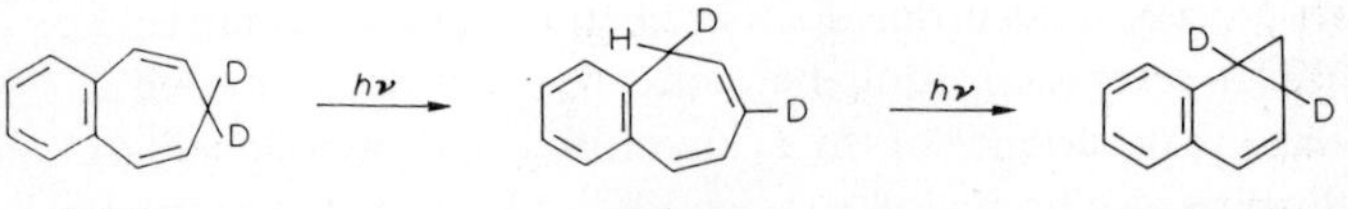

(107d) —$h\nu$→ (110d) —$h\nu$→ (109d′)

formation of (*110*) could not be observed in the irradiation of (*107*), the operation of a 1,3-hydrogen shift process was excluded. A reinvestigation [95] of the photochemistry of (*107*) as well as results in other systems [96] has cast some doubt upon the complete absence of a 1,3-shift in 3,4-benzotropilidenes. Thus, irradiation of carefully purified (*107*) showed at low conversion a ca. 10:1 ratio of (*109*) and (*110*). If only 10% of (*110*) or (*110d*) is formed in these irradiations, then even with a reasonable

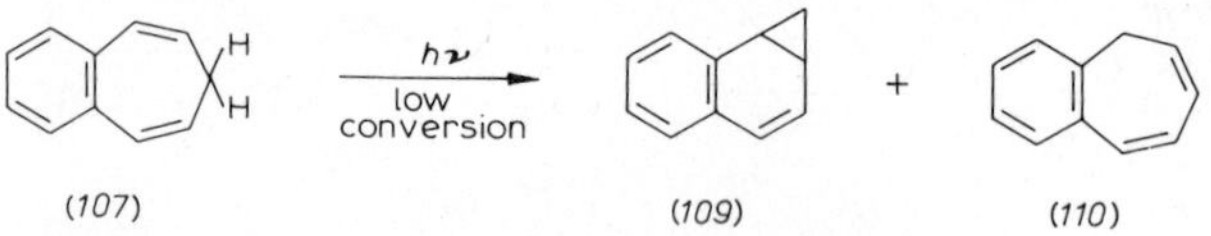

(107) —$h\nu$, low conversion→ (109) + (110)

isotope effect for the (*110d*) → (*109d*) conversion the amount of deuterium at the cyclopropyl–benzylic position might not be detectable within experimental error using ordinary n.m.r. integration methods.

The precise isotope effect for hydrogen migration in these and other systems, while of value in quantitative assessment of labeling studies, is not known with certainty. An indication that the ratio of hydrogen to deuterium migration may be appreciable derives from a study of the photochemistry of benzotropilidenes, (*111*) and (*111d*), as summarized below [97]. The validity of a precise number for the isotope effect of hydrogen versus deuterium migration is somewhat questionable since the corresponding rates of the reaction are unknown. However, since the isotope effect for ester mi-

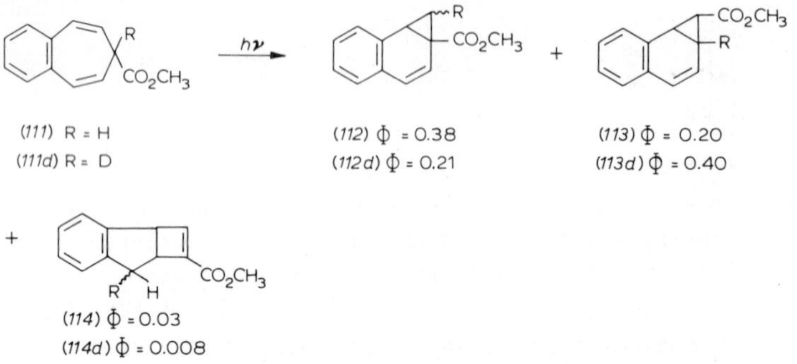

(111) R = H
(111d) R = D
—$h\nu$→
(112) Φ = 0.38
(112d) Φ = 0.21
+
(113) Φ = 0.20
(113d) Φ = 0.40
+
(114) Φ = 0.03
(114d) Φ = 0.008

gration might be expected to be small, if the rate of carbomethoxy migration is assumed to be equal in (*111*) and (*111d*), then k_H/k_D would be ~ 3.8. Interestingly, the overall quantum yields for photolysis of (*111*) and (*111d*) are identical within experimental error.

The production of (*114*) in this reaction was shown not to arise by further irradiation of initially formed (*112*) or (*113*). Furthermore, the lower quantum yield for production of (*114d*) was suggestive of an isotope effect in one of the steps of the (*111*) → (*114*) conversion. These facts, together with the labeling pattern in (*114d*), would be in agreement with a small percentage of 1,3-hydrogen shift in excited (*111*)

(*111*) —hν→ (*115*) —hν, Φ = 0.61→ (*114*)

to afford (*115*) which undergoes further secondary reaction to yield (*114*). Further support for this proposal derives from an independent study [97] which showed that excited (*115*) produces (*114*) with a quantum efficiency of 0.61. Using the quantum yields for the (*111*) → (*114*) (0.03) and (*115*) → (*114*) (0.61) conversions, a quantum efficiency for the (*111*) → (*115*) process is calculated as 0.10. Thus, this data would indicate a preference of 4:1 for 1,7- versus 1,3-hydrogen shift as compared to the 10:1 ratio observed for the parent 3,4-benzotropilidene.

A third system where facile 1,7-shifts have been observed is the 1,2-benzotropilidenes. Here again deuterium labeling of the parent hydrocarbon (*116d*) has provided the strongest support for the process [94]. Thus, irradiation of (*116d*) produces (*117d*) (65–95%) and (*118d*) (~ 0.5%), the (*116*) → (*117*) transformation occurring with a quantum efficiency of 0.63 [98]. It is noteworthy that the partitioning between a 1,7-hydrogen shift and an electrocyclic ring closure to the cyclobutene is quite

(*116 d*) —hν→ (*117 d″*) + (*118 d*)

sensitive to the nature of the 5-substituent as the data below illustrate [98]. Thus, in some systems (i.e. (*119c*)), the product of the 1,7-shift is only of minor importance.

(*119*) —hν→ (*120*) + (*121*)

X	ϕ_{120}	ϕ_{118}
(a) CH_3	0.37	0.09
(b) CN	0.20	0.35
(c) CO_2CH_3	0.09	0.61

B. 1,5-Shifts

In contrast to the numerous examples of photochemical 1,7-hydrogen shifts, reports of 1,5-hydrogen shifts in the excited state are rare. This certainly is a consequence of the restrictions imposed on the concertedness of such processes by the Woodward–Hoffmann Rules. Thus, while photochemical 1,7-hydrogen shifts can occur in a suprafacial sense, the analogous 1,5-hydrogen shifts are predicted to occur antarafacially. Indeed in systems conformationally disposed towards an antarafacial 1,5-hydrogen shift such processes have been observed [99,100].

In Roth's study [101a,b] of the photochemistry of 1,3,6-cyclooctatriene, (*122*), the formation of (*124*) was considered to arise from the initial isomerization of (*122*) to (*123*) followed by secondary photolysis of (*123*) to yield (*124*). The (*122*) → (*123*) conversion could have occurred by either a 1,3- or 1,5-hydrogen shift process. When the deuterium labeled material (*122d*) was irradiated, the product, (*124d*), showed a deuterium distribution consistent with the 1,5-photochemical hydrogen shift route as shown below. The excellent agreement between the observed labeling pattern and that

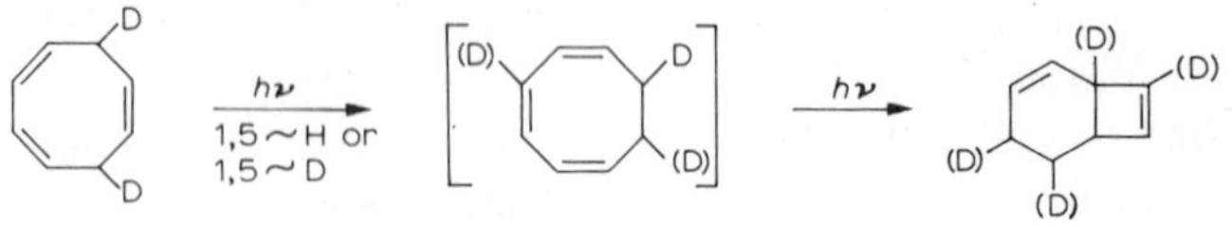

(*122d*) (*123d*) (*124d*)

calculated for the 1,5-shift mechanism (Table 7) seemed unusual since any isotope effect in the first step of the reaction would have led to a marked deviation from the statistical values assumed in the calculations. An alternate sequence which is also in agreement with the labeling results involves the di-π-methane rearrangement of (*122d*) to (*125d*) and (*126d*) (see p. 282). Since (*126d*) could not be detected in the former reaction due to its facile photoisomerization, a homo-1,7-shift in (*126d*) would afford (*123d'*) which could then photocyclize to (*124d'*). The apparent lack of an isotope effect in the rearrangement would be explained if (*126d*) showed a large degree of stereospecificity in migrating the group adjacent to the cyclopropane ring. Such a

TABLE 7

CALCULATED AND MEASURED HYDROGEN DISTRIBUTIONS FOR THE (*122d*) → (*124d*) CONVERSION

Distribution	H_2, H_3, H_7, H_8 (%)	H_1, H_6 (%)	H_4, H_5 (%)
Found	47.6	21.9	30.5
Calc. for 1, 5-shift	46.9	21.9	31.2
Calc, for 1, 3-shift	43.8	25.0	31.2
Calc. for 1, 7-homo shift	46.9	21.9	31.2

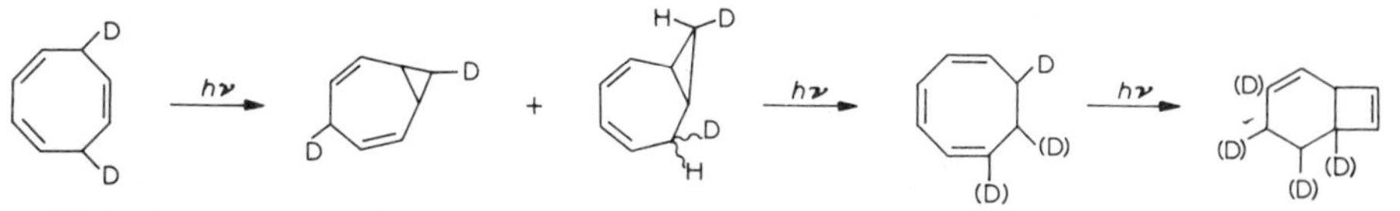

(122d) (125d) (126d) (123d′) (124d′)

specificity would not be unreasonable since a continuous overlap of the sigma bond migrating, the cyclopropane sigma bond, and the π-system in a Hückel transition state could involve stringent stereoelectronic requirements on the process. Furthermore, from results noted in the photochemistry of (*111*) and (*111d*), it would not be unreasonable to expect that while the rates of migration of hydrogen and deuterium would differ, the overall quantum efficiencies for the process could be identical. Since (*126d*) would be expected to be formed with about equal amounts of deuterium *cis* or *trans* to the cyclopropane, excellent agreement between the labeling observed and calculated would result (Table 7). While the author is aware of no previous precedent for the homo-1,7 shift, recent studies of the photochemistry of (*127*) by Burdett et al. [101b] are nicely accommodated by a highly specific 1,7-homo shift to yield (*128*) which subsequently cyclizes to (*129*).

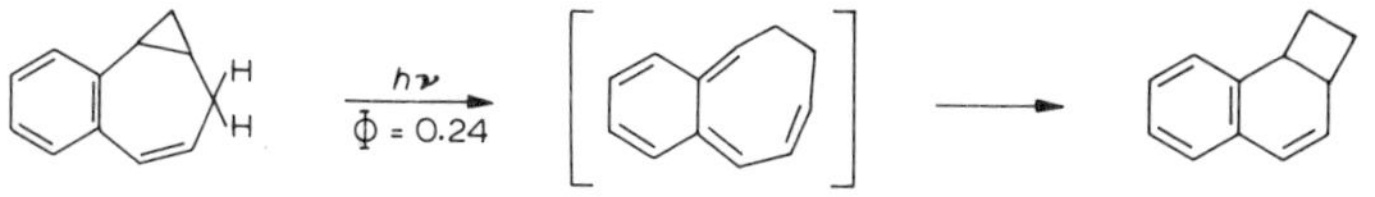

hν, Φ = 0.24

(127) (128) (129)

While the question of suprafacial 1,5-hydrogen shifts appears open, strong evidence from labeling studies is indicative of 1,5-carbon shifts in selected systems. Consider the case of the general benzonorcaradiene system below for which two different 1,5-carbon shifts are a priori possibilities. As noted earlier 1,2- and 3,4-benzotropilidenes

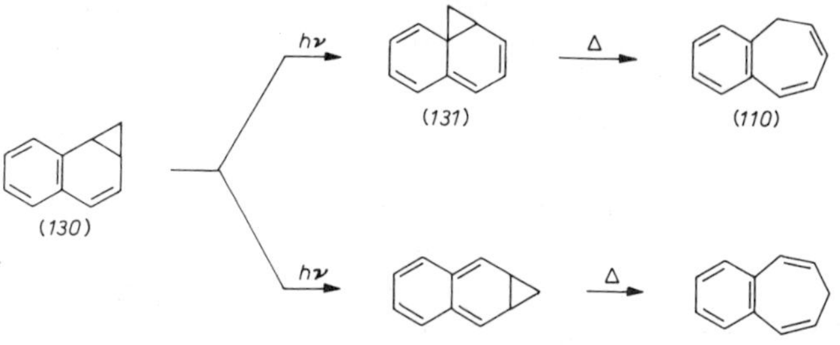

(130) (131) (110) (133) (107)

such as (*110*) and (*107*) are themselves quite photochemically reactive. Furthermore, (*107*) and (*110*) strongly absorb in the same regions of the ultraviolet spectrum as the starting material, (*130*). While low conversion irradiations at wavelengths in which the absorption of the starting material is maximized against products sometimes lead to the detection of the initially formed compound [97], this is quite often difficult even under optimum conditions.

In the case of the benzonorcaradienes, the problem is further compounded since the major products from irradiation of 3,4-benzotropilidenes are benzonorcaradienes while 1,2-benzotropilidenes often also yield benzonorcaradienes. A valuable technique in instances such as the above is to utilize deuterium labeling to support the intermediacy of undetected ground state intermediates. An early indication of photochemical 1,5-carbon shifts arose from a study of the photochemistry of (*134*) [102]. Irradiation of (*134*) results in slow conversion to a complex mixture of products. While (*135*) and (*136*) would be derived from photolysis of (*134*) to a carbomethoxy carbene and naphthalene, and (*137*) and (*138*) were products of 1,2-hydrogen shift in a ring opened biradical, the formation of the rearranged isomers, (*112*) and (*113*), was less apparent. It was originally proposed that (*139*), formed by a 1,5-carbon shift from (*134*), was a possible intermediate [102], and other work has supported the involvement of (*140*) in the reaction sequence as well [96].

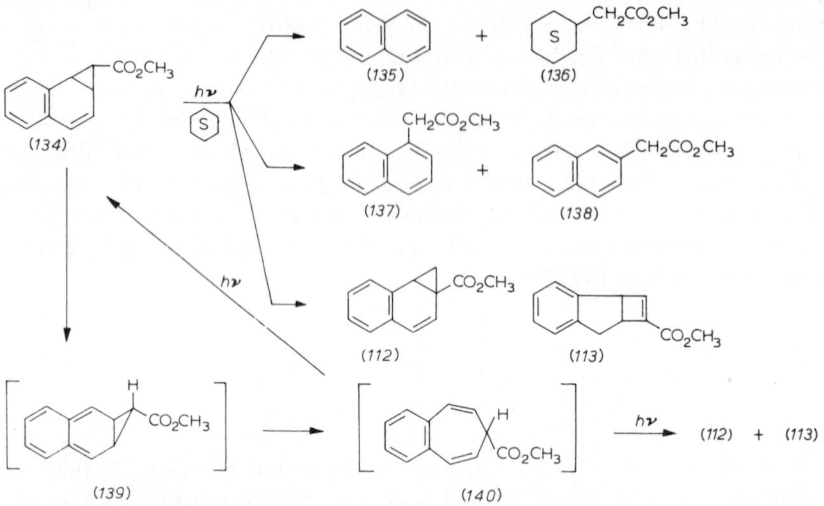

In later work Gruber and Pomerantz [103] studied the photochemistry of the benzonorcaradiene-7,7-d_2 and generalized the involvement of the 1,5-sigmatropic rearrangement. Thus, irradiation of (*130d*) besides affording fragmentation and hydrogen shift products, gives recovered (*130d*) in which the deuterium has been scrambled. The isomerization of (*130d*) to (*130d'*) and (*130d''*), together with the detection of small amounts of (*110*) and (*107*) in the irradiation of the undeuterated compound, supports the sequence shown below. Likewise, when the deuterated ester (*134d*) was studied, irradiation to 21% conversion resulted in deuterium scrambling to the extent noted below [104]. This result is consistent with the scheme noted above for (*130d*) irradiation, assuming that the ester migrates preferentially as opposed to deuterium in the 1,7-shift step. The comparable amount of scrambling to the two positions suggests little preference for the directionality of the initial 1,5-shift process.

(130 d) (131d) (110d) (130d′) (133d) (107d) (130d″)

(134d) (134 d′) (~80%) (7%) (9%)

In contrast to the two systems noted above, irradiation of (*112*) proceeds much more efficiently and cleanly producing only (*114*) at 300 nm [97]. The deuterium labeled compound (*112d*) afforded (*114d*) with exclusive labeling in the benzylic methylene position, consistent with preferential 1,5-sigmatropic shift to the aromatic ring carbon. Studies at 267 nm further support the mechanism implied by the deute-

(112d) (141d) (115 d) (114 d)

rium labeling results as at this wavelength the reaction of (*112*) to (*115*) can be studied independently of the (*115*) → (*114*) conversion. Detailed studies [97] in this system allowed the calculation of the quantum yield for the (*112*) → (*115*) reaction which was found to be 0.55. Hence, in this particular system the 1,5-shift is not only highly specific but also of good efficiency. It might be supposed that other substituted benzonorcaradienes would likewise show highly specific 1,5-shift processes.

While the labeling studies furnish strong support for the 1,5-shift processes in benzonorcaradienes, the question of inversion or retention of configuration at the migrating group should be resolved soon. Irradiation of optically active (*142*) does yield (*143*) which is optically active [105]. This work indicates that the (*142*) → (*143*) reaction does proceed in a highly stereospecific manner and from a knowledge of the absolute configurations of (*142*) and (*143*) the question of inversion or retention in the

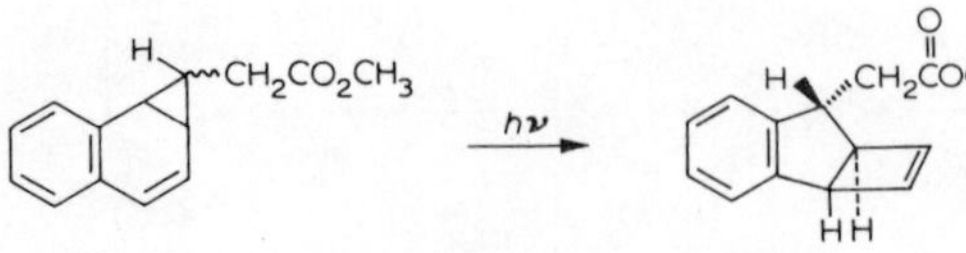

(142) 1:2 *endo-exo* $[\alpha]_D = +23.50$

(143) $[\alpha]_D^{25} = -117°$

migrating group can be answered. Mechanistic studies of the homofulvene to spiro-cyclopentadiene reaction are also consistent with a process involving a photochemical 1,5-sigmatropic shift with inversion of configuration [106,107].

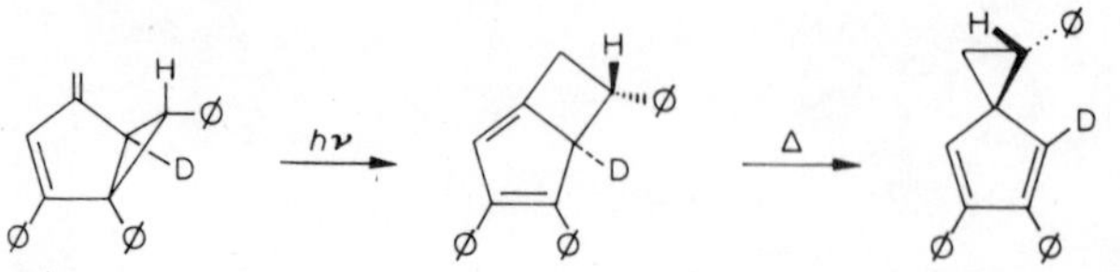

(144) (145) (146)

VI. DEUTERIUM LABELING IN PHOTOSENSITIZED NUCLEOPHILIC ADDITION REACTIONS OF CYCLIC OLEFINS

While Markovnikov addition of various adducts to ground state olefins has been known and studied for many years, it is only recently that light induced polar additions to olefin systems were recorded [108–110c]. In 1966 two groups [111a,b] noted that six- and seven-membered cyclic olefins undergo addition reactions when sensitized by benzene, toluene, or xylene. This particular reaction appears to be gener-

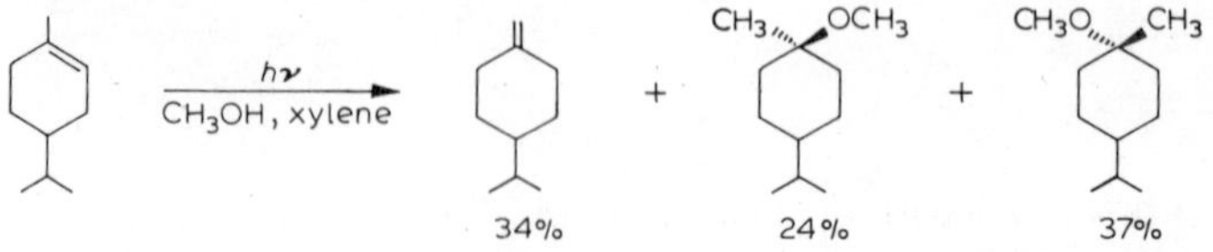

$h\nu$, CH_3OH, xylene

(147) → (148) 34% + (149) 24% + (150) 37%

al for the addition of water, alcohols, and carboxylic acids to six- and seven-membered cyclic olefins. However, the ease of addition is sensitive to the degree and type of olefin substitution as well as the acidity of the medium [112]. Acyclic, exocyclic, smaller and larger ring mono-olefins are generally much less reactive in these processes. While it was originally noted that cyclooctene was unreactive towards photosensitized addition in 1% sulfuric acid–methanol [112], a recent study has reported that cyclooctyl acetate is formed from irradiation of *cis*-cyclooctene in acetic acid–benzene (40:1) [113]. However, it is not clear whether the *cis*-cyclooctene is itself responsible for the reaction, as *trans*-cyclooctene, which may be formed in the reaction [114], undergoes addition of acetic acid in the dark to yield cyclooctyl acetate.

In view of the synthetic potential of performing addition reactions to simple olefins under very mild conditions, the mechanism and scope of these processes have been extensively investigated. Deuterium labeling studies have been especially useful in establishing the two-step nature of the addition and have furnished strong evidence for the intervention of cationic intermediates. Thus, photohydration of octalin (*151*) followed by washing the products with water, yielded a mixture of exocyclic olefin (*152*) and alcohols (*153*) and (*154*) [108]. All three products were shown to have deuterium in the equatorial position (> 90%) and to have essentially identical distributions of d_0, d_1, and d_2. These results strongly suggest a common intermediate for these three products. Interestingly, the hydration reactions of this and other olefins proceed faster

(*151*) $\xrightarrow[\text{xylene}]{h\nu,\ D_2O,\ DME}$ (*152*) 37% + (*153*) 30% + (*154*) 24%

in water than in deuterium oxide (for (*151*) the rate factor is about 7). The higher acidity of water versus deuterium oxide would suggest protonation as a key step in these reactions.

While the precise nature of the reactive species in these photoadditions has not been rigorously established, an attractive intermediate is a metastable *trans*-olefin. The formation of a mixture of equatorial and axial alcohols from (*151*) would be accounted for by protonation of a high energy ground state in the Markovnikov sense to produce a cation. The capture of cation by nucleophiles would yield a mixture of *cis*- and *trans*-addition products whose composition would be a function of the particular steric environment of the electron-deficient center. The high reactivity of the six- and seven-membered rings would then arise from the flexibility in the systems which allow distortion to the high energy ground state. The lower reactivity of exocyclic olefins and acyclic olefins would be due to the relatively free rotation about the olefin bond and thus non-attainment of a high energy ground state species. For smaller rigid cyclic systems, relaxation of the excited triplet to a distorted configuration would involve much more strain than in the six- and seven-membered rings; thus, these high energy systems often undergo radical reactions with solvents (vide infra).

Following the initial reports of photosensitized addition reactions, examples of photochemical methyl migration [115], double bond reduction [120], and fragmentation reactions of allylic alcohols [116–119] have appeared. These reactions may also be conveniently viewed as proceeding via cationic intermediates, and for the latter two processes deuterium labeling studies support this contention. Irradiation of cholesterol, (*155*), in *tert*-butyl alcohol–water with xylene as sensitizer yielded two products, (*156*) and (*157*) [116]. When the reaction was performed in deuterium oxide, and the exchangeable hydrogens washed out with water, (*156d*) and (*157d*) were observed with the indicated deuterium content. The high percentage of d_2 species in

(*157d*) suggested substantial double bond isomerization to the $\Delta^{4,5}$-position prior to its formation. While Waters and Witkop proposed a mechanism involving ring contraction followed by intramolecular capture of the primary carbonium ion yielding (*157*), recent studies seem more consistent with a fragmentation reaction of (*160*) to (*161*) followed by intramolecular oxetane formation. This latter mechanism is especially attractive since the unsaturated aldehyde, (*161*), has been detected in an independent study of the photosensitized reactions of (*159*) [118,119].

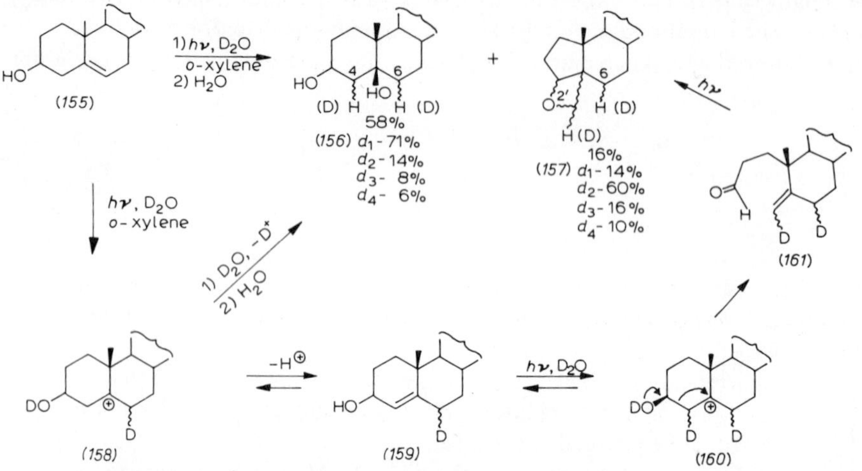

Finally, deuterium labeling has provided an interesting insight into the photoreduction of (*162*) in alcoholic media. Most photochemical reductions of carbon–carbon double bonds, while under vastly different conditions to those noted here, have been attributed to radical processes. However, reduction of 10-methyl-$\Delta^{1(9)}$-octalin in 2-deuterio-propan-2-ol with xylene as sensitizer afforded (*164d*), the position of the deuterium being established by synthesis [120]. On the basis of ground state radical chemistry, radical abstraction would have occurred at the carbon–hydrogen bond of the alcohol affording deuterium at C-1. Since the deuterium was located exclusively at the tertiary position, the authors considered this product as most reasonably arising by deuteride transfer from isopropoxide to cation (*163*).

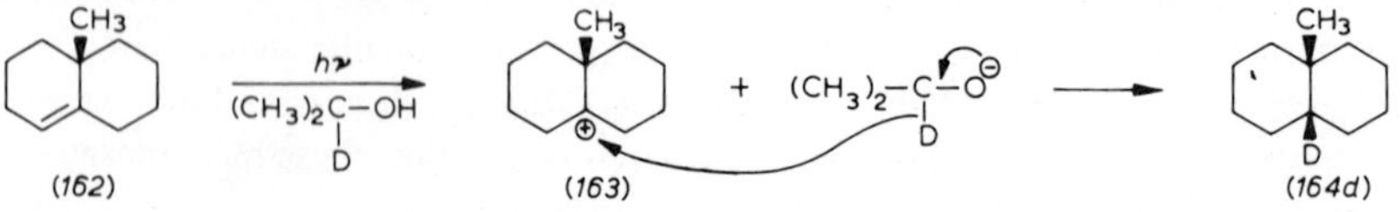

Earlier it was noted that smaller ring cyclic olefins (i.e. cyclopentene, 1-methylcyclopentene, 1-phenylcyclopentene) and rigid bicyclic systems (i.e., norbornene, 2-methylnorbornene, bicyclo[2.2.2]oct-2-ene) are generally unreactive towards photosensitized polar additions [112]. However, these systems often do afford isomerized olefins, reduction products, and solvent incorporated products upon sensitized irradiation. The latter products have been reasonably interpreted as arising from radical addition processes. Since both olefin isomerization and reduction were noted to occur from polar intermediates in the case of cyclohexyl and decalyl systems, the possibility of cation intermediates for related processes in cyclopentenes was studied. When 1-methylcyclopentene and 1-methylcyclohexene were simultaneously irradiated in methanol-*O-d*, the results noted below were obtained. Thus, while the cyclohexyl system showed substantial deuterium incorporation in both (*169d*) and recovered (*168d*),

(165) $\xrightarrow[\text{xylene}]{h\nu,\ CH_3OD}$ 50% (165) + 23% (166) + 12% (167)

(168) $\xrightarrow[\text{xylene}]{h\nu,\ CH_3OD}$ (168d) 17% (d_0 67%, d_1 28%, d_2 5%) + (169d) 21% (d_0 7%, d_1 79%, d_2 13%)

no deuterium incorporation was noted in either recovered (*165*) or the products, (*166*) and (*167*). The lack of deuterium incorporation in the sensitized photolysis of (*165*) indicates that deuteration of the reactive species is not occurring and the source of the rearrangement and reduction products does not involve a cationic intermediate. The formation of (*166*) and (*167*) from (*165*) most reasonably arise via radical processes.

While rigid cycloalkenes are usually quite unreactive towards photosensitized bimolecular addition, the reaction has been reported to occur intramolecularly in the norbornene (*170*) to afford (*171*) as the major product in unreported yield [121].

(170) HOCH$_2$ (D) $\xrightarrow{h\nu,\ \text{benzene}}$ (171) H (D), O—CH$_2$

When the deuterated alcohol was irradiated, (*171d*) was produced with the deuterium stereospecifically endo. As in the cases noted earlier, the reaction was substantially slower for the deuterated alcohol. The specific deuterium incorporation and the rate effect in going from the hydrogen to the deuterium compound perhaps suggest a similar mechanism to that occurring for the cyclohexenes and cycloheptenes. If this observation proves general, it would be possible to effect a variety of sensitized intramolecular additions in the five-membered ring systems.

VII. LABELING STUDIES OF THE DI-π-METHANE TO VINYLCYCLOPROPANE REARRANGEMENT

Exploratory photochemical studies on various organic systems in the early 1960's resulted in several reports of novel isomerizations. For one of these photochemical rearrangements, the di-π-methane reaction, deuterium labeling has not only been used to study the structural changes involved, but also to elucidate the energetics of various bonding processes in the reaction. One of the earliest examples of what is now termed the di-π-methane reaction was recorded in Roth's photochemical study of the cyclo-octatriene isomers [101a]. The irradiation of 1,3,6-cyclooctatriene (*172*) afforded bicyclo-[5.1.0]octa-2,5-diene (*174*) in 30% yield in addition to several other products. Strong evidence that a hydrogen shift was not involved in this reaction came from a study of the deuterium labeled compound (*172d*) which formed (*174d*) with the indicated labeling. The reaction was viewed as proceeding through biradical (*173d*) which

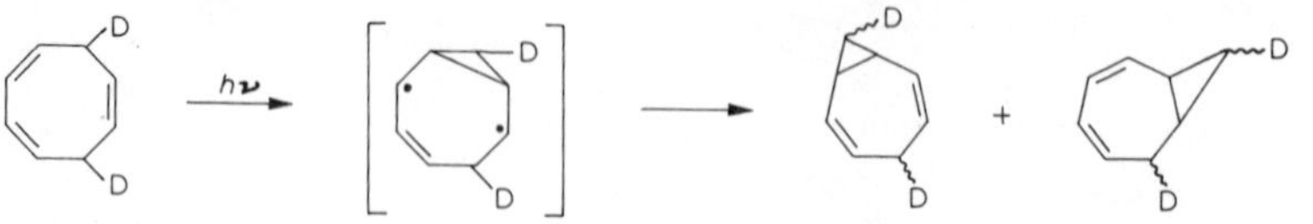

(172d) (173d) (174d) (175d)

could rearrange by two possible paths, yielding (*174d*) and (*175d*). The failure to isolate (*175d*) from the reaction was attributed to its further rearrangement under the reaction conditions. However, from more recent results (*175d*) would also be expected to be the minor process [123].

In the period from 1965–7 several other groups reported seemingly related transformations in a variety of systems [124a–f]; however, it was in connection with Zimmerman's labeling study of the sensitized reaction of barrelene (*176*) to produce semibullvalene (*179*) that the general nature of the earlier reactions was noted [125]. In this instance irradiation of the hexadeuteriobarrelene gave semibullvalene with the deuterium distribution shown below (● signifies hydrogen substitution). The observed labeling pattern supported a mechanism formally involving only two double bonds of the barrelene and excluded a simpler mechanism which had originally been considered.

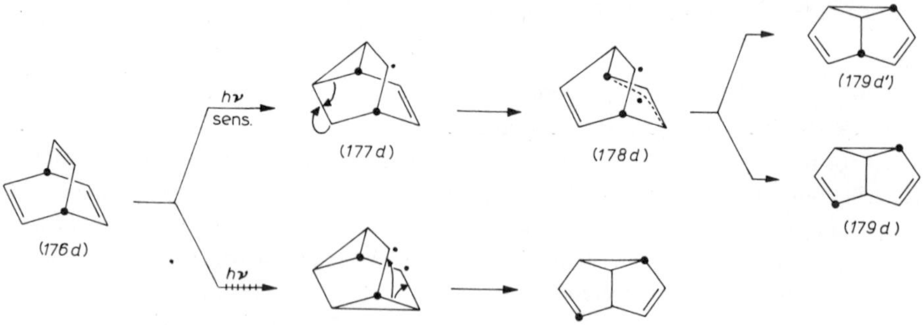

(176d) (177d) (178d) (179d') (179d) (180d) (179d)

(● signifies hydrogen substitution)

Thus, Zimmerman proposed that the reaction was of a general nature and formally involved conversion of a methano carbon flanked by two unsaturated linkages to a vinyl cyclopropane moiety (the di-π-methane to vinyl cyclopropane reaction). The process is conveniently pictured as proceeding initially via a three-membered ring formation followed by three-ring cleavage in a different sense and bond formation. The discrete molecular details on these individual steps depend upon the molecule reacting

(181) $\xrightarrow{h\nu}$ (182) → (183)

and the multiplicity of the excited state involved. In some instances the total process may be concerted [126b,c].

In the ensuing years, the general nature of the reaction, varying aspects of the mechanism [126a–c] and the synthetic utility of this process in both carbocyclic and heterocyclic [127a–e] systems have been studied. One point of general mechanistic interest has been the energetics of the alternate modes of bonding in reactions having more than one possible di-π-methane–vinylcyclopropane reaction. It is in this connection that Zimmerman has made extensive use of deuterium labeling studies in the barrelene systems. Direct irradiation of benzobarrelene [128] (*184*) produces benzocyclooctatetraene (*185*) while the triplet sensitized process produces primarily benzosemibullvalene (*186*). For each case the product may arise via two binding routes, initial bond formation between the two vinyl centers (vinyl–vinyl bridging) or between the olefinic group and the aromatic ring (benzo–vinyl bridging). As can be noted from Scheme II these alternative modes of bonding can be readily distinguished

(184) $\xrightarrow{h\nu}$ (185)

(184) $\xrightarrow[\text{sens.}]{h\nu}$ (186) 10 parts + (185) 1 part

by the distribution of label in the final products. Using benzobarrelene which was deuterated except for the bridgehead positions, it was shown that in the triplet state the di-π-methane product was produced exclusively via vinyl–vinyl bridging. The small amount of benzocyclooctatetraene (*185d*) formed in these sensitized irradiations was

(184d) $\xrightarrow[CH_3\overset{O}{\overset{\|}{C}}CH_3]{h\nu}$ (186d) + (185d) + (185d′)

(185d) $\xrightarrow{h\nu}$ (185d′)

(• signifies hydrogen substitution)

Bonding possibilities for benzobarrelene

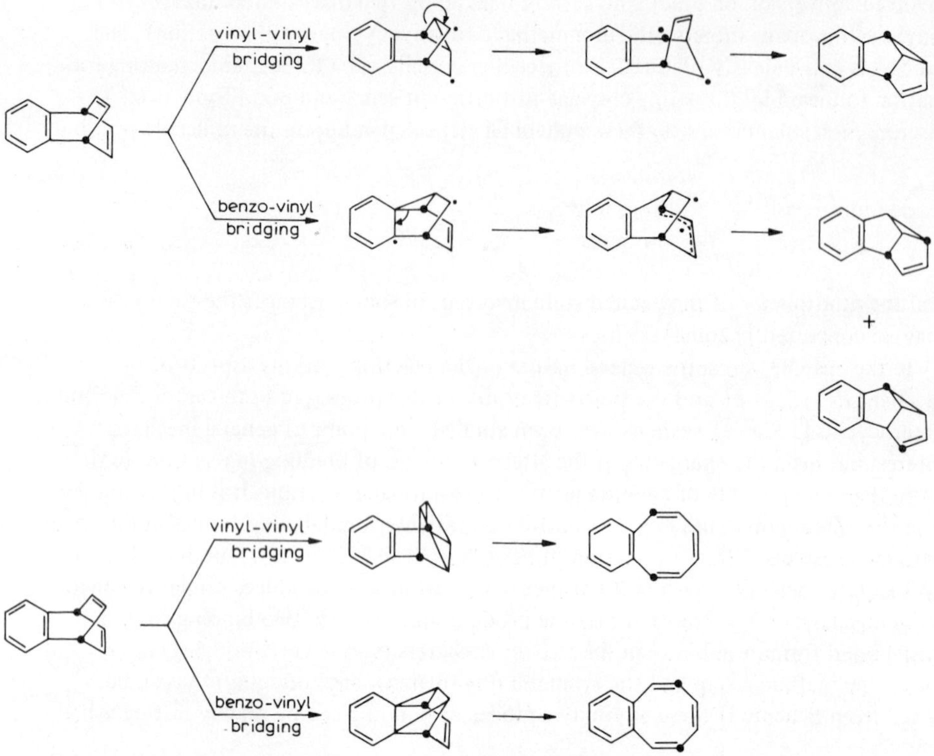

vinyl-vinyl bridging

benzo-vinyl bridging

+

vinyl-vinyl bridging

benzo-vinyl bridging

proposed to arise primarily from some direct excitation of the benzobarrelene with a lesser amount of (*185d'*) being formed from a secondary reaction of the benzosemibullvalene. An independent experiment showed that labeled semibullvalene (*186d*) rearranged to (*185d'*). In contrast to the sensitized irradiation, which showed entirely vinyl–vinyl bonding, the direct excitation of (*184d*) yielded cyclooctatetraene which arose primarily from benzo–vinyl bonding.

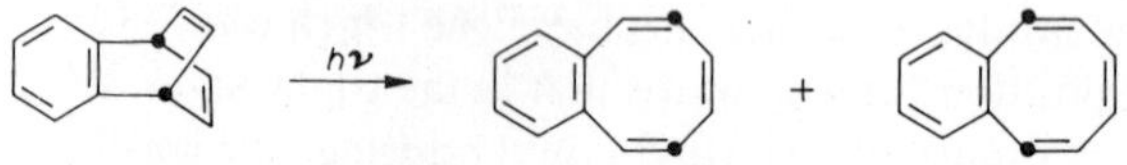

$h\nu$

(*184 d*) (*185 d*) 94 : 6 (*185 d'*)

In further efforts directed at elucidating the origin of the bonding preferences in these systems, studies on 1,2-naphthobarrelene [129], 2,3-naphthobarrelene [129], and benzo-2,3-naphthobarrelene [130] were reported. For the labeled 2,3-naphthobarrelene (*187d*) both direct and sensitized reaction led to (*188d*) in good yield (97% for $h\nu$ direct) via vinyl–vinyl bridging. The total absence of 2,3-naphthocyclooctatetraene

$h\nu$

$\Phi_{direct} = \Phi_{sens} = 0.45$

(*187d*) (*188d*)

(• signifies hydrogen substitution)

in the singlet state reaction was attributed to the necessity of destroying aromaticity simultaneously in both rings in the benzo–vinyl bridging process. In contrast, the 1,2-naphthobarrelene photochemistry was much more complicated. Direct irradiation to 50% conversion led to a nearly perfect mass balance and the formation of three products, (*190d*), (*191d*) and (*192d*). However, acetone sensitized irradiation of (*189d*) led to the equal quantities of (*193d*) and (*194d*) in 91% overall yield. The dissection of the labeling results in this system is more complicated since in addition to

$h\nu$

(1.00) (1.00) (0.42) (0.58) (1.00) (0.42) (0.58) (1.00)

(*189d*) (*190d*) 37% (*191d*) 6.5% (*192d*) 6.5%

Here the distribution of hydrogen label is given in parentheses. The number designations indicate the labeling pattern.

there being two types of vinyl–vinyl (α'–α' and β'–β') and naphtho–vinyl (α–α' or β–β') bridging processes, following each type of initial naphtho–vinyl bonding there

α' β' α β

$h\nu$, sens.

$\Phi = 0.50$

(*189d*) (*193d*) (*194d*)

are two ways in which the penultimate allylic radical may close. Comparison of the experimentally determined labeling results for the various possibilities indicated that the sensitized reaction proceeded entirely by α-naphtho–vinyl bonding. Strikingly, in the direct irradiation process, the formation of naphthosemibullvalenes, (*191d*) and (*192d*), was occurring 58% by the α-naphtho–vinyl bonding mechanism and 42% by α- and β-vinyl–vinyl bridging. As observed in benzobarrelene, the 1,2-naphthocyclooctatetraene, (*190d*), was formed with aromatic vinyl bonding.

In the last reported study in the series, either direct or sensitized irradiation of benzo-2,3-naphthobarrelene gave the benzo-2,3-naphthosemibullvalene. Irradiation of the deuterium labeled compound showed that the naphtho–vinyl bridging (*196d*) and benzo–vinyl bridging (*197d*) mechanisms were occurring to equal extents. Using the

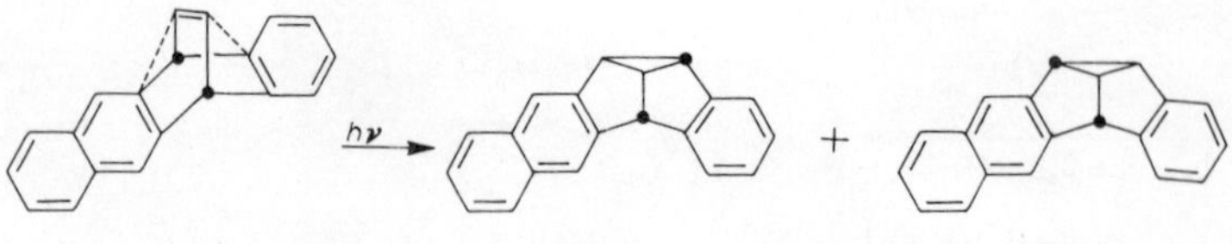

(195d) (196d) (197d)

(• signifies hydrogen substitution)

results in these barrelene derivatives, the observed pattern of triplet reactivity in the bridging process is: α-naphtho–vinyl > vinyl–vinyl > β-naphtho–vinyl ≃ benzo–vinyl. These results are not consistent with reactivity arising from that portion of the molecule having the highest degree of excitation energy. Had that been the case, the β-naphthobarrelene and benzo-2,3-naphthobarrelene would have utilized β-naphtho–vinyl bridging exclusively. Instead the authors propose that the direction of bonding is controlled by the energy of the triplet species along the reaction coordinate where the bridging groups are brought into proximity and overlap weakly. By approximating the triplet energy of the chromophores involved in bridging from known triplet energies of species approximating to the bonding, reasonable agreement with experimental results was obtained. Thus, using an α-vinyl naphthalene triplet [E_T = 56.3 kcal mol^{-1}] as a model for α-naphtho–vinyl bridging, a butadiene-like triplet [E_T = 53–61 kcal mol^{-1}] for vinyl–vinyl bridging, a β-vinyl naphthalene triplet [E_T = 59.5 kcal mol^{-1}] for β-naphtho–vinyl bridging, and a styryl triplet [E_T = 59.8 kcal mol^{-1}] for benzo–vinyl bridging, the observed order of bonding follows, provided the butadiene system is about 58 kcal mol^{-1}.

Finally, deuterium labeling was used to clarify the mechanism of the acetone sensitized photoisomerization of (*198*) to (*202*) [131] (benzobasketene is also formed in the reaction). Here two mechanisms were considered, a di-π-methane pathway and a photochemically allowed $\pi^2_a + \sigma^2_a$ cyclo-addition. In this instance deuterium labeling established that the di-π-methane route was followed.

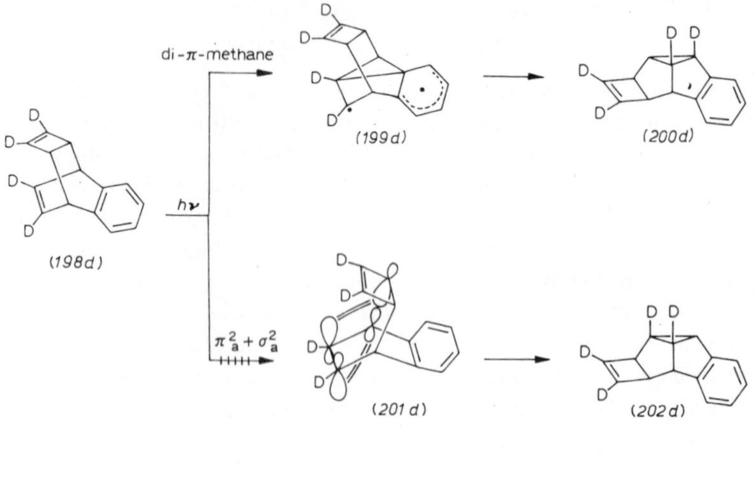

di-π-methane

hν

$\pi^2_a + \sigma^2_a$

(198d) (199d) (200d) (201d) (202d)

While the di-π-methane to vinyl cyclopropane conversion is a quite general one, in certain systems other processes supersede the a priori di-π-methane reaction. Unfortunately, the final products from these reactions are often not distinguishable from the di-π-methane products except by labeling; thus, some reservation must be exercised in attributing formal di-π-methane products as arising via the general mechanism described here. While 3,4-benzotropilidenes (i.e. (*111*)) possess the requisite moiety for di-π-methane reaction, labeling establishes that this route is not followed (see p. 270). Furthermore, in certain cases substitution can divert a system from the di-π-methane reaction to other processes. In one instance replacement of the methyl groups at C-3 by hydrogen changes the mechanism of the photolysis. Thus, irradiation of (*203*) affords the expected di-π-methane product. However, for the molecule lacking methyl substitution, two products were observed, one of which formally arises via the di-π-

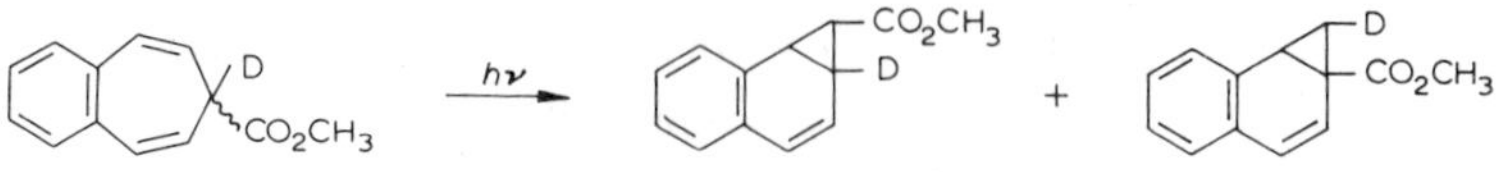

(111) (113) (112)

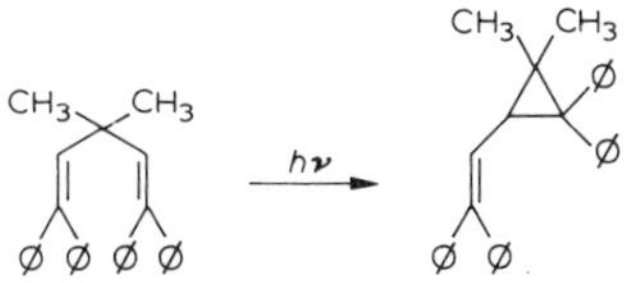

(203) (204)

methane reaction [132]. Deuterium labeling indicates that (*206d*) does not arise from a di-π-methane process; instead, both (*206d*) and (*207d*) are reasonably interpreted as being formed from biradical (*208d*), produced by 1,2-hydrogen in excited (*205d*).

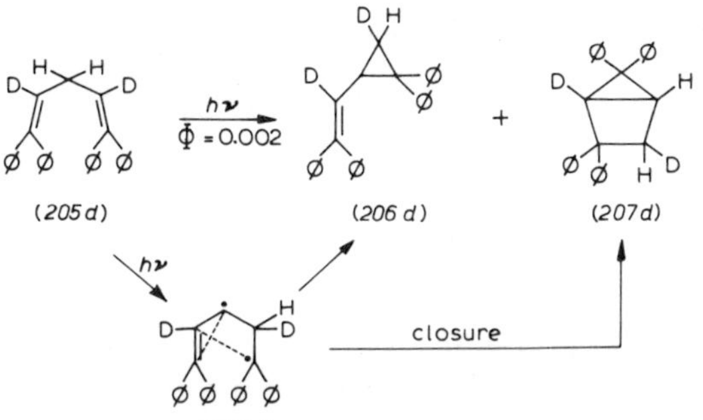

(205d) (206d) (207d)

(208d)

VIII. SUMMARY

As the introductory portion of this chapter illustrates, substitution of deuterium for hydrogen may have a profound influence on the rates of decay for both the excited singlet and triplet states. While this certainly complicates detailed interpretation of isotope effects in many photochemical systems, it does not exclude their meaningful utilization in mechanistic photochemistry. The discussion of deuterium isotope studies in the subsequent pages was chosen to illustrate the general uses of the technique and was not intended to be exhaustive in its coverage. The utilization of deuterium as a label in establishing the gross molecular changes in a photochemical rearrangement will certainly continue to be the major use in organic photochemistry. However, as our understanding of deuterium substitution on excited state decay processes is advanced, undoubtedly, deuterium isotope effect studies will afford perspectives into the more intimate details of photochemical rearrangements. While somewhat unrelated to this discussion, it is important to re-emphasize the potential synthetic utility of photochemical reactions in the synthesis of specifically deuterated molecules [133]. Finally, the author hopes this brief review will be of value to those investigators contemplating deuterium labeling studies in organic photochemistry.

REFERENCES

1a C.A. Hutchison and B.W. Mangum, J. Chem. Phys., 32 (1960) 1261.
1b M.R. Wright, R.P. Frosch and G.W. Robinson, J. Chem. Phys., 33 (1960) 934.
1c S.G. Hadley, H.E. Rast and R.A. Keller, J. Chem. Phys., 39 (1963) 705.
1d R.E. Kellogg and R.P. Schwenker, J. Chem. Phys., 41 (1964) 2860.
2 D. Bryce-Smith (Ed.), Photochemistry, Vol. 1, The Chemical Society, London, 1972, pp. 4–30.
2a D. Bryce-Smith (Ed.), Photochemistry, Vol. 3, The Chemical Society, London, 1972, pp. 43–62.
3 E.W. Schlag, S. Schneider and S.F. Fischer, Annual Review of Physical Chemistry, Annual Reviews, Inc., Palo Alto, California, 1971, pp. 465–521.
4 J. Birks, Photophysics of Aromatic Molecules, Wiley-Interscience, New York, 1970, Chaps. 4–6.
5a J. Jortner, Pure Appl. Chem., 24 (1970) 165.
5b J. Jortner, S.A. Rice and R.M. Hochstrasser, in W.A. Noyes, G.S. Hammond and J.N. Pitts, Jr., (Eds.), Advances in Photochemistry, Vol. 7, Interscience, New York, 1969, p. 149.
6 W. Siebrand, in A.B. Zahlan (Ed.), The Triplet State, University Press, Cambridge, 1967, pp. 31–45.
7a P.M. Johnson and L. Ziegler, J. Chem. Phys., 56 (1972) 2169.
7b P.M. Johnson and M.C. Studer, Chem. Phys. Lett., 18 (1973) 341.
8 R. Li and E.C. Lim, J. Chem. Phys., 57 (1972) 605.
9 S. Fischer and E.C. Lim., Chem. Phys. Lett., 14 (1972) 40.
10 For an elementary discussion see: N.J. Turro, Molecular Photochemistry, W.A. Benjamin, New York, 1965, pp. 67–70.
11 S. Fischer and S. Schneider, Chem. Phys. Lett., 10 (1971) 392.
12a R.F. Borkman and D.R. Kearns, J. Chem. Phys., 44 (1966) 945.
12b R.E. Rebbert and P. Ausloss, J. Amer. Chem. Soc., 87 (1965) 5569.
13 A.A. Lamola, J. Chem. Phys., 47 (1967) 4810.
14a N.J. Turro and R. Engel, J. Amer. Chem. Soc., 90 (1968) 2989.

14b N.J. Turro and R. Engel, Mol. Photochem., 1 (1969) 359.
14c R.F. Borkman, Chem. Phys. Lett., 9 (1971) 77.
15 N.C. Yang, S.L. Murov and T.C. Shieh, Chem. Phys. Lett., 3 (1969) 6.
16 T.E. Marlin and A.H. Kalantar, J. Chem. Phys., 48 (1968) 4996.
17 T.E. Marlin and A.H. Kalantar, Chem. Phys. Lett., 1 (1968) 623.
18 J.D. Simpson, W.H. Offen and J.G. Burr, Chem. Phys. Lett., 2 (1968) 383.
19 R.J. Watts and S.J. Strickler, J. Chem. Phys., 49 (1968) 3867.
20a J.D. Laposa and H. Singh, Chem. Phys. Lett., 4 (1969) 288.
20b J. Saltiel, J.T. D'Agostino, W.G. Herkstroeter, G. Saint-Ruf and N.P. Buiv-Hoi, J. Amer. Chem. Soc., 95 (1973) 2543.
21a S.H. Lin, Trans. Faraday Soc., 66 (1970) 1879.
21b W. Siebrand, Chem. Phys. Lett., 6 (1970) 192.
21c B.R. Henry and W. Siebrand, J. Chem. Phys., 54 (1971) 1072; Chem. Phys. Lett., 3 (1969) 327; 7 (1970) 533.
21d B. Scharf, Chem. Phys. Lett., 14 (1972) 475.
21e S.H. Lin, Mol. Phys., 21 (1971) 853.
22 J.P. Simons and A.L. Smith, Chem. Phys. Lett., 16 (1972) 536.
23 For example, see R.D. McQuigg and J.G. Calvert, J. Amer. Chem. Soc., 91 (1969) 1590.
24a B. Sharf and R. Selbey, Chem. Phys. Lett., 5 (1970) 314.
24b B. Sharf, J. Chem. Phys., 55 (1971) 320.
25 E.C. Lim and J.D. Laposa, J. Chem. Phys., 41 (1964) 3257.
26 J.D. Laposa, E.C. Lim and R.E. Kellogg, J. Chem. Phys., 42 (1965) 3025.
27 C.L. Ermolaev and E.B. Sveshnikova, Opt. Spektrosk., 16 (1964) 320.
28 I.B. Berlman, Handbook of Fluorescence Spectra of Aromatic Molecules, Academic Press, New York, 1965.
29a G.D. Johnson, L.M. Logan and I.G. Ross, J. Mol. Spectrosc., 14 (1964) 198.
29b A.E.W. Knight and B.K. Selinger, Chem. Phys. Lett., 12 (1971) 419.
30 P.F. Jones and S. Siegel, Chem. Phys. Lett., 2 (1968) 486.
31 C.M. Breuer and E.K.C. Lee, J. Chem. Phys., 51 (1969) 3615.
32 B.K. Selinger and W.R. Ware, J. Chem. Phys., 53 (1970) 3160.
33 J.O. Uy and E.C. Lim, Chem. Phys. Lett., 7 (1970) 306.
34 E.C. Lim and H.R. Bhattacharjee, Chem. Phys. Lett., 9 (1971) 249.
35 G.M. Breuer and E.K.C. Lee, J. Phys. Chem., 75 (1971) 989.
36 L. Stryer, J. Amer. Chem. Soc., 88 (1966) 5708 and references cited therein.
37 Th. Förster, Chem. Phys. Lett., 17 (1972) 309.
38 Th. Förster and R. Rokos, Chem. Phys. Lett., 1 (1967) 279 and references cited therein.
39 J. Eisinger and G. Navon, J. Chem. Phys., 50 (1969) 2069 and references cited therein.
40 S.S. Lehrer, J. Amer. Chem. Soc., 92 (1970) 3459 and references cited therein.
41 G.O. Schenck and R. Steinmetz, Bull. Soc. Chim. Belg., 71 (1962) 781.
42 J. Saltiel, K.R. Neuberger and M. Wrighton, J. Amer. Chem. Soc., 91 (1968) 3658.
43 N.C. Yang, J.I. Cohen and A. Shani, J. Amer. Chem. Soc., 90 (1968) 3264.
44 For another viewpoint, see S.M. Japar, M. Pomerantz and E.W. Abrahamson, Chem. Phys. Lett., 2 (1968) 137.
45 For leading references see R.A. Caldwell, G.W. Sovocool and R.R. Gajewski, J. Amer. Chem. Soc., 95 (1973) 2549.
46 R.A. Caldwell and G.W. Sovocool, J. Amer, Chem. Soc., 90 (1968) 7138.
47 R.A. Caldwell, G.W. Sovocool and R.J. Peresie, J. Amer. Chem. Soc., 93 (1971) 779; 95 (1973) 1496.
48 R.A. Caldwell, J. Amer. Chem. Soc., 92 (1970) 1439.
49 R.A. Caldwell and S.P. James, J. Amer. Chem. Soc., 91 (1969) 5184.

50 R.A. Caldwell and R.P. Gajewski, J. Amer. Chem. Soc., 93 (1971) 532.
51 M.W. Schmidt and E.K.C. Lee, J. Amer. Chem. Soc., 90 (1968) 5919; 92 (1970) 3579.
52 M. Matsuoka and M. Szwarz, J. Amer. Chem. Soc., 83 (1961) 1261.
53 S. Seltzer, J. Amer. Chem. Soc., 83 (1961) 1861.
54 R. Hoffmann, S. Swaminathan, B.G. Odell and R. Gleiter, J. Amer. Chem. Soc., 92 (1970) 7091.
55 D.R. Arnold. Advan. Photochem., 6 (1968) 301.
56 W.M. Moore, G.S. Hammond and R.P. Foss, J. Amer. Chem. Soc., 83 (1961) 2789.
57 W.M. Moore and M.D. Ketchum, J. Phys. Chem., 68 (1964) 214.
58 F.D. Lewis and J.G. Magyar, J. Org. Chem., 37 (1972) 2102.
59a D.S. Kendall and P.A. Leermakers, J. Amer. Chem. Soc., 88 (1966) 2766.
59b S.G. Cohen and S. Aktipis, Tetrahedron Lett., (1965) 579.
59c W.C. Agosta and A.B. Smith, J. Amer. Chem. Soc., 93 (1971) 5513.
59d S. Wolff, W.L. Schreiber, A.B. Smith and W.C. Agosta, J. Amer. Chem. Soc., 94 (1972) 7797.
60 E. König, H. Musso and U.I. Zahorszky, Angew. Chem., 84 (1972) 33; Angew. Chem. Int. Ed. Engl., 11 (1972) 45.
61 M. Halwer, J. Amer. Chem. Soc., 73 (1951) 4870.
62 W.M. Moore and C. Baylor, J. Amer. Chem. Soc., 88 (1966) 5677.
63 C.H. DePuy, H.L. Jones and W.M. Moore, J. Amer. Chem. Soc., 95 (1973) 477.
64 P.J. Wagner and G.S. Hammond, J. Amer. Chem. Soc., 87 (1964) 4009.
65 T.J. Dougherty, J. Amer. Chem. Soc., 87 (1964) 4011.
66 For leading references see P.J. Wagner, P.A. Kelso and R.G. Zepp, J. Amer. Chem. Soc., 94 (1972) 7480.
67 R. Srinivasan, J. Amer. Chem. Soc., 81 (1959) 5061.
68 G.R. McMillan, J.G. Calvert and J.N. Pitts, J. Amer. Chem. Soc., 86 (1964) 3602.
69 R.P. Borkowski and P. Ausloss, J. Phys. Chem., 65 (1961) 2257.
70 D.R. Coulson and N.C. Yang, J. Amer. Chem. Soc., 88 (1966) 4511.
71a A. Padwa and W. Bergmark, Tetrahedron Lett., (1968) 5795.
71b C.P. Casey and R.A. Boggs, J. Amer. Chem. Soc., 94 (1972) 6457.
72 C. Djerassi and B. Zeeh, Chem. Ind. (London), (1967) 358.
73 F.D. Lewis, J. Amer. Chem. Soc., 92 (1970) 5602.
74 T.R. Darling and N.J. Turro, J. Amer. Chem. Soc., 94 (1972) 4366.
75 A. Padwa and R. Gruber, J. Amer. Chem. Soc., 92 (1970) 107.
76 A. Padwa and W. Eisenhardt, J. Amer. Chem. Soc., 93 (1971) 1400.
77 For a discussion of the intramolecular reactions of α-, β-unsaturated ketones see:
77a H.E. Zimmerman, Science, 153 (1966) 837.
77b K. Schaffner, in W.A. Noyes, G.S. Hammond and J.N. Pitts, Jr. (Eds), Advances in Photochemistry, Vol. 4, Interscience, New York, 1966, p. 81.
77c P.J. Kropp, in O.L. Chapman (Ed.), Organic Photochemistry, Vol. 1, M. Dekker, New York, 1967.
78 N.C. Yang and M.J. Jorgenson, Tetrahedron Lett., (1964) 1203.
79 R. Noyori, H. Inoue and M. Kato, J. Amer. Chem. Soc., 92 (1970) 6699.
80 R.Y. Levina, V.N. Kostin, and P.A. Gembitskiĭ, Zh. Obshch. Khim., 29 (1959) 2456. However, see reference 78 for somewhat different observations.
81a H. Wehrli, R. Wenger, K. Schaffner and O. Jeger, Helv. Chim. Acta, 46 (1963) 678.
81b S. Kuwata and K. Schaffner, Helv. Chim. Acta, 52 (1969) 173.
81c D. Bellus, D.R. Kearns and K. Schaffner, Helv. Chim. Acta, 52 (1969) 971.
82 P.W. Jennings, Dissertation, University of Utah, 1965
83 Y. Yamada, H. Uda and K. Nakanishi, Chem. Commun., (1966) 423.
84 J. Gloor, K. Schaffner and O. Jeger, Helv. Chim. Acta, 54 (1971) 1864.

85 J.L. Ruhlen and P.A. Leermakers, J. Amer. Chem. Soc., 89 (1967) 4944.
86 W. Herz and M.G. Nair, J. Amer. Chem. Soc., 89 (1967) 5474.
87a W.L. Schreiber and W.C. Agosta, J. Amer. Chem. Soc., 93 (1971) 3814.
87b W.C. Agosta and A.B. Smith, J. Amer. Chem. Soc., 93 (1971) 5513.
87c S. Wolff, W.L. Schreiber, A.B. Smith and W.C. Agosta, J. Amer. Chem. Soc., 94 (1972) 7797.
88 M.J. Jorgenson, J. Amer. Chem. Soc., 91 (1969) 198.
89 W.E. Doering and P.P. Gaspar, J. Amer. Chem. Soc., 85 (1963) 3043.
90 W.R. Roth, Angew. Chem., 75 (1963) 921; Angew. Chem. Int. Ed. Engl., 2 (1963) 688.
91 A.P. ter Borg and H. Kloosterziel, Rec. Trav. Chim. Pays-Bas, 84 (1965) 241.
92 A.P. ter Borg, E. Razenberg and H. Kloosterziel, Chem. Commun., (1967) 1210.
93 For leading references see:
93a A.R. Brember, A.A. Gorman, R.L. Leyfand and J.B. Sheridan, Tetrahedron Lett., (1970) 2511.
93b L.B. Jones and V.K. Jones, J. Amer. Chem. Soc., 90 (1968) 1540.
94 M. Pomerantz and G.W. Gruber, J. Amer. Chem. Soc., 93 (1971) 6615.
95 K.A. Burdett, D.H. Yates and J.S. Swenton, Tetrahedron Lett., (1973) 783.
96 J.S. Swenton and D.M. Madigan, Tetrahedron, 28 (1972) 2703.
97 D.M. Madigan and J.S. Swenton, J. Amer. Chem. Soc., 93 (1971) 6316.
98 K.A. Burdett and J.S. Swenton, unpublished results.
99a E.F. Kiefer and J.Y. Fukunaga, Tetrahedron Lett., (1969) 993.
99b E.F. Kiefer and C.H. Tanna, J. Amer. Chem. Soc., 91 (1969) 4478.
100 W.G. Dauben, C.D. Poulter and C. Suter, J. Amer. Chem. Soc., 92 (1970) 7408.
101a W.R. Roth and B. Peltzer, Justus Liebigs Ann. Chem., 685 (1965) 56.
101b K.A. Burdett, T.J. Ikeler and J.S. Swenton, J. Amer. Chem. Soc., 95 (1973) 2702.
102 J.S. Swenton and A.J. Krubsack, J. Amer. Chem. Soc., 91 (1969) 786.
103 G.W. Gruber and M. Pomerantz, J. Amer. Chem. Soc., 91 (1969) 4004.
104 P.D. Rosso, D.M. Madigan and J.S. Swenton, unpublished results.
105 M. Kato, M. Kawamura, Y. Okamoto and T. Miwa, Tetrahedron Lett., (1972) 1171.
106 N.K. Hamer and M. Stubbs, Chem. Commun., (1970) 1013.
107 For an alternate mechanistic viewpoint see: H.E. Zimmerman, D.F. Juers, J.M. McCall and B. Schröder, J. Amer. Chem. Soc., 92 (1970) 3474.
108 J.A. Marshall, Accounts Chem. Res., 2 (1969) 33.
109a P.J. Kropp, J. Org. Chem., 35 (1970) 2435 and references cited therein.
109b P.J. Kropp, Pure Appl. Chem., 24 (1970) 585.
110 For leading references concerning nucleophilic additions to dienes see:
110a C.C. Leznoff and G. Just. Can. J. Chem., 42 (1964) 2801.
110b G. Bauslaugh, G. Just and E. Lee-Ruff., Can. J. Chem., 42 (1966) 2837.
110c J.A. Barltrop and H.E. Browning, Chem. Commun., (1968) 1481.
111a P.J. Kropp, J. Amer. Chem. Soc., 88 (1966) 4091.
111b J.A. Marshall and R.D. Carroll, J. Amer. Chem. Soc., 88 (1966) 4092.
112. P.J. Kropp, J. Amer. Chem. Soc., 91 (1969) 5783.
113 H. Kato and M. Kawanisi, Tetrahedron Lett., (1970) 865.
114 J.S. Swenton, J. Org. Chem., 34 (1969) 3217.
115 J.A. Marshall and A.R. Hochstettler, Chem. Commun., (1967) 732.
116 J.A. Waters and B. Witkop, J. Org. Chem., 34 (1969) 3774.
117 P.J. Kropp and H.J. Krauss, J. Amer. Chem. Soc., 91 (1969) 7466.
118 D. Guinard and R. Beugelmans, Tetrahedron Lett., (1970) 1705.
119 See also J.A. Marshall and J.P. Arrington, J. Org. Chem., 36 (1971) 214.
120 J.A. Marshall and A.R. Hochstetter, Chem. Commun., (1968) 296.
121 P.J. Kropp and H.J. Krauss, J. Amer. Chem. Soc., 91 (1969) 7466.
122 H.E. Zimmerman and A.C. Pratt, J. Amer. Chem. Soc., 92 (1970) 1407.
123 H.E. Zimmerman and A. Baum, J. Amer. Chem. Soc., 93 (1971) 3646.

124a E. Ciganek, J. Amer. Chem. Soc., 88 (1966) 2882.
124b J.R. Edman, J. Amer. Chem. Soc., 88 (1966) 3454.
124c W.R. Roth and B. Peltzer, Angew. Chem., 76 (1964) 378.
124d J. Zirner and S. Winstein, Proc. Chem. Soc. London, (1964) 235.
124e G.W. Griffin, J. Covell, R.C. Petterson, R.M. Dodson and G. Klose, J. Amer. Chem. Soc., 87 (1965) 1410.
124f M. Jones and L.T. Scott, J. Amer. Chem. Soc., 89 (1967) 150.
125 H.E. Zimmerman, R.W. Binkley, R.S. Givens, G.L. Grunewald and M.A. Sherwin, J. Amer. Chem. Soc., 89 (1967) 3932; 91 (1969) 3316.
126 For leading references see:
126a J.S. Swenton, J.A. Hyatt, A.L. Crumrine and T.J. Walker, J. Amer. Chem. Soc., 93 (1971) 4808.
126b H.E. Zimmerman, P. Baeckstrom, T. Johnson and D.W. Kurtz, J. Amer. Chem. Soc., 94 (1972) 5504.
126c P.S. Mariano and J. Ko, J. Amer. Chem. Soc., 94 (1972) 1766.
127 For leading references see:
127a H. Hart, R.K. Murray and G.D. Appleyard, Tetrahedron Lett., (1969) 4785.
127b L.A. Paquette and R.H. Meisinger, Tetrahedron Lett., (1970) 1479.
127c L.A. Paquette, J.R. Malpass and G.R. Krow, J. Amer. Chem. Soc., 92 (1970) 1980.
127d T.J. Katz, J.C. Carnaham, G.M. Clarke and N. Acton, J. Amer. Chem. Soc., 92 (1970) 734.
127e P.D. Rosso, J. Oberdier and J.S. Swenton, Tetrahedron Lett., (1971) 3947.
128 H.E. Zimmerman, R.S. Givens and R.M. Pagni, J. Amer. Chem. Soc., 90 (1968) 6096.
129 H.E. Zimmerman and C.O. Bender, J. Amer. Chem. Soc., 92 (1970) 4366.
130 H.E. Zimmerman and M. Viriot-Villaume, J. Amer. Chem. Soc., 95 (1973) 1274.
131 I. Murata and Y. Sugihara, Tetrahedron Lett., (1972) 3785.
132 H.E. Zimmerman and J.A. Pincock, J. Amer. Chem. Soc., 94 (1972) 6208.
133 A.F. Thomas, Deuterium Labeling in Organic Chemistry, Meredith Corporation, New York, 1971.

SUBJECT INDEX